P9-DVQ-230
02970374

Physical Illness and Depression in Older Adults

THE PLENUM SERIES IN SOCIAL/CLINICAL PSYCHOLOGY

Series Editor: C. R. Snyder

University of Kansas
Lawrence, Kansas

Current Volumes in the Series:

ADVANCED PERSONALITY
Edited by David F. Barone, Michel Hersen, and Vincent B. Van Hasselt

AGGRESSION
Biological, Developmental, and Social Perspectives
Edited by Seymour Feshbach and Jolanta Zagrodzka

AVERSIVE INTERPERSONAL BEHAVIORS
Edited by Robin M. Kowalski

COERCION AND AGGRESSIVE COMMUNITY TREATMENT
A New Frontier in Mental Health Law
Edited by Deborah L. Dennis and John Monahan

HANDBOOK OF SOCIAL COMPARISON
Theory and Research
Edited by Jerry Suls and Ladd Wheeler

HUMOR
The Psychology of Living Buoyantly
Herbert M. Lefcourt

THE IMPORTANCE OF PSYCHOLOGICAL TRAITS
A Cross-Cultural Study
John E. Williams, Robert C. Satterwhite, and José L. Saiz

PERSONAL CONTROL IN ACTION
Cognitive and Motivational Mechanisms
Edited by Miroslaw Kofta, Gifford Weary, and Grzegorz Sedek

PHYSICAL ILLNESS AND DEPRESSION IN OLDER ADULTS
A Handbook of Theory, Research, and Practice
Edited by Gail M. Williamson, David R. Shaffer, and Patricia A. Parmelee

THE REVISED NEO PERSONALITY INVENTORY
Clinical and Research Applications
Ralph L. Piedmont

SOCIAL COGNITIVE PSYCHOLOGY
History and Current Domains
David F. Barone, James E. Maddux, and C. R. Snyder

SOURCEBOOK OF SOCIAL SUPPORT AND PERSONALITY
Edited by Gregory R. Pierce, Brian Lakey, Irwin G. Sarason, and Barbara R. Sarason

---

A Continuation Order Plan is available for this series. A continuation order will bring delivery of each new volume immediately upon publication. Volumes are billed only upon actual shipment. For further information please contact the publisher.

Physical Illness and Depression in Older Adults

A Handbook of Theory, Research, and Practice

Edited by

Gail M. Williamson
David R. Shaffer

University of Georgia
Athens, Georgia

and

Patricia A. Parmelee

Genesis Health Ventures
Kennett Square, Pennsylvania

Kluwer Academic / Plenum Publishers
New York Boston Dordrecht London Moscow

Library of Congress Cataloging-in-Publication Data

Physical illness and depression in older adults: a handbook of theory, research, and practice/[edited by Gail M. Williamson, David R. Shaffer, and Patricia A. Parmelee
p. ; cm — (The Plenum series in social/clinical psychology)
Includes bibliographical references and index.
ISBN 0-306-46269-9
1. Depression in old age—Handbooks, manuals, etc. 2. Aged—Health and hygiene—Handbooks, manuals, etc. 3. Aged—Mental health—Handbooks, manuals, etc. 4. Geriatrics—Handbooks, manuals, etc. I. Williamson, Gail M. II. Shaffer, David R. (David Reed), 1946– III. Parmelee, Patricia A. IV. Series.
[DNLM]: 1. Depressive Disorder—Aged. 2. Geriatrics. 3. Risk Factors. WM 171 P5775 2000]
RC537.5 .P48 2000
618.97'68527—dc21 00-035714

ISBN: 0-306-46269-9

233 Spring Street, New York, N.Y. 10013

http://www.wkap.nl/

10 9 8 7 6 5 4 3 2 1

A C.I.P. record for this book is available from the Library of Congress

Printed in the United States of America

To NELL MCEWIN,
mother of Gail Williamson,
and quintessential role model of combining femininity
with independence, achievement,
and overcoming adversity
to set one's sights on nothing less than the best

Contributors

Katherine L. Applegate, Department of Psychiatry, Ohio State University College of Medicine, Columbus, Ohio 43210

Jamila Bookwala, Department of Psychology, Pennsylvania State University, Abington, Pennsylvania 19001

Martha L. Bruce, Department of Psychiatry, Weill Medical College of Cornell University, White Plains, New York 10605

Eric D. Caine, Department of Psychiatry, University of Rochester Medical Center, Rochester, New York 14642

Jen Cheavens, Department of Psychology, University of Kansas, Lawrence, Kansas 66045

Mary Amanda Dew, Department of Psychiatry, Western Psychiatric Institute and Clinic, University of Pittsburgh, Pittsburgh, Pennsylvania 15213

Ronald Glaser, Department of Medical Microbiology and Immunology, Ohio State University College of Medicine, Columbus, Ohio 43210

Barry Gurland, Columbia University Stroud Center, New York, New York 10032

Sidney Katz, Columbia University Stroud Center, New York, New York 10032

Janice K. Kiecolt-Glaser, Department of Psychiatry, Ohio State University College of Medicine, Columbus, Ohio 43210

Ellen J. Klausner, Department of Psychiatry, Weill Medical College of Cornell University, White Plains, New York 10605

Charles E. Lance, Department of Psychology, University of Georgia, Athens, Georgia 30602

M. Powell Lawton, Polisher Research Institute, Philadelphia Geriatric Center, 5301 York Road, Philadelphia, Pennsylvania 19141

Jeffrey M. Lyness, Department of Psychiatry, University of Rochester Medical Center, Rochester, New York 14642

Adam W. Meade, Department of Psychology, University of Georgia, Athens, Georgia 30602

L. Stephen Miller, Department of Psychology, University of Georgia, Athens, Georgia 30602

Mark D. Miller, Department of Psychiatry, University of Pittsburgh School of Medicine, Pittsburgh, Pennsylvania 15213

Benoit H. Mulsant, Department of Psychiatry, Western Psychiatric Institute and Clinic, University of Pittsburgh, Pittsburgh, Pennsylvania 15213

Patricia A. Parmelee, Managed Care Division, Genesis Health Ventures, 148 West State Street, Kennett Square, Pennsylvania 19348

Zachary M. Pine, Geriatric Division, Department of Medicine, University of California at San Francisco, San Francisco, California 94143

Bruce G. Pollock, Department of Psychiatry, Western Psychiatric Institute and Clinic, University of Pittsburgh, Pittsburgh, Pennsylvania 15213

John W. Reich, Department of Psychology, Arizona State University, Tempe, Arizona 85287

Charles F. Reynolds III, Department of Psychiatry, Western Psychiatric Institute and Clinic, University of Pittsburgh, Pittsburgh, Pennsylvania 15213

Bruce Rollman, Department of Psychiatry, University of Pittsburgh School of Medicine, Pittsburgh, Pennsylvania 15213

Ramesh Sairam, Department of Psychiatry, University of Texas Southwestern Medical Center, Dallas, Texas 75235

Herbert C. Schulberg, Department of Psychiatry, Weill Medical College of Cornell University, White Plains, New York 10605

Amy S. Schultz, Department of Psychology, Arizona State University, Tempe, Arizona 85287

Richard Schulz, Department of Psychiatry, University of Pittsburgh School of Medicine, Pittsburgh, Pennsylvania 15213

David R. Shaffer, Department of Psychology, University of Georgia, Athens, Georgia 30602

C. R. Snyder, Department of Psychology, University of Kansas, Lawrence, Kansas 66045

Myron F. Weiner, Department of Psychiatry, University of Texas Southwestern Medical Center, Dallas, Texas 75235

Gail M. Williamson, Department of Psychology, University of Georgia, Athens, Georgia, 30602

Jennifer L. Yee, Center for Social and Urban Research, University of Pittsburgh, Pittsburgh, Pennsylvania 15213

Alex J. Zautra, Department of Psychology, Arizona State University, Tempe, Arizona 85287

Foreword

Aging is inevitable—A "psychological recession" is not . . .

As I go about my daily life, I read and hear about the sometimes scary things that are happening to other people. As the saying goes, bad news sells newspapers. But I usually can take some solace in reasoning that this bad stuff assuredly will *not* occur in my life. After reading this book, however, one message has gotten through—I cannot dismiss "those" older people described in the various chapters as being dissimilar to me. After all, "old person" is a term that *can be applied to me in a few more years*. On this point, I once heard the following rhetorical question applied to the prejudice actions of the TV character Archie Bunker: "What would he say about "those" Puerto Ricans, if, on his next birthday, he knew that he would become a Puerto Rican?" As to aging, we best pay close attention because we soon will be "those" elders.

This is why the alarming facts of this book—that our elders often are experiencing elevated levels of physical illnesses and depression—grabbed me by the proverbial throat. So, too, should these facts alert America's baby boomers, because the talk of illness and depression certainly takes some of the shine off their ensuing "golden years." As the physical illnesses and dark shadows of depression creep into many elders' lives, they are rendered far from painless, less than happy, and generally not too "golden." I call this a "psychological recession," where large numbers of elders have lost their good health, as well as their positive outlooks. To compound matters, a psychological recession involves the elder's implicit and sometimes explicit comparison of the present circumstances with those of his or her younger days.

Physical illnesses and depression are serious: anyone who has experienced them can testify to such an assertion. They take away life's pleasures, rob simple joys, and leave us wondering about our purpose in being. And such physical problems and depression do not impact just the one person who "has" them. They spread, and soon friends and loved ones also are engulfed by them. The "downer" grows—and so we have a psychological recession for all those involved.

What can we do to better understand and help the waves of aging people who are mired in such psychological recession? This question portends answers for the present elders, as well as the waves of future elders—people like you and me. For this reason, this book both alarms and excites me. The alarm I already have sounded. As to the excitement, in the following pages we are treated to new insights about understanding and lifting the veils from the illness and despondency problems. The thinking in this volume is catalytic and fresh, producing a reaction akin to that of solving a problem. After thoroughly describing the depth and breadth of the physical illness and depression problems faced by elders, the present researchers share their insights and solutions. They tell us what we now know, and provide an agenda for future research and interventions. Thus, the editors and authors of this volume paint a realistic view about how the forces of physical illness and depression can be understood *and overcome*.

In conclusion, there is bad and good news on these pages. The bad news is that physical illness and depression may touch us in the senior years. The good news is that these threats to our well-being can be understood and, in many, if not most instances, changed for the better. As such, aging is inevitable; a psychological recession is not.

C. R. Snyder
Lawrence, Kansas

Preface

The population is aging. Baby boomers are approaching retirement age, and individuals over age 85 make up the single fastest-growing segment of the population. Our society is faced with quickly finding answers to questions about longevity and quality of life that, because they pertained to so few people, were not matters of urgency just a few decades ago.

Historically, diagnosis and treatment of depression in elderly people was hampered by mythical thinking. Depression was viewed as an unavoidable consequence of the losses (e.g., declining health, retirement, and bereavement) that naturally accompany advancing age. Epidemiological studies exploded this myth by indicating that severe depression was not as widespread a problem in older adults as was commonly believed.

Still, even clinically diagnosable depression typically went untreated. Among the reasons were the difficulty of diagnosing depression when it is accompanied by physical illness and beliefs that available treatments for depression were not suitable for and would not be accepted by elderly persons. In recent years, we have seen some major breakthroughs. First, researchers and practitioners took note of the fact that, although major depression is relatively uncommon in elderly adults, it is not necessary for depressed affect to meet clinical diagnostic criteria to negatively impact quality of life. Second, substantial strides have been made toward better methods for identifying depressive symptomatology in this segment of the population. Third, research efforts have clearly shown that the association between physical illness and depression is bidirectional and complicated by numerous biological, social, and psychological factors. Finally, advances in pharmacological and behavioral therapies have produced more suitable interventions, and current treatment strategies have been proven to be both effective for and accepted by depressed older adults.

Professionals working at the intersection of physical illness and depression are an interdisciplinary group that can model (albeit less than perfectly) cross-discipline cooperation. From its inception, the idea for this handbook was to bring together in one place a comprehensive sample of up-to-date work from various disciplines in an effort to stimulate further theoretically based and intervention-oriented research.

In addition to introductory and final summary chapters, this handbook consists of three major sections. The first considers risk factors for the development of depression in older adults. How are disability, chronic pain, and vascular disease, all of which increase in incidence with old age, related to depression? What about circumstances of elderly persons' ordinary, everyday lives? How detrimental is the age-related likelihood of providing care to a chronically ill family member?

The second section graphically illustrates the need for complex models when studying associations between physical illness and depression. What is the impact of depression on the immune systems of older adults? How do depression and personal evaluation of one's quality of life influence end-of-life attitudes and behaviors? Does losing valued activities contribute to the relation between illness and depression? How can we best evaluate change over time in complex longitudinal models?

The third section considers critical diagnostic and treatment issues. What are the difficulties for primary care physicians, who are the most frequent contacts with the medical system for older adults, in recognizing and treating comorbid illness and depression? How can major depressive disorder be identified in Alzheimer's disease patients? Are pharmacotherapy and individual psychotherapy effective in older adults? What about group therapy? Must we be certain of causal influences before instituting treatment?

The editors of this volume strongly advocate an integrative interdisciplinary approach to answering the myriad questions posed by efforts to study and treat depression in older persons. We are all products of basic training in experimental social psychology who, at various points in our careers, made either partial or complete transitions into applied research with heavy emphasis on health psychology, particularly in the context of the older adult population. Although we each realized early on that a multidisciplinary approach was the only rational way to deal with the complexity of physical illness and depression in older adults, we brought with us the respect for, and insistence upon, theoretical and methodological rigor—whether the specific study under consideration was experimental, correlational, or some combination of both. Interdisciplinary as they are, without exception, our contributing authors meet these stringent criteria. They uniformly fall into the category of researchers and practitioners operating on the cutting edge of inquiry in the domain of physical illness and depression in the rapidly increasing proportion of the population that consists of elderly people. We are indebted to each and every one of them for taking time from their hectic schedules to contribute to this book. We also are gratified that they felt such an effort was important enough to do so.

Many individuals provided a great deal of support for this endeavor, often through their enthusiasm for and encouragement of the direction our own idiosyncratic interests seemed to be taking us. Among those influential persons devoted to the enterprise of scientific inquiry, regardless of personal theoretical or methodological orientation, are Margaret S. Clark, Clyde Hendrick, and M. Powell Lawton. For their encouragement and support for taking the "dangerous" step, we are eternally grateful.

Finally, Rick Snyder and Eliot Werner deserve particular thanks for their guidance when we needed it but, more especially, for their faith in our ability to carry out this project. Their willingness to let us do it on our own, their tolerance, and their senses of humor both lightened and enlightened this enterprise.

Gail M. Williamson
David R. Shaffer
Patricia A. Parmelee

Contents

1

Physical Illness and Depression in Older Adults

An Introduction

DAVID R. SHAFFER

In early 1990, my father was diagnosed with prostate cancer and told that, although he would probably live for many years, he would never be cured of this life-threatening condition. By the end of that summer, this formerly cheerful and optimistic man had become listless, moody, and hard to live with—and, in his own words, "perpetually blue." When I visited in August and asked why he felt so depressed, he had obviously considered his situation, responding immediately, listing several reasons. He said he either could not do, or no longer felt like doing, many of things he had always enjoyed. To him, that meant he was not the same Herb Shaffer he had always seen himself as being. Moreover, the drugs he was taking made him irritable with everyone, and even though he knew this was not "him," he could not seem to help it. Dad was also distressed that his doctor had dismissed his depressed affect as "temporary" and unworthy of treatment, although he claimed (later) to have gained some relief from his symptoms by taking sizable doses of an herbal treatment, St. John's wort.

At the time of my visit, I was an experimental social psychologist, still several years away from acting on any research interests in health-related issues. Yet the themes that my father expressed would all make sense to social psychologists. A serious physical illness had resulted in restriction of valued activities, loss of control and sense of self, and perhaps even helplessness-induced despair. In fact, my father's state caused this social psychologist to seriously ponder both the factors that might contribute to, and the inevitability of any links between, physical illness and depression in older adults.

DAVID R. SHAFFER • Department of Psychology, University of Georgia, Athens, Georgia 30602.

Physical Illness and Depression in Older Adults: A Handbook of Theory, Research, and Practice, edited by Gail M. Williamson, David R. Shaffer, and Patricia A. Parmelee. Kluwer Academic/Plenum Publishers, New York, 2000.

HISTORICAL CONSIDERATIONS

Only 20 years ago, my father's open acknowledgment of an adverse and pervasive affective reaction stemming from a physical illness might have been considered atypical. Indeed, the physical illnesses and social–economic difficulties that elderly adults often face had led many health care professionals in the 1970s and 1980s to conclude that negative affect was a *normal* and probably inconsequential reaction to such problems, an attitude that older patients often shared. Unfortunately, this stance conspired to make depression among older adults an underdiagnosed and undertreated disease (Lebowitz et al., 1997).

Following a decade of progress in understanding the incidence, diagnosis, and treatment of depression in older adults, the National Institutes of Health held the 1991 NIH Consensus Development Conference on the Diagnosis and Treatment of Depression in Late Life (NIH Consensus Statement, 1991). The objectives of this team of biomedical and behavioral scientists, surgeons, and other health care professionals were to resolve questions surrounding what was known about the "epidemiology, pathogenesis, pathophysiology, prevention, and treatment of depression in the elderly and to alert both the professional and lay public to the seriousness of depression in late life, to its manifestations and useful treatments, and to areas needing further study" (p. 1).

The Consensus Panel concluded that depressive illness is widespread among older adults and a serious public health concern. Moreover, late-onset depression can be diagnosed and differentiated from normal aging, and one hallmark of depression in older people is its comorbidity with medical illness. The panel stressed that a variety of treatments have been found to be safe and efficacious for older adults and that depressed elderly patients should be treated vigorously and monitored for a potentially recurring condition that is likely to require long-term intervention.

As a result of this conference and the research it has stimulated, considerable progress has been made in understanding the etiology, diagnosis, treatment, and design of care provision systems for depression in late life. In light of these developments, NIH staff, in collaboration with experts in the field, have recently issued a second consensus statement to update the original conclusions drawn by the Consensus Panel of 1991 (Lebowitz et al., 1997). Among the important new findings highlighted in this Consensus update are the following:

1. The association of late-onset depression with brain abnormalities and vascular disease.
2. Transactional relationships among physical illness, depression, and functional disability.
3. The clinical significance of *subsydromal depression* in older adults, that is, levels of depressive symptomatology that are associated with increased risk of major depression, physical disability, medical illness, and high use of health care services but that do not meet current *Diagnostic and Statistical Manual of Mental Disorders* (DSM-IV; American Psychiatric Association, 1994) criteria for major depression or dysthymia.
4. The clinical utility of new antidepressant treatments, including selective serotonin reuptake inhibitors (SSRIs) and standardized psychotherapies.

5. The *necessity* for long-term treatment of late-life depression.
6. The need for increased sensitivity to the identification of depression among older patients by their primary care providers.
7. A clear relationship between depression and suicide among older adults.

PRIMARY OBJECTIVES OF THIS VOLUME

Of course, one limitation of consensus statements such as that just referenced above is that they present conclusions while offering little more than abbreviated reference lists as support for their pronouncements. Moreover, references cited by multidisciplinary panels, necessarily multidisciplinary, are thus spread across a large number of scientific journals, many of which may be difficult to access for readers interested in physical illness and depression among the elderly.

Given that our population is growing progressively older, with more and more individuals living longer with chronic illnesses and disabilities, it is of increasing importance for the lay public and professionals alike to understand the interrelationship of depression, disability, and physical illness, the mechanisms underlying these interrelations, and their implications for diagnosis and treatment. Simply stated, our primary objective in preparing this volume was to summarize and interpret a carefully selected portion of the knowledge about physical illness and depression that has come to light over the past 20 years.

To pursue this objective, we assembled an impressive array of nationally and internationally recognized experts in gerontology, geriatric psychiatry, and health psychology who were asked to provide cutting-edge summaries in their areas of expertise on the interface between physical illness/disability and depression in older adults. Chapter authors have reviewed and interpreted the state of research in their own areas of expertise, identified deficiencies in existing knowledge, offered practice and policy implications, and suggested important directions for future research. Thus, we have endeavored to assemble a current state-of-the-art "handbook" for gerontologists, geriatricians, primary care physicians, nurses, geriatric psychiatrists, rehabilitation specialists, health psychologists, and social workers—one that incorporates a variety of theoretical perspectives and contemporary reviews of the literature, with an emphasis on application of theory and research to issues of diagnosis and treatment.

ORGANIZATION AND OVERVIEW OF THE VOLUME

The handbook is organized into four broad sections (or parts) that address the major etiological, diagnostic, and treatment issues and conclusions raised by the Consensus Panels on the Diagnosis and Treatment of Depression in Late Life (Lebowitz et al., 1997; NIH Consensus Statement, 1991).

Part I, "Risk Factors," focuses on medical, physiological, psychological, and psychosocial factors that place one at risk for experiencing depression in late life, while also examining the bidirectional (or transactional) patterns of association between physical morbidities, functional disabilities, and late-onset depression. In Chapter 2, Martha L. Bruce reviews epidemiological and clinical data demonstrating both that disability increases risk of depression and that depressive symptoms and disorders increase the risk of subsequent functional decline. Bruce addresses the potential psychological, physiological, cognitive, and social mechanisms that may underlie this bidirectional relationship, noting that identification of these mechanisms is essential should we hope to forestall or prevent what might otherwise be a spiral of decline in vulnerable older adults.

More detailed examinations of particular risk factors begin with Chapter 3, in which Jeffrey M. Lyness and Eric D. Caine review the literature on the association between late-onset depression and such cardiovascular risk factors (CVRFs) as coronary artery disease, diabetes mellitus, hypertension, and atrial fibrillation. The authors critically evaluate two pathobiological models specifying that CVRFs may contribute to depression in either of two ways: (1) through brain damage via small vessel cerebrovascular disease, or (2) by producing functional alterations in neurotransmitter systems underlying depressive pathogenesis. They caution, however, that correlational data consistent with the pathobiological models can be taken as support for bidirectional viewpoints specifying that depression is a risk factor for cardiovascular disease and death, and they acknowledge the likelihood that a variety of psychosocial factors may mediate or moderate the relationship between CVRFs and depression among older adults.

A psychological risk factor is the primary concern in Chapter 4. Here, Gail M. Williamson discusses a mechanism that appears to qualify the well-known (but often modest) association between the subjective experiences of illness-induced pain and depression *in persons of all ages*. Both cross-sectional and longitudinal data converge to suggest that pain and depression are most strongly associated for those individuals whose condition has led to a particular kind of functional impairment—restriction of routine daily activities. This finding has clear implications for intervention. Additionally, the observation that pain, functional disability, and depression are less strongly interrelated among patients—be they children or adults—who have experienced pain the longest implies that pain sufferers may adapt to pain over time, becoming less distressed by it or by the routine activities it restricts.

In Chapter 5, Alex J. Zautra, Amy S. Schultz, and John W. Reich examine small, everyday stressors as potentially important psychosocial correlates of depression among older adults and small, everyday positive events as potential contributors to recovery from depressive symptoms. After describing the development of the Inventory of Small Life Events (ISLE) scale, Zautra and associates present data from their Life Events and Aging Project (LEAP) to indicate that everyday life events (both negative *and* positive experiences) play a central role in the maintenance and/or alleviation of depressive symptoms in physically disabled but not conjugally bereaved older adults. A major implication is that interventions that prove most fruitful for depressed elders may depend very crucially on the type of major life stress that precipitated their symptomatology.

In Chapter 6, Jamila Bookwala, Jennifer L. Yee, and Richard Schulz conclude Part I by calling our attention to a potent psychosocial contributor to late-onset depressive symptomatology—the experience of providing care to a chronically ill family member. Their review of the literature convincingly indicates that, compared to matched noncaregiver control samples or community norms, providing primary care to either a demented or nondemented elder places one at risk for poorer mental health outcomes (i.e., anxiety disorders, depression). Moreover, caregiving is reliably associated with poorer self-reported health, and there is evidence for links between caregiving and objective measures of poor health (e.g., physical disabilities, depressed immune function, and adverse physiological outcomes; see also Chapter 7). Bookwala et al. examine known sociodemographic, personological, and social correlates of psychiatric and physical morbidity among caregivers, while pointing to interpretive ambiguities in our current knowledge base and suggesting strategies for addressing them in future research.

Part II, "Conditioning Variables and Outcomes", concentrates on variables that appear to intervene, and thus qualify, relations among depression, functional impairment, and physical health in older adults. In recent years, two ways of looking at depression and its relation to physical morbidity and mortality have emerged. Depression might be viewed as (1) a biopsychological state that "causes" physical morbidity, or (2) an indicator of somatic resources; that is, a meter of one's ability to control existing disease and/or fight pathogens to prevent disease. These alternative views of depression raise a number of questions for intervention and treatment of depression in older populations.

Katherine L. Applegate, Janet K. Kiecolt-Glaser, and Ronald Glaser address both viewpoints in Chapter 7, as they summarize research suggesting that depression and other forms of life stress (e.g., caregiving) pose greater risk of undermining the physical health of older, rather than younger, adults by further impairing an *aging* immune system. Both endrocrinological and behavioral variables are highlighted as possible mediators of relations between stress, depression, and impaired immune functioning in older populations, and implications for research and clinical practice are discussed.

In Chapter 8, M. Powell Lawton examines relations among physical illness, depression, and end-of-life attitudes and behaviors. His review of the literature reveals that depression is neither inevitable in the terminal phase of life nor a necessary motivation for avoiding life-sustaining treatment or ending one's life. And although quality of life often erodes with declining health, Lawton proposes that many positive, non-health-related aspects of life strongly influence the judgments of seriously ill elders, depressed or otherwise, when thinking about life prolonging treatment. Lawton introduces a promising intervening variable, Valuation of Life (VOL), describes its measurement, and illustrates its utility in research demonstrating that VOL mediates the impact of both positive and negative aspects of one's life (including illness and depression) on end-of-life attitudes. The chapter concludes with an insightful list of unanswered questions to guide future research on end-of-life issues.

Interestingly, one of the positive factors that Lawton cities as an important contributor to life-sustaining motivation is the ability to carry out valued routine activities. In Chapter 9, Gail M. Williamson and David R. Shaffer evaluate their *activity restriction model*—a viewpoint specifying that restriction of routine activities by a major life stressor is a powerful intervening variable that plays a central role in psychological adjustment. Williamson and Shaffer conceptualize illnesses, disability, and even role-related burdens of providing care for an ill elder as major life stressors; depression, and such other affective reactions as resentment, as possible reactions to these stressors; and activity restriction as a *mediator* of association between health-related stressors and emotional well-being. Activity restriction is consistently shown to mediate associations between a variety of illness-related stressors and depressed affect (in both patients and caregivers), even after other known contributors to depression in older adults are controlled. Moreover, the impact of nonillness-related variables (e.g., social support; financial resources; public self-consciousness) on depressed affect is shown to be qualified by patient and caregiver levels of activity restriction. Implications of these findings for future research and intervention are discussed.

Charles E. Lance, Adam W. Meade, and Gail M. Williamson conclude Part II with a methodological contribution in Chapter 10. The authors begin by reviewing the current "state of the practice" of analyzing longitudinal data to detect developmental change. They then critique current practices, comparing them with criteria defining a more idealized approach to the measurement of longitudinal change. Lance et al. provide a general introduction to one analytic technique, *latent growth modeling* (LGM), that meets most of these criteria, and follow up by applying LGM to an existing data set, comparing the information provided to that gleaned from more traditional analytic procedures. The interpretive differences that arise imply that there is much to be gained by measuring longitudinal change through newer "state of the art" statistical techniques.

Part III, "Diagnosis and Treatment," addresses the multiple challenges that primary care physicians and mental health care professionals face in properly diagnosing and treating depression and its concomitants in older adults. In Chapter 11, Herbert C. Schulberg, Richard Schulz, Mark D. Miller, and Bruce Rollman point out that primary care physicians, the very practitioners to whom older adults most often turn for help with mood disturbances, often perceive their clinical skills as inadequate for the psychiatric task they face. The chapter explores complexities of medical and depressive comorbidity in the primary care sector, along with strategies for physicians faced with differential diagnostic and treatment decisions.

Complexities introduced by the comorbidy between major depressive disorder (MDD) and Alzheimer's disease (AD) are the primary focus of Chapter 12. Here, Myron F. Weiner and Ramesh Sairam review the pertinent literature, concluding that MDD in late life is a risk factor for the development of AD, but that MDD is overestimated by both clinicians and caregivers as a complication of AD. Weiner and Sairam describe the factors that contribute to overdiagnosis of MDD among AD patients and offer new criteria for use by clinicians in order to lessen the likelihood of an inappropriate diagnosis. Also included in the chapter is a discussion of procedures for and benefits of treating depression among AD patients who have been certified as clinically depressed.

The focus shifts from diagnostic to treatment issues in the next two chapters. In Chapter 13, Charles F. Reynolds III, Mark D. Miller, Benoit H. Mulsant, Mary Amanda Dew, and Bruce G. Pollack adopt a medical model, viewing geriatric depression as a chronic, relapsing illness that requires active antidepressant medication to relieve symptoms, followed by long-term continuation therapy. For the clinically oriented reader, the authors describe in considerable detail how they carry out pharmacological treatment of geriatric depression. For the research-oriented reader, results of a 10-year study combining medication and interpersonal psychotherapy (IPT) are presented, suggesting that an approach combining pharmacotherapy with IPT may be optimal for achieving and maintaining wellness in elderly adults, particularly those above the age of 70. The authors conclude by suggesting strategies for pursuing much-needed cost-effectiveness studies of combined treatments versus monotherapies as interventions for depression in later life.

Suicidal ideation (and behavior) is not uncommon among depressed older adults (Lebowitz et al., 1997), particularly those who express a persistent sense of hopelessness. Unfortunately, psychotherapy, pharmacotherapy, and their combination do not always prove to be successful in alleviating hopeless thinking and/or eventual suicidal behaviors in depressed elders. In Chapter 14, Ellen J. Klausner, C. R. Snyder, and Jen Cheavens introduce an innovative group intervention for treating late-life depression—one based on C. R. Snyder's Hope Theory and the authors' goal-focused model of hope. The intervention was tailored to reduce functional disabilities and other symptoms of major depression among patients who had not achieved remission with medication or other types of psychotherapy, and special attention was paid to patients' persistent symptoms of hopelessness. The authors outline Snyder's Hope Theory and describe how it was used in group therapy to help patients identify and achieve practical, individualized goals. Compared to equally depressed older outpatients in an alternative group-therapeutic (control) procedure known to be somewhat effective in reducing depressive symptomatology, outpatients in the new, hope-inspired Goal-Focused Group Psychotherapy (GFGP) displayed clear decreases in anxiety, hopelessness, and functional disability, as well as improvements in social-interactive behaviors. The success of this small-scale study of depressed community-dwelling elders prompts the authors to standardize their GFGP for use in clinical trials with a broader range of patients in a variety of treatment settings.

In Chapter 15, Barry Gurland, Sidney Katz, and Zachary M. Pine conclude Part III with a provocative essay on the limitations of information on cause and effect that can be derived from longitudinal studies, with particular emphasis on ambiguities arising from varying views of depression as a cause or an effect of physical illness/disability. The authors conclude that interrelationships among physical illness, disabilities, and depression are best viewed as a whole-person phenomenon, explainable only as a complex web of interacting attributes; that is, a reciprocally interactive subjective–objective complex influenced by body, mind, values, living and nonliving environments, and life experiences in context (including treatment). They argue that it is counterproductive to search or to advocate for a "prime" cause and a "primary" effect in the physical disorder–depression relationship, for these two agents converge as irreducible wholes. Similarly, the authors view moderators of this complex interaction as parts of the whole that cannot be located readily on a path between cause and effect. This is clearly a provocative essay, but one that is far from nihilistic. Gurland et al. argue convincingly that practitioners can and should be able to live with tolerable uncertainty about cause and effect with regard to the physical illness/disability/depression complex in older adults, and they propose strategies for treatment in the face of causal uncertainty.

Part IV concludes the volume by summarizing major themes and principles that have been stressed throughout. In Chapter 16, L. Stephen Miller skillfully reviews and synthesizes our rapidly expanding knowledge of the complex interplay between physical illness, disability, and depression in older populations and offers his views of the implications of this knowledge base for future research, diagnosis, intervention, and public policy.

POSTSCRIPT

On a personal note, I have come away from my editorial role in compiling this volume having learned far more than I expected to learn about physical illness and depression in late life and having had many of my preconceived notions disconfirmed. As C. R. Snyder notes in the Foreword, there is both good and bad news in this book, and working on it has been a real growth experience for its editorial team. We now offer the volume, with the hope that its readers will experience the same benefits.

REFERENCES

American Psychiatric Association. (1994). *Diagnostic and statistical manual of mental disorders* (4th ed.). Washington, DC: Author.

NIH Consensus Development Conference Consensus Statement. (1991, November 4–6). *Diagnosis and treatment of depression in late life*.

Lebowitz, B. D., Pearson, J. L., Schneider, L. S., Reynolds, C. F., Alexopoulos, G. S., Bruce, M. L., Conwell, Y., Katz, I. R., Meyers, B. S., Morrison, M. F., Mossey, J., Niederehe, G., & Parmelee, P. (1997). Diagnosis and treatment of depression in late life: Consensus statement update. *Journal of the American Medical Association, 278,* 1186–1190.

I

Risk Factors

2

Depression and Disability

MARTHA L. BRUCE

In 1991, the National Institutes of Health (NIH) Consensus Development Panel on Depression in Late Life noted that depression causes "suffering . . . and burdens families and institutions providing care for the elderly by disabling those who might otherwise be able-bodied" (NIH Consensus Panel, 1992). Disability itself has been implicated in the risk of late-life depression (Bruce & Hoff, 1994; Gurland, Wilder, & Berkman, 1988; Katz, 1996). These two relatively straightforward observations have profound implications for the lives of older adults and the strategies needed to maintain and enhance their quality of life. Evidence of the contribution of depression to disability and, conversely, of disability to depression, suggests that each could initiate a declining spiral itself associated with a range of other negative outcomes, including institutionalization and death.

The purpose of this chapter is to review evidence suggesting a reciprocal relationship between depression and disability, and, in reviewing potential explanations for this relationship, to suggest approaches to investigate this relationship. The chapter begins by reviewing the major terms (depression and disability), as their definitions also have an impact on the findings.

BACKGROUND AND TERMS

Depression

In everyday usage, the term *depression* refers to a wide range of affect, from transient sadness to persistent symptoms to a clinical diagnosis. This chapter refers to research in which the term is more narrowly defined, usually referring either to moderate to

MARTHA L. BRUCE • Department of Psychiatry, Weill Medical College of Cornell University, White Plains, New York 10605.

Physical Illness and Depression in Older Adults: A Handbook of Theory, Research, and Practice, edited by Gail M. Williamson, David R. Shaffer, and Patricia A. Parmelee. Kluwer Academic/Plenum Publishers, New York, 2000.

severe depressive symptoms or to depressive disorders as determined using diagnostic criteria such as the fourth edition of the Diagnostic and Statistical Manual of Mental Disorders (DSM-IV; American Psychiatric Association, 1994) or the International Statistical Classification of Diseases and Other Health Problems (ICD-10; World Health Organization, 1992). Understanding what the term *depression* means and how it is measured is critical for interpreting the data from any specific study as well as for building a comprehensive understanding of the relationship between depression and disability.

The essential feature of a major depressive diagnosis, as defined by DSM-IV, is a period of at least 2 weeks during which there is depressed mood or the loss of interest/pleasure in nearly all activities. To meet full criteria for an episode of major depression, individuals must also concurrently experience symptoms, also lasting 2 weeks or more, from at least four out of a list of seven symptom groups: (1) changes in weight or appetite; (2) changes in sleep; (3) changes in psychomotor activity; (4) decreased energy; (5) feelings of worthlessness or guilt; (6) difficulty thinking, concentrating, or making decisions; and (7) recurrent thoughts of death or suicidal ideation, plans, or attempts. Furthermore, the symptoms must not be solely the physiological effects of substance use (i.e., drugs, alcohol, or medication) or of a medical condition. In addition, these symptoms must be severe, meaning they are associated with clinically significant distress and/or with impairment in social, occupational, or other types of functioning.

Alternative indicators of depression generally differ from the diagnosis of major depression by being less restrictive concerning one or more of the DSM-IV criteria. Other diagnostic approaches, for example, differ by reducing the number of required symptom groups (e.g., the DSM-IV proposed minor depression) or by eliminating the necessity of judging the etiology of the symptoms (i.e., "inclusive" measures of major depression accept symptoms possibly resulting from medical illness or medication use; Koenig, George, Peterson, & Pieper, 1997). Symptom scales, for example, the Center for Epidemiologic Studies—Depression Scale (CES-D; Radloff, 1977), the Geriatric Depression Scale (GDS), Yesavage, Brink, Rose, & Lum, 1983, and the Beck Depression Inventory—Primary Care (BDI-PC), Beck, Guth, Steer, & Ball, 1997, and the Hopkins Symptom Checklist (HSCL), Derogatis, Lopman, Rickels, Uhlenhuth, & Covi, 1974) produce continuous measures of depression ranging from little or no symptomatology to considerable symptomatology. Symptom scales also tend to be inclusive, so that the etiology of each symptom is not evaluated. In addition, most symptom scales do not assess severity in terms of symptoms causing distress or impairment (e.g., the CES-D), and some do not evaluate symptom duration so that current symptoms are counted without assessing how long they have lasted or how frequently they occur (e.g., the GDS).

Each of these seemingly subtle variations in measures poses potentially large methodological challenges for the study of disability and depression in old age. Diagnostic dichotomies, such as those derived from DSM-IV, are vulnerable to classification error, especially around the milder cases, sometimes resulting in biased estimates in relationship to another variable. Depression measures that include symptoms possibly due to medical conditions may overestimate the relationship of depression to disability due to the same medical conditions (Cohen-Cole & Stoudemire, 1987). Older cohorts of adults, in whom the prevalence of physical disability is the highest (Fried &

Guralnick, 1997), tend to be reluctant to report symptoms of depression, often attribute symptoms to causes other than depression, or are less likely than younger adults to acknowledge distress associated with these symptoms (Tweed, Blazer, & Ciarlo, 1992). Furthermore, role impairment due to depression could be misattributed to disability (or vice versa). Each of these points is discussed in greater detail later in this chapter.

Disability

Perhaps even more than depression, *disability* has numerous definitions. The scope of the term varies from a more narrow emphasis on basic tasks, such as the ability to feed oneself or to move around one's home, to a wide range of physical, social, recreational, and employment activities. Although the different major disability models offer conflicting terms and definitions (Johnson & Wolinsky, 1993; Nagi, 1976; Pope & Tarlov, 1991; World Health Organization, 1980), each shares the notion that disability is part of a process influenced by biological and social factors. An influential example is the World Health Organization's *International Classification of Impairment, Disabilities, and Handicaps* (ICIDH; World Health Organization, 1980). The ICIDH differentiates three constructs: (1) *impairment,* defined as any loss or abnormality of psychological, physiological, or anatomical structure of function (e.g., cough, skin problem, fainting, breathlessness, eyesight problem, and specific medical problems); (2) *disability,* defined as a restriction or lack, resulting from an impairment, in the ability to perform an activity in the manner considered normal for a human being (e.g., activities of daily living [ADLs], mobility,); and (3) *handicap,* defined as a disadvantage for an individual resulting from ill health compared to what is normal for someone of the same age, sex, and background. Although the WHO is currently revising the ICIDH, the older classification reflects the use of the term *disability* in most research to date.

Even within the ICIDH and similar schema, the concept of disability can be broadly constructed. In this chapter, disability refers primarily to the subset of functional problems in performing fundamental activities related to community living, specifically, ADLs such as bathing, dressing, and eating; *instrumental activities of daily living* (IADLs) such as preparing meals, caring for the home, maintaining finances; and *mobility* such as walking and climbing stairs (Spector, 1996). Because most models of disability posit a causal, if not hierarchical, relationship between these so-called "physical" disabilities and disability in other domains such as social function, role function, and employment function, a more thorough understanding of the relationship between physical disability and depression will ultimately contribute to understanding the relationship of depression to these other functional outcomes. Study of social, role, or employment functioning is further complicated by their conceptual overlap with other constructs having large research literatures, for example, social support, social networks, role theory, activity theory, and productive activities.

Like depression, disability can be measured as a classification variable or as continuous variable. Measures of ADL disability, for example, often count the number of ADL limitations but because so few people in the general population of elderly adults report any ADL disability, the continuous variable is usually highly skewed and a dichotomous (any/none) version is often reported. Other approaches to measuring disability have resulted in a broader spread. An example comes from the Medical Outcomes Study—Short Form (Stewart, Hays, & Ware, 1988), in which the physical function subscale (PF-10) assessed several behaviors ranging from ADL activities to strenuous exercise. The PF-10 yields a score from 0 (serious limitations, equivalent to ADL disability) to 100 (high activity).

Traditionally, self-report measures of disability ask individuals to report their purported behavior or what they believe they are able to do. More recently, performance tests, such as the 6- minute walk (Guayatt, Walter, & Norman, 1987), the Timed Up and Go (Podsiadlo & Richardson, 1991), and the Hand Grip (Fried, Ettinger, Lind, Newman, & Gardin, 1994) have been developed as objective measures of disability or, as sometimes conceptualized, indicators of performance capacity.

REVIEW OF CROSS-SECTIONAL AND LONGITUDINAL FINDINGS

Cross-Sectional

Empirical research documenting the cross-sectional association between depression and disability is surprisingly recent, with only a few studies investigating the topic prior to the 1980s. Evidence of the association is found in various types of samples. Particularly important is evidence from population-based samples, usually defined by geographic parameters, because this kind of sample is less likely to be biased by the presence of either disability or depression. In other types of samples, for example, nursing home residents, primary care patients, or psychiatric patients, disability or depression may well influence the likelihood of a person being in that service setting. Inability to care for oneself (i.e., a disability) is a primary reason an older person enters a nursing home (Foley et al., 1992); suffering from depressive symptoms is one reason a person goes to a general medical or mental health provider (Leaf et al., 1988).

The problem is not so much the variation in the prevalence of disability or depression across these different settings but that the relationship between disability and depression at any one point in time and over time may also vary. Evidence of this problem was observed in population-based data from the New Haven site of the Epidemiologic Catchment Area study (Lyness et al., 1996). In the elderly subsample of these data ($N = 2{,}576$), major depression was positively associated with functional disability after controlling for demographic factors and medical morbidity. The association varied, however, when the sample was subdivided by whether subjects had used mental health or medical services. Major depression was associated with *less* physical disability among individuals who used mental health services (adjusted odds ratio [OR] = 0.38), with *more* disability among individuals who used medical but not mental health services (adjusted OR = 1.85), and was not associated with disability among those who used neither source of care (adjusted OR = 0.83). These different relationships between disability and depression reflect in part the extent to which presence of disability, depression, or both, influences the decision whether or not to seek care, when to seek care, and where that care is sought.

Since so much of the elderly population uses medical services, it follows that population-based studies also support the association between depression and physical disability. The relationship has been reported in studies assessing depressive symptoms (Berkman et al., 1993; Craig & Van Natta, 1983; Berkman et al. 1986; Kennedy et al., 1989; Ormel, Kempen, Brilman, Beekman, & Van Sonderen, 1997), usually adjusting for medical and cognitive health. Population-based studies of elderly adults using diagnostic criteria for depression similarly report an association with disability measures (Beekman, Deeg, Braam, Smit, & Van Tilburg, 1997; Forsell, Jorm, & Winblad, 1994; Henderson et al., 1993; Prince, Harwood, Blizard, Thomas, & Mann, 1997;).

Although population-based studies offer the least biased estimates of the relationship between depression and disability, other kinds of samples are also useful. The higher prevalence of depression, disability, or both, in some settings provides additional statistical power for investigating the relationships. The role of these settings in case identification of both conditions is also useful for planning and evaluating interventions. Studies in primary care are relevant, therefore, as the vast majority of older adults use primary care services and most depression as well as disability are treated in primary care settings. The Medical Outcomes Study provided strong evidence of the negative impact of depression on functioning; patients diagnosed with major depression reported poorer physical function, measured by the PF-10 described earlier, than well patients or patients with any of several medical conditions, including hypertension, diabetes, or arthritis (Wells et al. 1989). These findings have been replicated in more recent studies and with an array of indicators, including specific measures of ADL disability (e.g., Jaffe, Froom, & Galambos, 1994; Williams, Kerber, Mulrow, Medina, & Aguilar, 1995). Nursing home residents and home care patients are also useful populations to study because the prevalence of ADL disability, so low in the general population, is very high in these groups, providing substantial statistical power (Fried & Guralnik, 1997).

Other studies have focused on individuals identified either by depression, disabilities, or specific medical conditions. In an epidemiologic sample of people with disabilities (Turner & Noh, 1988; Turner & Beiser 1990), symptoms of depression were associated with degree of disability. Similarly, in studies of patients with major depression, disability was associated with severity of depression (Alexopoulos et al., 1996; Lyness, Caine, Conwell, King, & Cox, 1993). The relationship of depressive symptoms or disorders to functional status has also been documented in patients with specific medical conditions, for example, rheumatoid arthritis (McFarlane & Brooks, 1988), spinal cord injuries (Schulz & Decker, 1985) and Parkinson's Disease (Brown, MacCarthy, Gotham, Der, & Marsden, 1988; Cole et al., 1996).

Longitudinal

Findings from cross-sectional studies beg the question of the extent to which the association observed between depression and disability at one point in time is a function of disability's influence on depression or vice versa. Data from longitudinal studies in which participants are assessed at multiple time points, provide support for both hypotheses.

Several recent studies support the hypothesis that depression affects disability status over time. One study (Bruce, Seeman, Merrill, & Blazer, 1994) examined the predictive effect of depressive symptoms on onset ADL disability using data from the MacArthur Study of Successful Aging, a subset of the high (physical and cognitive) functioning elderly assessed by the Established Populations for the Epidemiologic Studies of the Elderly (EPESE) population–samples in three communities. A strength of this sample for this kind of analysis was the large number of baseline health assessments, including not only self-reported medical conditions but also objective measures such as lung function, blood pressure, and body mass. The EPESE study was also one of the first epidemiological studies to collect performance indicators of physical ability (e.g., walking across a room, semitandem stand, tandem stand, walk, seat-to-rise). Controlling for all these factors, baseline depressive symptoms (assessed with the HSCL depression subscale) predicted onset ADL disability at 2.5-year follow-up in both males and females. Similar findings have been reported using a diagnostic measure of depression in the 12-year follow-up study of the Baltimore Epidemiologic Catchment Area (ECA) sample (Armenian, Pratt, Gallo, & Eaton, 1998). In these data, individuals with major depression at baseline were over four times more likely to report incident ADL disability at follow-up, controlling for medical illness and sociodemographic factors.

Penninx and colleagues (1998) examined data from one of the EPESE communities to assess whether depressive symptoms at baseline affect decline in physical performance measures at 4-year follow-up. In their analysis of the Iowa EPESE data (1988), baseline measures of depressive symptoms were associated with subsequent decline in performance as measured at 4-year follow-up both among the total sample and among disability-free individuals at baseline. To the extent that measuring performance is less influenced by concurrent mood, cognition, or personality than self-reported physical function, a topic discussed later in this chapter, these data are particularly strong evidence of the link between depressive symptoms and subsequent physical decline.

Strawbridge, Shema, Balfour, Higby, and Kaplan (1998) used a different strategy in their analyses of prospective data collected over four decades as part of the Alameda County study. Their analyses targeted the broader concept of frailty as an outcome, defining frailty by four domains of measures, including physical function, nutritional function (e.g., loss of appetite and unexplained weight loss), cognitive function, and sensory problems. In these data, depressives symptoms contributed to the prediction of frailty, although the overlap between aspects of the frailty measures (e.g., nutritional function) and depressive symptoms undermines the usefulness of this study in understanding the relationship between depression and disability.

Several prospective studies have reported the converse effect—that disability is associated with an increased risk of depression. The results have been found using a variety of indicators of depression, from symptom scores to incident of major depressive disorder (Bruce & Hoff, 1994; Henderson et al., 1997; Kennedy, Kelman, & Thomas, 1990), and have been applied to both onset and course of depressive symptoms (Bruce & Hoff, 1994; Wells, Burnam, Rogers, Hays, & Camp, 1992). In one example, Kennedy and colleagues (1990) found in a community sample of elderly adults that increases in disability levels over time predicted emergence of depressive symptoms. Using data from the New Haven site of the ECA study, Bruce and Hoff (1994) reported a marginally significant ($p < .08$) effect of baseline ADL disability on first-onset major depression at 1-year follow-up, controlling for sociodemographic, physical health, social isolation, psychiatric comorbidity, and homebound status. Perhaps more important, however, the authors reported that another baseline indicator of disability, subjects being confined to their bed or chair over the prior 2-week period, significantly and strongly (OR = 5.03) predicted first onset major depression, controlling for ADL disability and the other covariates.

POTENTIAL EXPLANATIONS

Why do the data suggest that depression and disability contribute to the risk of each other over time? In considering the potential explanation for these findings, this section first poses several methodological artifacts to the findings and then reviews some of the hypothesized substantive reasons.

Methodological Artifacts

Any hypothesis testing of the relationship between depression and disability must first ensure that the results will not be an artifact of methodological problems. These studies are most vulnerable to three types of problems: (1) selection bias, (2) confounding, and (3) measurement error.

The importance of selection bias was discussed earlier in the context of data sources, with the main point being that both depression and disability affect help seeking in different kinds of care settings, so that the relationships between the two observed in one setting may differ from those in other settings. Selection bias may not affect the internal validity of longitudinal analyses of the relationship of depression and disability over time, although the external validity in terms of generalizability to other types of individuals would need to be demonstrated.

Confounding occurs when two variables are statistically associated with each other, not because they are causally related but because each has an independent relationship to a third, common variable. In terms of the observed relationship between depression and disability, medical illnesses are particularly relevant as potential confounding variables, since previous research has documented a causal association between some form of medical illness and each of the other two. To date, studies linking depression and disability vary substantially in the amount of available data related to medical status, ranging from self-reports of medical conditions to medical record reviews to clinical measures of lung function, blood pressure, and the like.

One way to reduce confounding is to control for variables representing medical status or other potential confounding variables in statistical analyses. Thoughtful interpretation of this kind of analysis is important, however, as simply controlling for a third variable may obscure more complex causal paths. Berkman and colleagues (1986) made this point when reporting a relationship between medical illness and essentially each symptom of depression. They argue, at least in terms of their cross-sectional study, that because medical illness was related to nonsomatic symptoms (e.g., depressed mood) as well as somatic symptoms (e.g., loss of appetite) merely adjusting for medical morbidity may be "overcontrolling" and obscure potential intricate causal relationships among depression, medical illness, and other variables such as disability.

In some cases, what is labeled as "confounding" can also be addressed by better measurement. The extent to which the observed relationship between depression and disability is an artifact of measurement refers primarily to the concern that the assessment of depression is affected by disability status or, conversely, that the assessment of disability is affected by depression.

Disability may affect the assessment of depression in several ways. As noted, DSM-IV specifically distinguishes between symptoms due to "mental" versus "physical" causes, warning not to include symptoms that are clearly due to a general medical condition. Differentiating "mental" from "physical" symptoms is especially difficult for disorders such as depression, with high levels of medical comorbidity, and for the study of psychiatric disorders in the medically ill (Katz, 1996). There is no gold standard, laboratory test, or methodology generally accepted by the field for distinguishing symptoms of depression from those associated with medical illness. Cohen-Cole and Stoudemire (1987) differentiate four ways to approach making the diagnosis: (1) *etiological,* when symptoms count toward the diagnosis of depression only if they are not "caused" by physical illness, which is the approach stipulated by DSM-IV and the decision rule for the assessment tools such as the Structured Clinical Interview for Axis I DSM-IV Disorders (SCID; Spitzer, Gibbon, & Williams, 1995), although neither explains how to accomplish this task; (2) *inclusive,* when the etiological criterion is ignored so that symptoms of depression are counted whether or not they might be attributable to a primary physical problem; this approach increases sensitivity at the expense of specificity; (3) *substitutive,* when additional psychological, affective, or cognitive symptoms are substituted for somatic symptoms in making the diagnosis (e.g., Clark, Cavanaugh, & Gibbons, 1983); and (4) *exclusive,* when somatic items are eliminated from the existing criteria and the diagnosis is made on the basis of nonsomatic symptoms. The strategy chosen has profound implications for the estimated rates of disorder, especially in medically ill populations. For example, in a sample of elderly medical inpatients, Koenig and colleagues (1997) report a twofold difference (from 10.4% to 20.7%) in the prevalence rate of major depression depending upon which of the strategies is used. As important, in cross-sectional data, the approach to diagnosis presumably affects the nature of observed relationships with disability, with a greater association expected with inclusive measures compared to the other three approaches.

A second problem in the assessment of DSM-IV criteria for major depression is the requirement that "symptoms cause clinically significant distress or impairment in social, occupational or other important areas of functioning" (American Psychiatric Association, 1994, p. 320). In very old adults with disabilities, finding evidence of impairment above and beyond the limitations placed on a life already circumscribed by physical disability can be challenging to an accurate diagnosis of depression. Indeed, in our own study of elderly patients receiving visiting nurse care for medical problems (Bruce et al., 1998), many of our patients with depressive symptoms initially state that their symptoms do not cause impairment, as they are already limited (e.g., bedbound or chairbound) by the illness. With further probing, some patients admit that they would otherwise telephone their friends, participate in their hobbies, maintain their financial records, or listen to music were it not for their depressive symptoms.

More generally, assessment of whether depressive symptoms cause impairment can be affected both positively and negatively by the presence of physical disabilities, because disabilities, can modify the scope or nature of a person's roles. On the one hand, role reduction resulting from physical limitations may be inaccurately attributed to depression, so that the severity of depression will be artificially inflated. Conversely, physical disabilities may make it more difficult to identify evidence of role impairment. One example comes from Boyd, Weissman, Thompson, and Myers (1982) who compared depression assessed using the CES-D (Radloff, 1977) and Schedule of Affective Disorders and Schizophrenia (SADS/RDC; Spitzer & Endicott, 1978) in 482 community-dwelling adults. The authors describe an 85-year-old woman who scored very high on the CES-D but did not make criteria for major depression using the SADS. The woman lived alone, had almost no social contact, and was homebound, in large part due to physical disabilities. She, in essence, had no family, work, or other social role. So while the woman affirmed almost all the SADS depression items, the items could not be scored positive because she had neither sought help and, having no role to impair, reported no role impairment. In essence, the woman's physical disabilities made it harder for researchers to evaluate the severity of the depression.

Finally, the presence of disability complicates the assessment of specific depressive symptoms, for example, worthlessness. In our visiting nurse patients, many of whom have recently experienced new-onset ADL or IADL disability, one of the more difficult symptoms to assess reliably is feelings of worthlessness. Many patients express frustration, anger, or demoralization with being unable to take care of themselves and being newly reliant on others. Some of these patients also question their ability to contribute to their family and household. But fewer of these patients equate their diminished function as a devaluation of their worth as a person, a symptom of depression.

Conversely, depressed mood may affect the assessment of disability. Some evidence indicates that depressed mood affects not only functional behavior but also an individual's perceptions, and therefore reporting, of his or her own disability level. Comparisons of self-report versus informant data, for example, suggest that depression may contribute to an overestimation of behavioral disability in some adults, although the research evidence is equivocal (Morgado et al., 1991; Kuriansky et al., 1976; Little, 1986). Informant questionnaires, direct observation, or other means of verifying self-report information may improve the reliability of disability data.

Theoretical Explanations

Although the empirical data may reflect methodological problems, the notion that disability and depression may contribute to the risk of each other also finds strong theoretical support. Several hypotheses have been suggested.

Disability as a Risk Factor for Depression. Most potential explanations of why disability may be risk factor for depression fall into two types (see Gurland et al., 1988). The first argues that disability is a prodromal indicator of underlying physical, cognitive, or emotional dysfunction leading to the onset of depression. In its extreme form, the argument is that disability is the first observable evidence of a depression. In a more general version of this explanation, disability is indicative of an underlying mechanism, itself needing further identification, that leads to depression. Especially important to testing this kind of hypothesis is initially ruling out the possibility of poor measurement or confounding.

Alternatively, several investigators have conceptualized disability as a stressful condition that increases the risk for depression (Gurland et al., 1988; Schulz & Williamson, 1993; Turner & Noh, 1988, Williamson & Shaffer, Chapter 9, this volume). From the perspective of stress research, it may be useful to differentiate new or aggravated disability as a life event from the potential chronic stress associated with ongoing disability. The potential stress associated with new disability is compatible with both major theories of how events affect the risk of depression (McLean & Link,

1994). In one view, life events, such as the onset of disability, are disruptive experiences requiring readjustment or behavioral change in the day to day living of life. The disruption and adjustment associated with new disability are the mechanisms increasing the risk of depression (Holmes & Rahe, 1976; Dohrenwend, Krasnoff, Askensay, & Dohrenwend, 1978). In the other view, it is the meaning of life events and the emotional response to that meaning that increase the risk of depression. In terms of new disability, reactions such as feeling worthless or hopeless may contribute to depression. Finally, others have conceptualized disability as a chronic strain because, by its nature, it involves ongoing challenges to accomplishing daily tasks (Turner & Noh, 1988).

Depression as a Risk Factor for Disability. Although most conceptual models of disability, even recently published ones (Judge, Schechtman, Cress, et al., 1996), do not include depression, a relatively large number of theoretical explanations for why depression would affect disability have been offered (Armenian et al., 1998, Gurland et al., 1988; Penninx et al., 1998). In general, testing these explanations demands more specificity concerning what aspect of depression predicts functional decline than is available in most empirical analyses to date, a topic discussed in the next section. They include the hypothesis that specific depressive symptoms, such as fatigue, sleep disturbance, and appetite disruption, have direct debilitating effects that result over time in functional decline and disability. Alternatively, the debilitating effects of depressive symptoms may work indirectly in that depressive symptoms result in intermediary behaviors that then affect functional status. For example, depressive symptoms of hopelessness or anhedonia may reduce motivation, leading to poor health behaviors (e.g., poor nutrition, smoking, lack of exercise) or impeding use of health services lead to inadequate appropriate preventive and screening care, lack of compliance to medical treatments, or poorly maintained rehabilitation regimens. Cognitive symptoms of depression may erode a person's ability to understand what tasks are needed or how to approach accomplishing these tasks.

Other hypotheses address the contribution of psychological distress to neural, hormonal, and/or immunological alterations resulting in susceptibility to disease and decline in physical health. Similarly, depression may in other ways increase vulnerability to other risk factors, such as medical illness and life events, for disability (Turner & Noh, 1988).

FUTURE DIRECTIONS

Generally speaking, research to date has demonstrated the existence of a relationship between depression and disability without telling us much about the mechanism driving this relationship. In several respects, the data lack enough detail to test the hypotheses offered by the theoretical literature. Arguably, then, we need to augment our current approach to this task. The next section suggests several strategies that could be useful: (1) studies that further specify the components or aspects of disability and depression that are related; (2) analytic approaches to longitudinal data that model the changes in both depression and disability over time; (3) studies that identify factors mediating the relationship between depression and disability; and (4) interventions studies that attempt to modify the relationship.

Specificity of Measurement

As described earlier, the association between depression and disability has been documented using a variety of measures in a variety of samples. This variation suggests that the relationship is robust and generalizable. On the other hand, this seeming variety obscures how limited these measures are. Most studies have used aggregate indicators of disability, depression, or both. Many of the measures are dichotomous (i.e., depressed or not, disabled or not) while the continuous measures aggregate across a single dimension, usually number of symptoms.

On the other hand, many of the hypotheses suggest that specific aspects of depression are linked to disability and vice versa. Disaggregation of measures is essential to test some of these hypotheses, making it possible to identify what aspects of disability are related to changes in depression, and what aspects of depression are related to changes in disability. For example, Steffens and colleagues (1999), using cross-sectional data on depressed patients, reported an association of depression severity with IADL disability but not ADL disability. These data might suggest that the effect of depression is greater for more discretionary activities, lending support for the hypothesis that depression affects disability through undermining motivation (Bruce, 1999). In contrast, however, longitudinal EPESE data indicating that depressive symptoms have an effect on physical performance measures—measures presumably less affected by motivation—have been interpreted as evidence that the motivation hypothesis is insufficient (Penninx et al., 1998), although motivation remains one potential factor in a series of links between depression at one point in time and reduced function a year or more later.

Depression. With few exceptions, most studies of depression and disability investigate only a single dimension of depression (i.e., severity as measured by number of symptoms or symptom groups), without examining the role of other forms of heterogeneity in the risk for and outcomes of disability. Reynolds, Lebowitz, and Kupfer (1993) identified several other sources of heterogeneity particularly relevant to geriatric patients: (1) *types* of specific symptoms or symptom groups (e.g., psychomotor retardation, hopelessness) consistent with several hypotheses that link specific types of depressive symptoms to depression; (2) *persistence* (i.e., whether the episode is transient and self-limiting vs. persistent and potentially benefitting from intervention), (3) whether this is a *first versus recurrent* episode of depression, and if a first episode of major depression, whether there have been previous episodes of subsyndromal depression (Lyness, Pearson, Lebowitz,& Kupfer, 1994); (4) *age of onset*; (5) *severity* in terms of magnitude of intensity of symptoms; (6) *duration* of the current episode; (7) *anxiety* symptoms; and (8) *neurocognitive function* (e.g., executive dysfunction). While some of these latter categories might be conceptualized as independent, comorbid conditions as opposed to features of the depression, some evidence suggests that certain depressions in late life are differentiated by their relationship to underlying vascular illness and characterized by neuropsychological dysfunction (Alexopoulos et al., 1999).

Disability. Although disability is often operationalized as a unidimensional construct, the research literature points to a number of somewhat overlapping features potentially relevant to disability's relationship to depression (Fried & Guralnik, 1997): (1) *Severity* itself is multidimensional, including both *intensity* (i.e., the degree of difficulty) and *extensivity* (i.e., the number of activities in which a person is limited); (2) *duration*, specifically, whether disability is acute (i.e., occurring recently) or chronic; (3) *course*, for chronic conditions, whether the level of severity is stable or changing; and (4) *use of assistance* in terms of persons or mechanical aids (e.g., walkers).

Fried and colleagues (1994) have introduced another dimension that they label *subclinical* disability. In their samples, a large proportion of elderly adults did not report disability (defined as difficulty in performing ADL, IADL, or mobility activities) but compensated their behavior from prior practice in order to accomplish these activities. Compensatory behavior usually reflected a change in either the frequency or method of doing an activity. For example, a woman might not report difficulty in grocery shopping but, upon further questioning, report going to the store less frequently than usual or leaning on the grocery cart for support. In some specific activities, as many as 30% of the group who reported no difficulty met these criteria for subclinical disability.

Another detectable difference is between *perceived* and *behavioral* disability. These terms differentiate a person who perceives him- or herself as being able to perform an activity (i.e., *can* do) from whether the person actually performs those activities (i.e., *does* do). The reasons for differences between perceived and behavioral disability may be informative. In our preliminary data on homecare nursing patients, a number of respondents reported that they do not do activities that they believe they are capable of doing because it takes too much effort, the tasks seemed too difficult, they feet embarrassed, or just do not feel like it—all reasons that might be affected by depressed mood or other symptoms of depression.

In some cases, the difference between what a person does do and what a person can do is a result of social roles (e.g., a married man may never prepare his own meals, or a married woman may never do the bills) or the absence of an environmental challenge (e.g., the apartment building has an elevator, or the bedroom is on the main floor of the house), leading to the question of whether or not a person *has* to do an activity. One of the defining differences between ADL and IADL activities is that ADLs are considered necessary activities that all people would do for themselves if they could. Accordingly, one possible reason for the finding discussed earlier (Steffens et al., 1999) that severity of depression is associated with IADL disability, but not ADL disability, in cognitively intact, depressed patients may be that depression undermines the effort needed to do only those activities that are judged discretionary, such as shopping or cleaning, in comparison to ADL activities, such as eating or toileting.

Part of the challenge for researchers interested in measuring disability is determining the extent to which social roles, environmental options, or decisions to forego an activity are actually independent of disability. People purposely move to single-floor homes in order to avoid stairs or other environmental challenges. Moreover, several of our older homecare patients report not doing housework or budgets (e.g., "my daughter took over that task several years ago), suggesting that the reason the social roles changed was that these individuals were no longer able or willing to do the task.

Longitudinal Modeling

Taken together, the longitudinal findings suggest a spiral overtime, whereby depression increases the risk of disability, disability increases the risk of more depression, and each condition continues to reinforce the other. A spiral model would suggest that once a person enters into the spiral, either through depression or disability, a trajectory of decline is somewhat inevitable. Whether this model accurately reflects the depression–disability process has not been adequately tested, especially with community-based data. Nor is this model compatible with the simple observation that not all people who get depressed also become disabled, and that not all people who have disabilities become depressed, suggesting that the spiral model may at best only partially fit the depression–disability process or, indeed, may fit well for only specific subgroups of people.

Needed to test the spiral hypothesis is concurrent analysis of multiple and, ideally, frequent assessments of both disability and depression over time. One of the better examples of this kind of analysis is an early study conducted by Aneshensel and colleagues (1984). Their data came from a population-based sample of adults with baseline plus three sets of follow-up interviews (4-month intervals). The strength of this study is the use of structural equation (LISREL) modeling of several assessments conducted at short time intervals. The findings have limited usefulness for our current purpose, since the study did not directly address the question of disability but, rather, defined a general illness construct by combining reported bed-days, activity-restricted days, and a count of health conditions. In the full sample, as well in the older ages (≥ age 45), illness was related to an immediate increase in depressive symptoms, while depression was related to an increase in subsequent illness.

Because of the low prevalence of more specific measures of disability (i.e., ADL disability) and diagnostic measures of depression, the feasibility of testing the spiral hypothesis with this kind of analysis in population-based samples is uncertain. Analyses of other types of samples are useful, as noted earlier, when the prevalence of these conditions is high; findings from such samples, however, lend only some support to the spiral hypothesis. For example, von Korff, Ormel, Katon, and Lin (1992) analyzed the course of depressive symptoms and disability in a sample of depressed (moderate-to-severe symptoms) primary care patients, ages 18–75, who used an excessive number of health services. Disability was determined by two measures: bed-days and a 5-item score (ranging from ADL-related disabilities to limitation in social activities or hobbies). Their data suggest a synchrony between severity of depressive symptoms and disability such that increases or decreases in depression are mirrored by similar changes in disability, also suggesting that repeated measures are needed at very short intervals to identify the underlying causal relationship between depression and disability.

Oxman and Hull (1997) also conducted a structural equation analysis of the depression–disability relationship, using data from heart surgery patients. In this sample, the relationships between depression and disability changed over time, so that ADL impairment at 1-month postsurgery increased the risk of depression at 6 months, but postsurgery depression did not affect ADL levels at 6 months, suggesting that in these patients, any effect of depressive symptoms or functional status is transient and weak, especially in comparison to the profound changes in medical burden they are experiencing.

Mediating Variables

More frequent and detailed assessments would also make it possible to identify factors that link changes between depression and disability. Several of the hypothesized explanations for the relationship between disability and depression implicate such a mediating variable. These hypothesized variables could be measured and their mediating relationship explicitly tested. Some of the proposed mediators are relatively easy to measure, such as self-neglect, inactivity, poor health habits (e.g., smoking, malnutrition), and underuse of preventive and early intervention screening services. Each of these intermediary outcomes could be measured and its link to depression and disability estimated over time. Other hypothesized mediating factors, such as a deterioration in a person's ability to cope or other vulnerability to life events, are more difficult, but not impossible, to measure.

As measures of depression and disability become more specific, so can the test of intermediary factors, allowing for the investigation of complex models. Rigorous testing of these mediators requires longitudinal data and statistical modeling discussed earlier.

Modifying Variables

Finally, one way to identify the key factors of a system is to disrupt it. This strategy can take the form of either identifying modifying variables (e.g., interaction terms) in observational data (see Williamson & Shaffer, Chapter 9, this volume, for examples) or conducting intervention studies designed to change the natural course of depression and disability. The point here is that intervention studies are not only useful for identifying how to treat a problem but also for learning about the underlying mechanisms that drive the problem. A variety of approaches could be useful, including intervention studies that solely target treatment of depression, those focused on reduction of disability, and those specially designed interventions for people with both depression and disability.

The interventions themselves may involve traditional treatments, so that changes in disability might be observed in depressed patients receiving antidepressant treatment, or changes in depression might be observed in persons with disabilities undergoing physical rehabilitation. This approach has been seen more frequently in recent years with a growing number of clinical trials for depression, including functional outcomes (Borson et al., 1992; Heiligenstin et al., 1995; Mazumdar et al., 1996; Small et al., 1996). For example, a placebo controlled study demonstrated an association between 6 weeks of fluoxetine treatment and improvement on scales measuring physical functioning (Heligenstein et al., 1995). This approach is also feasible in effectiveness studies. One study randomized discharged elderly, acute, hospitalized patients with moderate, but not severe, depressive symptoms to usual care versus interpersonal counseling; in this case, the intervention affected depressive symptoms and self-rated health but not physical or social functioning (Mossey, Knott, Higgins, & Talerico, 1996). Comparisons of findings among different intervention studies need to account for the same kinds of factors relevant to observational studies such as the source of subjects, the severity of clinical status, and the types of measures, as well as the nature and duration of the intervention, in order to ensure that observed differences in outcomes are a function of the intervention rather than differences in the samples or methodologies.

Conversely, studies might impact a therapy traditionally used in one domain for use in the other to test the effect of physical rehabilitation on changing the course of depression in a sample of depressed persons, or to test the effect of depression therapies in improving the function of persons with disabilities. One example of this kind of strategy is a study conducted by Singh and colleagues (1998), who randomized a community sample of elderly adults meeting DSM-IV criteria for major depression ($N = 20$). The intervention, a 10-week exercise training program, was compared to a health education control condition. The intervention significantly improved depressive status (as measured using several symptom scales). Although the intervention was associated with improvement in a number of other domains (e.g., vitality, bodily pain, role function, emotional function, and social function), no effect was observed for ADL or IADL disability. The authors interpret these findings as evidence of a biological effect of depression rather than a purely cognitive-behavioral one. The data might also suggest that these biological as well as cognitive changes may in time affect physical disability, but that the process is slower than could be observed in 10 weeks.

Finally and conversely, controlled treatment trials offer a way of obtaining data on the relationship between depression and disability in which the potentially confounding effect of treatment is minimized and/or carefully described. The frequent assessment periods often characterizing such trials are also useful for charting change in depression and disability relative to each other in time.

To summarize this section, all the "future directions" offered in this section share a common need to increase the specificity of our investigations of the relationships between depression and disability in terms of both measurement and analysis. As such detail already exists within the fields of both depression and disability research, this goal may well best be accomplished by an interdisciplinary approach that draws on the methodological and conceptual strengths of both areas of research.

CONCLUDING COMMENTS

In some lights, the connection between depression and disability seems almost commonsensical and obvious. But empirical data on the nature of this relationship suggest that the links between the two are surprisingly subtle and complex. This chapter has recommended a number of analytic strategies that could be useful to clarifying these relationships. Given the heterogeneity of both depression and disability, it is likely that simple solutions will be insufficient. In a very real sense, this complexity offers the potential that not one but a variety of strategies may be useful in reducing the burden of depression and disability in the lives of older adults.

ACKNOWLEDGMENTS: Supported by NIMH Grants Nos. K02 MH01634 and R01 MH56482.

REFERENCES

Alexopoulos, G. A., Bruce, M. I., Kalayum, B., Silbersweig, D., & Stern E. (1999). Vascular depression: A new view of late life depression. *Dialogues in Clinical Neuroscience, 1,* 68–80.

Alexopoulos, G. A., Vrontou, C., Kakuma, T., Meyers, B., Young, R., Klausner, E., & Clarkin, J. F. (1996). Disability in geriatric depression. *American Journal of Psychiatry, 153,* 877–885.

American Psychiatric Association. (1994). *Diagnostic and statistical manual of mental disorders* (4th Ed.). Washington, DC: Author.

Aneshensel, C. S., Frerichs, R. R., & Huba, G. S. (1984). Depression and physical illness: A multiwave nonrecursive model. *Journal of Health and Social Behavior, 25,* 350–371.

Armenian, H. K., Pratt, L. A., Gallo, J., & Eaton, W. W. (1998). Psychopathology as a predictor of disability: A population-based follow-up study in Baltimore, Maryland. *American Journal of Epidemiology, 148,* 269–275.

Beck, A. T., Guth, D., Steer, R. A., & Ball, R. (1997). Screening for major depression disorders in medical inpatients with the Beck Depression Inventory for Primary Care. *Behavioral Research Therapy,* 785–791.

Beekman, A. T. F., Deeg, D. J. H., Braam, A. W., Smit, J. H., & Van Tilburg, W. (1997). Consequences of major and minor depression in later life: A study of disability, well-being, and service utilization. *Psychological Medicine, 27,* 1397–1409.

Berkman, L. F., Berkman, C. S., Kasl, S., Freeman, D. H., Jr., Leo, L., Ostfeld, A. M., Cornoi-Huntley, J., & Brody, J. A. (1986). Depressive symptoms in relation to physical health and functioning in the elderly. *American Journal of Epidemiology, 124,* 372–388.

Borson, S., McDonald, G., Gayle, T., Deffeback, M., Lakshminarayan, S., & VanTuinen, C. (1992). Improvement in mood, physical symptoms, and function with nortriptyline for depression in patients with chronic obstructive pulmonary disease. *Psychosomatics, 33,* 190–201.

Boyd, J. H., Weissman, M. M., Thompson, W. D., & Myers, J. K. (1982). Screening for depression in a community sample. *Archives of General Psychiatry, 39,* 1195–1200.

Brown, R. G., MacCarthy, B., Gotham, A. M., Der, G. J., & Marsden, C.D. (1988). Depression and disability in Parkinson's disease: A follow-up of 132 cases. *Psychological Medicine, 18,* 49–55.

Bruce, M. L. (1999). Depression and disability. *American Journal of Geriatric Psychiatry, 7,* 1–4.

Bruce, M. L., & Hoff, R. A. (1994). Social and physical health risk factors for first onset major depressive disorder in a community sample. *Social Psychiatry and Psychiatric Epidemiology, 29,* 165–171.

Bruce, M. L., Meyers, B. S., Alexopoulos, G. S., Raue, P. J., Brown, E. J., & McAvay, G. (1998, July). *Prevalence and short-term course of major depression in elderly homecare nursing patients.* Paper presented at the NIMH International Conference on Mental Health Problems in the General Health Care Sector. Baltimore, MD.

Bruce, M. L., Seeman, T. E., Merrill, S. S., & Blazer, D. G. (1994). The impact of depressive symptomatology on physical disability: MacArthur studies of successful aging. *American Journal of Public Health, 84,* 1796–1799.

Clark, D .C., Cavanaugh, S., & Gibbons, R.D. (1983). The core symptoms of depression in medical and psychiatric patients. *Journal of Nervous and Mental Disease, 171,* 705–713.

Cohen-Cole, S. A., & Stoudemire, A. (1987). Major depression and physical illness: Special considerations in diagnosis and biological treatment. *Psychiatric Clinics of North America, 10,* 1–17.

Cole, S. A., Woodard, J. L., Juncos, J. L., Kogos, J. L., Youngstrom, E. A., & Watts, R. L. (1996). Depression and disability in Parkinson's disease. *Journal of Neuropsychiatry and Clinical Neurosciences, 8,* 20–5.

Craig, T. J., & Van Natta, P. A. (1983). Disability and depressive symptoms in two communities. *American Journal of Psychiatry, 140,* 598–601.

Derogatis, L. R., Lopman, R. S., Rickels, K., Uhlenhuth, E. H., & Covi, L. (1974). The Hopkins Symptom Checklist (HSCL): A measure of primary symptom dimensions. In P. Pinchot (Ed.), *Psychological Measurements in Psychopharmacology: Modern Problems in Pharmacopsychiatry, 7,* 79–110.

Dohrenwend, B. S., Krasnoff, L., Askensay, A. R., & Dohrenwend, B. P. (1978). Exemplification of a method for scaling life events: The PERI life events scale. *Journal of Health and Social Behavior, 19,* 205–229.

Foley, D. J., Ostfeld, A. M., Branch, L. G., Wallace, R. B., McGloin, J. M., & Cornoi-Huntley, J. C. (1992). The risk of nursing home admission in three communities. *Journal of Aging and Health, 4,* 155–173.

Forsell, Y., Jorm, A. F., & Winblad, B. (1994). Association of age, sex, cognitive dysfunction, and disability with major depressive symptoms in an elderly sample. *American Journal of Psychiatry, 151,* 1600–1604.

Fried, L. P., Ettinger, W. H., Lind, B., Newman, A. B., & Gardin, J. (1994). Physical disability in older adults: A physiological approach. *Journal of Clinical Epidemiology, 47,* 747–760.

Fried, L. P., & Guralnik, J. M. (1997). Disability in older adults: Evidence regarding significance, etiology, and risk. *Journal of the American Geriatric Society, 45,* 92–100.

Guayatt, G., Walter, S., & Norman, G. (1987). Measuring change over time: Assessing the usefulness of evalutative instruments. *Journal of Chronic Disease, 40,* 171–178.

Gurland, B. J., Wilder, D. E., & Berkman, C.(1988). Depression and disability in the elderly: Reciprocal relations and changes with age. *International Journal of Geriatric Psychiatry, 3,* 163–179.

Heiligenstin, J. H., Ware, J. E., Beusterien, K. M., Roback, P. J., Andrejasich, C., & Tollefson, G. D. (1995). Acute effects of fluoxetine versus placebo on functional health and well-being in late-life depression. *International Psychogeriatrics, 7,* 125–137.

Henderson, A. S., Jorm, A., F., MacKinnon, A., Christensen, H., Scott, L.R., Korten, A.E., & Doyle, C. (1993). The prevalence of depressive disorders and the distribution of depressive symptoms in later life: A survey using Draft ICD-10 and DSM-III-R. *Psychological Medicine, 23,* 719–729.

Henderson, A. S., Korten, A. E., Jacomb, P. A., MacKinnon, A. J., Jorm, A. F., Christensen, H., & Rodgers, B. (1997). The course of depression in the elderly: A longitudinal community-based study in Australia. *Psychological Medicine, 27,* 119–129.

Holmes, T. H., & Rahe, R. H. (1976). The social readjustment rating scale. *Journal of Psychosomatic Research, 11,* 213–218.

Jaffe, A., Froom, J., & Galambos, N. (1994). Minor depression and functional impairment. *Archives of Family Medicine, 3,* 1081–1086.

Johnson, R. J., & Wolinsky, F. D. (1993). The structure of health status among older adults: Disease, disability, functional limitation, and perceived health. *Journal of Health and Social Behavior, 34,* 105–121.

Judge, J. O., Schechtman, K., Cress, E., & the FICSIT Group. (1996). The relationship between physical performance measures and independence in instrumental activities of daily living. *Journal of the American Geriatric Society, 44,* 1332–1341.

Katz, I. R. (1996). On the inseparability of mental and physical health in aged persons: Lessons from depression and medical comorbidity. *American Journal of Geriatric Psychiatry, 4,* 1–6.

Kennedy, G. L., Kelman, H. R., & Thomas, C. (1990). The emergence of depressive symptoms in late life: The importance of declining health and increasing disability. *Journal of Community Health, 5,* 93–104.

Kennedy, G. J., Kelman, H. R., Thomas, C., Wisniewski, W., Metz, H., & Bijur, P. E. (1989). Hierarchy of characteristics associated with depressive symptoms in an urban elderly sample. *American Journal of Psychiatry, 146,* 220–225.

Koenig, H. G., Cohen, H. J., Blazer, D. G., Krishman, K. R., & Silbert, T. E. (1993). Profile of depressive symptoms in younger and older medical patients with major depression. *Journal of the American Geriatric Society, 41,* 1169–1176.

Koenig, H. G., George, L. K., Peterson, B. L., & Pieper, C. F. (1997). Depression in medically ill hospitalized older adults: Prevalence, characteristics, and course of symptoms according to six diagnostic schemes. *American Journal of Psychiatry, 154,* 1376–1383.

Kuriansky, J. B., Gurland, B. J., & Fleiss, J. L. (1976). The assessment of self-care in geriatric psychiatric patients by objective and subjective methods. *Journal of Clinical Psychology, 32,* 95–102.

Leaf, P. J., Bruce, M.L., Tischler, G. L., Freeman, D. H., Weissman, M. M., & Myers, J. K. (1988). Factors affecting the utilization of specialty and general medical mental health services. *Medical Care, 26,* 9–26.

Little, A. G., Hemsley, D. R., Volanas, P. J., & Bergmann, K. (1986). The relationship between alternative assessments of self-care ability in the elderly. *British Journal of Clinical Psychology, 25,* 51–59.

Lyness, J. M., Bruce, M. L., Koenig, H. C., Parmelee, P. A., Schulz, R., Lawton, M. P., & Reynolds, C. F. (1996). Depression and medical illness in late life: Report of a symposium. *Journal of the American Geriatric Society, 44,* 198–203.

Lyness, J. M., Caine, E. D., Conwell, Y., King, D. A., & Cox, C. (1993). Depressive symptoms, medical illness, and functional status in depressed psychiatric patients. *American Journal of Psychiatry, 150,* 910–915.

Lyness, J. M., Pearson, J. L., Lebowitz, B. D., & Kupfer, D. J. (1994). Age at onset of late-life depression: A research agenda report of a MacArthur Foundation–NIMH workshop. *American Journal of Geriatric Psychiatry, 2,* 4–8.

Mazumdar, S., Reynolds, C. F., Houck, P. R., Frank, E., Dew, M. A., & Kupfer, D. J. (1996). Quality of life in elderly patients with recurrent major depression: Factor analysis of the General Life Functioning Scale. *Psychiatry Research, 63,* 183–190.

McFarlane, A. C., & Brooks, P. M. (1988). Determinants of disability in rheumatoid arthritis. *British Journal of Rheumatology, 27,* 7–14.

McLean, D. E., & Link, B. G. (1994). Unraveling complexity: Strategies to refine concepts, measures, and research designs in the study of life events and mental health. In W.R. Avison & I.H. Gotlib (Eds.), *Stress and mental health* (pp. 15-42). New York: Plenum Press.

Morgado, A., Smith, M., Lecrubier, Y., & Widlocher, D. (1991). Depressed subjects unwittingly overreport poor social adjustment which they reappraise when recovered. *Journal of Nervous and Mental Disease, 179,* 614–619.

Mossey, J. M., Knott, K. A., Higgins, M., &Talerico, K. (1996). Effectiveness of a psychosocial intervention, interpersonal counseling, for subdysthymic depression in medically ill elderly. *Journals of Gerontology, 51,* 172–178.

Nagi, S. (1976). An epidemiology of disability among adults in the United States. *Milbank Memorial Fund Quarterly, 54,* 439–468.

NIH Consensus Development Panel on Depression in Late Life. (1992). Diagnosis and treatment of depression in late life. *Journal of the American Medical Association, 268,* 1018–1024.

Ormel, J., Kempen, G. J. M., Brilman, E., Beekman, A. T. F., & Van Sonderen, E. (1997). Chronic medical conditions and mental health in older people: Disability and psychosocial resources mediate specific mental health effects. *Psychological Medicine, 27,* 1065–1077.

Oxman, T. E., & Hull, J. G. (1997). Social support, depression, and activities of daily living in older heart surgery patients. *Journals of Gerontology, 52,* 1–15.

Penninx, B. W., Guralnick, J. M., Ferrucci, L., Simonsick, E. M., Dee, D. J. H., & Wallace, R. C. (1998). Depressive symptoms and physical decline in community-dwelling older persons. *Journal of the American Medical Association, 279,* 1720–1726.

Podsiadlo, D., & Richardson, S. (1991). The timed "Up and Go": A test of basic functional mobility for frail elderly persons. *Journal of the American Geriatric Society, 39,* 142–148.

Pope, A.M., & Tarlov, A.R. (Eds). (1991). *Disability in America: Toward a national agenda for prevention.* Washington, DC: National Academy Press.

Prince, M. J., Harwood, R. H., Blizard, R. A., Thomas, A., & Mann, A, H. (1997). Impairment, disability and handicap as risk factors for depression in old age: The Gospel Oak Project V. *Psychological Medicine, 27,* 311–321.

Radloff, L. (1977). The CES-D Scale: A self-report depression scale for research in the general population. *Applied Psychological Measurement, 1,* 385–397.

Reynolds, C. F., Lebowitz, B. D., & Kupfer, D. J. (1993). Recommendations for scientific reports on depression in late life. *American Journal of Geriatric Psychiatry, 1,* 231–233.

Schulz, R., & Decker, S. (1985). Long-term adjustment to physical disability: The role of social support, perceived control, and self-blame. *Journal of Personality and Social Psychology, 48,* 1162–1172.

Schulz, R., & Williamson, G. M. (1993). Psychosocial and behavioral dimensions of physical frailty. *Journals of Gerontology, 48,* 39–43.

Singh, N. A., Clements, K. M., & Fiatarone, M. A. (1997). A randomized controlled trial of progressive resistance training in depressed elders. *Journals of Gerontology. Series A, Biological Sciences and Medical Sciences, 52*(1), 27–35.

Small, G. W., Birkett, M., Meyers, B. S., Koran, L., Bystritsky, S., Nemeroff, C.B., & the Fluoxetine Collaborative Study Group. (1996). Impact of physical illness and quality of life and antidepressant response in geriatric major depression. *Journal of the American Geriatric Society, 44,* 1220–1225.

Spector, W. D. (1996). Functional disability scales. In B. Spiker (Ed.), *Quality of life and pharmacoeconomics in clinical trials* (pp. 133–143). Philadelphia: Lippincott–Raven.

Spitzer, R. L., & Endicott, J. (1978). *Schedule for Affective Disorders and Schizophrenia.* New York: Biometrics Research Division, New York State Psychiatric Institute.

Spitzer, R. L., Gibbon, M., & Williams, J. B. (1995). *Structured Clinical Interview for Axis I DSM-IV Disorders (SCID).* Washington, DC: American Psychiatric Association.

Steffens, D. C., Hays, J. C., & Krishman, K. R. (1999). Disability in geriatric depression. *American Journal of Geriatric Psychiatry, 7,* 34–40.

Stewart, A. L., Hays, R. D., & Ware, J. E. (1988). The MOS short-form General Health Survey. *Medical Care, 26,* 724–745.

Strawbridge, W. J., Shema, S. J., Balfour, J. L., Higby, H. R., & Kaplan, G. A. (1998). Antecedents of frailty over three decades in an older cohort. *Journals of Gerontology, 53,* 9–16.

Turner, R. J., & Beiser, M. (1990). Major depression and depressive symptomatology among the physically disabled: Assessing the role of chronic stress. *Journal of Nervous and Mental Disease, 178,* 343–350.

Turner, R., & Noh, S. (1988). Physical disability and depression: A longitudinal analysis. *Journal of Health and Social Behavior, 29,* 23–37.

Tweed, D. L., Blazer, D. G., & Ciarlo, J. A. (1992). Psychiatric epidemiology in elderly populations. In R. B. Wallace & R. F. Woolson (Eds.), *The epidemiologic study of the elderly* (pp. 213–233). New York: Oxford University Press.

Von Korff, M., Ormel, J., Katon, W., & Lin, E. H. B. (1992). Disability and depression among high utilizers of health care. *Archives of General Psychiatry, 49,* 91–100.

Wells, K. B., Burnam, M. A., Rogers, W., Hays, R., & Camp, P. (1992). The course of depression in adult outpatients: Results from the Medical Outcomes Study. *Archives of General Psychiatry, 49,* 788–794.

Wells, K. B., Stewart, A., Hays, R. D., Burnam, A., Rogers, W., Daniels, M., Berry, M. S., Greenfield, S., & Ware, J. (1989). The functioning and well-being of depressed patients. *Journal of the American Medical Association, 262,* 914–919.

Williams, J. W., Kerber, C. A., Mulrow, C. D., Medina, A., & Aguilar, C. (1995). Depressive disorders in primary care: Prevalence, functional disability and identification. *Journal of General Internal Medicine, 10,* 7–12.

World Health Organization. (1980). *International classification of impairments, disabilities, and handicaps.* Geneva, Switzerland: Author.

World Health Organization. (1992). *International statistical classification of diseases and other health problems.* Geneva, Switzerland: Author.

Yesavage, J. A., Brink, T. L., Rose, T. L., & Lum, O. (1983). Development and validation of a geriatric depression screening scale: A preliminary report. *Journal of Psychiatric Research, 17,* 37–49.

3

Vascular Disease and Depression

Models of the Interplay between Psychopathology and Medical Comorbidity

JEFFREY M. LYNESS and ERIC D. CAINE

INTRODUCTION

The Syndromic Approach to Psychopathology

Modern psychopathology has for the most part taken a syndromic approach to conceptualizing psychiatric conditions (Caine & Lyness, 2000). That is, specific disorders are defined by clusters of symptoms and signs. Such an approach offers the advantage of heightened diagnostic interrater reliability. Putatively, it also offers the opportunity to achieve better diagnostic validity by operationally defining clinical states that are more biologically homogeneous, thereby allowing us to elucidate the underlying pathogenesis(es). Yet considerable evidence suggests that syndromically defined entities are often *not* homogeneous; rather, biological heterogeneity is the rule rather than the exception (Caine & Joynt, 1986). Indeed, a single syndrome may have many possible causes, while a single etiology may manifest with myriad psychiatric symptoms or syndromes.

Partly because of such heterogeneity, correlative studies have largely failed to yield useful insights into the relevant mechanisms of psychopathology. Potentially more productive are theory-driven approaches. A conceptual model of disease process provides a framework within which empirical evidence (which may include results from correlative as well as interventional investigations) can be organized into that which does or does not provide support for its utility. Ultimately, accumulated empirical data may lead to acceptance, rejection, or modification of a model.

JEFFREY M. LYNESS and ERIC D. CAINE • Department of Psychiatry, University of Rochester Medical Center, Rochester, New York 14642.

Physical Illness and Depression in Older Adults: A Handbook of Theory, Research, and Practice, edited by Gail M. Williamson, David R. Shaffer, and Patricia A. Parmelee. Kluwer Academic/Plenum Publishers, New York, 2000.

Informed by these issues, we consider approaches to the study of cerebrovascular risk factors and depression in later life. Two theoretical models will be discussed and evaluated in light of available empirical support. What follows serves, we hope, as both a review of specific content areas and an illustrative paradigm for critically assessing biopsychosocial explanations of the interplay of physical illnesses and depression in older adults.

Medical Illness as a Risk Factor for Depression

The public health importance of depressive conditions in later life is enormous (National Institutions of Mental Health [NIH] Consensus Development Panel on Depression in Late Life, 1992), as the reader of this volume will have long since taken as an article of empirically supported faith. Yet despite continued intensive investigation into the clinical phenomenology, neurobiology, course, and treatment response of later-life depression, its etiologies remain elusive (Caine, Lyness, & King, 1993).

Depression in younger adults is associated with genetic factors, developmental factors including stressful early life events, limited cognitive problem-solving styles, and other aspects of an individual that may be subsumed under concepts of "personality" (only a small part of which is captured by diagnostic entities such as personality disorders). By contrast, acquired factors play a much greater role in mood pathology in older adults (Caine et al., 1993). Consistent with the theme of this volume, medical illnesses are the most consistently identified concomitants of major depression in older persons and the most powerful predictors of the course and outcome of depressive conditions (Caine et al., 1993; Katz, 1996). Risk factors by implication are related to mechanisms of disease (Evans, 1978). Therefore, study of the relationships between medical illnesses and depression may yield insights into the etiology(ies) or pathogenesis(es) of depression. Medical illnesses may contribute to the onset, recurrence, or course of depression via biological, psychological, or psychosocial routes.

"Secondariness"

Medical illnesses can directly alter brain physiology in such a fashion as to produce depressive symptoms; when this link is recognized clinically, the resultant depressive condition is known as "secondary" in DSM-IV parlance ("organic mood disorder" in previous DSM editions and during decades of clinical usage). We have noted elsewhere (Lyness & Caine, 1997) the principles by which one might establish "secondariness" (i.e., whether the disease in question does in fact contribute *pathobiologically*). The most obvious suspects for such a causal relationship are neurological diseases, wherein the putative etiology of the depression is a condition that itself arises in the brain. Typically, these disorders produce structural alterations that affect the function of regions thought to mediate the expression of psychopathology. Indeed, there is considerable evidence that specific neurological disorders can cause depression, examples including stroke (Robinson, 1997), multiple sclerosis (Meyerson, Richard, & Schiffer, 1997), and neurodegenerative conditions such as Huntington's disease, Parkinson's disease, or Alzheimer's disease (Meyerson et al., 1997; Porsteinsson, Tariot, & Schneider, 1997).

Systemic conditions also may affect brain physiology and lead to a secondary depression (Popkin & Andrews, 1997), presumably through neurotoxic, metabolic, or hormonal effects that ultimately lead to altered brain function. Although definitive empirical support for secondariness is lacking surprisingly often, such conditions almost certainly include drug-induced states, both from intoxication and from withdrawal (Smith & Atkinson, 1997), endocrinopathies such as Cushing's syndrome or hypothyroidism, and end-stage renal or hepatic disease (Popkin & Andrews, 1997).

Of course, the distinction between "neurological" and "systemic" conditions often is arbitrary. For example, abnormal findings on neuroimaging scans of depressed patients (to be reviewed later) suggest the possibility that later-life depression might be due to cerebrovascular disease. Yet cerebrovascular pathology itself may be related to the expression or consequences of systemic conditions. We now turn our attention to such systemic disorders, known as cerebrovascular risk factors, followed by consideration of two models by which these conditions might influence brain function and contribute to depression.

CEREBROVASCULAR RISK FACTORS AND DEPRESSION

Definition of Cerebrovascular Risk Factors

Cerebrovascular risk factors (CVRFs) are a group of systemic conditions that are risk factors for stroke, a manifestation of large vessel brain disease. The CVRFs include coronary artery disease, diabetes mellitus, hypertension, cigarette smoking, atrial fibrillation, and left ventricular hypertrophy (American Heart Association, 1990; Dyken et al., 1985). CVRFs' relationships to depression in later life warrant careful consideration for several reasons. They are common conditions that, except for smoking, have greater prevalence with older age (Moritz & Ostfeld, 1990). Cardiovascular disease (defined for these purposes as history of myocardial infarction, angina pectoris, coronary insufficiency, intermittent claudication, or congestive heart failure) and diabetes mellitus are among the leading causes of death in the elderly, while the other CVRFs also are strongly associated with mortality (National Center for Health Statistics, 1992). As well, CVRFs are associated with considerable morbidity, including decline in functional status (Bush, Miller, Criqui, & Barret-Connor, 1990).

Association with Depression

While conceptually appealing, published clinical studies leave one uncertain about the association of CVRFs with depression. Baldwin and Tomenson (1995) noted an association of CVRFs with late-onset depression among a group of older persons with major depression. Wells, Rogers, Burnam, and Camp (1993) reported a study of outpatients from general medical and mental health practices in which myocardial infarction was associated with poorer 2-year outcome of depression. As reviewed by Hayward (1995), several older studies found associations between depression and cardiovascular disease, hypertension, or hypercholesterolemia. However, interpretation of most of these studies must be tempered by methodological issues, including retrospective design, samples from long-term psychiatric institutions, or depression assessments using solely self-report scales or nonstandardized diagnostic schemes.

Hayward concluded, "studies evaluating the risk of cardiovascular disease in the psychiatrically ill yield mixed results" (p. 135).

A number of studies have failed to find a positive association between CVRFs and depression. Schleifer and colleagues (1989) did not find an association of cardiac illness severity with depression diagnosis in cardiac care unit inpatients suffering acute myocardial infarction. Nor did Greenwald et al. (1996) detect significant differences in the prevalences of hypertension, coronary artery disease, diabetes, or smoking in older depressed patients as compared with normal control subjects. Kumar et al. (1997) similarly failed to find a significant difference between cumulative CVRF severity in their depressed subjects and normal controls. Our group (Lyness et al., 1998a) studied older inpatients with major depression, reporting that cumulative CVRF severity was not significantly associated with depressive severity, nor was it greater in these depressives (or in a later-onset depression subgroup) as compared with normal controls. In our subjects, certain individual CVRFs (cardiovascular disease, diabetes, and atrial fibrillation) were more prevalent among the depressives than the controls, but others (antihypertensive treatment, systolic blood pressure, cigarette smoking, and left ventricular hypertrophy) were not. Only diabetes and atrial fibrillation were more prevalent after controlling for age, gender, and education. We extended this work by conducting similar analyses among a group of older patients recruited from primary care practices (Lyness et al., 1998b). After controlling for demographic variables, neither cumulative CVRF severity nor individual CVRFs were significantly associated with depression diagnosis or depressive symptom severity.

Two studies have noted an association of diabetes mellitus with depression (Littlefield, Rodin, Murray, & Craven, 1990; Lustman, Griffith, Gavard, & Clouse, 1992) but did not focus specifically on older patients. Similarly, there is considerable evidence for an association between cigarette smoking and major depression in psychiatric populations (Kendler et al., 1993), but data are scarce regarding both nonpsychiatric subject groups and the elderly (cf. the CVRF studies summarized earlier).

Causal Routes

It is important to recognize that even if a clear association between CVRFs and depression exists, such an association may reflect any of several mechanisms. First, there may be no direct etiological relationship. Any apparent association between CVRFs and depression may be due to each having an independent relationship with an unknown additional variable.

Another possibility is that CVRFs contribute to depression by biological effects that lead to brain dysfunction. We later consider two models of this type. The first one involves the disruption of functional neuroanatomy by small vessel ischemic brain disease; the second entails cytokine-mediated effects on the brain by the inflammatory processes inherent in systemic atherosclerosis.

A third set of possibilities for apparent causal associations must be recognized as well. Although this chapter focuses on pathobiological mechanisms, we cannot emphasize strongly enough that CVRFs may contribute to depression by psychological or psychosocial routes. One example involves the role of functional disability. Disability is a complex, multidetermined concept of fundamental import to clinicians working with older adults (Williams, 1990; Guralnik & Simonsick, 1993). Interpretation of reports of disability as a mediator of medical illnesses' effects on depression must be tempered by recognition of its "confound" with depression, that is, depression itself may affect the reporting of disability (Sinclair et al., submitted). Despite this, there is substantial evidence that disability is associated with depression (Broadhead, Blazer, George, & Tse, 1990; Kennedy, Kelman, & Thomas, 1990; Wells et al., 1989) as well as with CVRFs (Bush et al., 1990). Psychological models can be invoked to explain how external disability may lead to the changes in mental experiences that are part of depression. For example, disability may be viewed as representing a breakdown in primary control processes (Schulz, Heckhausen, & O'Brien, 1994). Control theory has the added advantage of empirical support for its relevance to later life and to negative affect and cognitive style (Heckhausen & Schulz, 1995). Further evidence for the usefulness of disability models of depression comes from the work of Williamson and Schulz (e.g., 1992, 1995), who found that activity restriction mediated the relationship between pain and depressive symptoms in outpatients suffering a range of medical illnesses. Other groups have not confirmed a mediating role for disability between pain and depression in nursing home residents (Parmelee, Katz, & Lawton, 1991), but interestingly, Parmelee, Katz, and Lawton (1992) found that disability (along with other factors) mediated depression's association with mortality. Few studies have examined the relationships among disability, depression, and CVRFs. Schleifer et al. (1989) noted that depression comorbid with poorer medical course predicted work disability in subjects with myocardial infarction. Lyness et al. (1998b) found that disability was independently associated with depressive symptoms and syndromes in primary care elderly, while CVRFs were not. The need is clear for studies assessing the extent to which disability mediates CVRFs' effects on depression (Williamson & Shaffer, Chapter 9, this volume) and for comparisons of the magnitude of this effect with that of other mediators.

Personality is an additional "nonneurological" route meriting attention. Several investigations have reported personality abnormalities in older patients with major depression (Abrams, Rosendahl, Card, & Alexopoulos, 1994; Schneider, Zemansky, Bender, & Sloane, 1992). In younger subjects, personality trait neuroticism was associated with poor depressive outcome (for a review, see Duberstein, Seidlitz, Lyness, & Conwell, 1999). Since neuroticism is broadly definable as emotional vulnerability to stress, it is plausible that it can moderate the relationship between CVRFs and depression. This is supported by Costa and McCrae's work (1987) demonstrating that neuroticism is associated with somatic symptoms. As well, our group (Lyness, Duberstein, King, Cox, & Caine, 1998c) found partial support for an interaction between neuroticism and overall organ system burden as predictors of depression in a cross-sectional study of primary care elderly. In a related study, we found that neuroticism was an independent predictor of depression while CVRFs were not (Lyness et al., 1998c).

Pain is a symptom with neurological, psychological, and psychosocial dimensions that may influence the relationship between CVRFs and depression. Pain is a frequent concomitant of CVRFs such as angina pectoris, intermittent claudication, and diabetic neuropathy. It has long been recognized that depressive symptoms can be a consequence (some would view them simply as a manifestation) of chronic pain; conversely, pain complaints may intensify during periods of increased depression. Pain was associated with depressive symptoms, independent of organ system burden, in long-term care (Cohen-Mansfield & Marx, 1993; Parmelee et al., 1991) and community-residing older outpatient populations (Williamson & Schulz, 1992, 1995). As noted earlier, these studies differed as to the mediating role of functional disability. These relationships have not been studied specifically in relation to CVRFs.

THE "STRUCTURAL" CEREBROVASCULAR DISEASE MODEL OF DEPRESSION

The Model

Neuronal Ischemia/Cell Death

Cardiovascular Disorders → Small Vessel Cerebrovascular Disease → Depression

This model, termed by some the "vascular depression" hypothesis, has received increased attention recently from researchers who work in the area of geriatric depression. Thus far, it may be the most carefully outlined biomedical theory of later-life depression, although data to test definitively its utility remain to be collected or published. The model and its supporting evidence have been reviewed elsewhere (Alexopoulos et al., 1997a). In this model, otherwise clinically occult cerebrovascular disease may "predispose, precipitate, or perpetuate some geriatric depressive syndromes" (p. 915) (Alexopoulos et al., 1997a).

Evidence for the Model

Evidence in favor of this model includes the following findings: As noted earlier, later-life depression, particularly that of older age of onset, is more related to acquired factors such as medical illnesses than to genetics or early life experiences. Older depressives, as compared to both younger depressives and age-matched nondepressed controls, demonstrate both a breadth and severity of neuropsychological abnormalities that may reflect underlying brain dysfunction (King, Cox, Lyness, & Caine, 1995; King, Cox, Lyness, Conwell, & Caine, 1998). There may be an interaction between age and depression, such that older age and more severe depressive symptoms are associated with greater neuropsychological deficits (King et al., 1995). Cognitive deficits also may predict a poorer outcome of depression (King, Caine, Conwell, & Cox, 1991).

In addition to neuropsychological evidence for brain dysfunction in later-life depression, neuroimaging techniques have shown an association of depression with abnormalities that are suggestive of cerebrovascular disease. Structural neuroimaging studies (both computed tomography and magnetic resonance imaging [MRI]) have noted increased cortical and central atrophy in older depressed patients as compared with controls, which is consistent with brain tissue damage (Rabins, Pearlson, Aylward, Kumar, & Dowell, 1991; Zubenko et al., 1990). Most MRI studies have shown greater diffuse hyperintensities on T_2-weighted images (in which regions with greater water content show as brighter, or more white, on the scans) in older depressives as compared with age-matched controls (see reviews by Alexopoulos et al., 1997a; Krishnan, 1991). While these hyperintensities are nonspecific in origin, studies of (nondepressed) older and mixed-age patients have shown that they are associated with CVRFs (Awad, Spetzler, Hodak, Award, & Carey, 1986). They also are correlated with postmortem neuropathological findings of arteriosclerosis and gliosis, consistent with small vessel ischemic disease (Pantoni & Garcia, 1997). These results have led to the suggestion that the MRI hyperintensities, which are indicative of increased water content, reflect increased permeability "leakiness" of the vessels, presumably from atherosclerosis.

One might expect that functional neuroimaging techniques may help shed light on the pathophysiological significance of the MRI hyperintensities. To date, they have only rarely been used with older depressed subjects; however, available data suggest that older depressives have more widespread and diffuse hypometabolism (reflecting decreased neuronal activity) than age-matched controls or younger depressives (Upadhyaya, Abou-Saleh, Wilson, Grime, & Critchley, 1990), again consistent with a disease process such as cerebrovascular injury.

Another line of evidence supporting the cerebrovascular theory of depression are the accumulated data that clinically manifest cerebrovascular disease (i.e., strokes probably cause or contribute to depressive symptoms and syndromes). Data to support this assertion have been summarized recently by Robinson (1997), who makes the following points: There is an association between strokes and depression; the prevalence of depression in stroke populations is greater than that found in patients with comparably debilitating non-central nervous system diseases, suggesting that the depression in stroke may be neurologically mediated and not solely a psychological reaction to disability; depression is associated with both stroke lesion size and location (e.g., left greater than right hemisphere; anterior pole in the left hemisphere vs. posterior pole in the right hemisphere); and the prevalence of depression is higher in vascular dementia than in dementia due to Alzheimer.

As well, patients with later-onset depression and CVRFs may have a distinctive clinical profile. Krishnan et al. (1997) classified patients into "vascular" and "nonvascular" depression groups, based on MRI criteria, and noted that the vascular group was older, and had a later age of onset of depression and a lower frequency of psychosis. Alexopoulos et al. (1997b) found that older patients with depression of later age of onset also suffer hypertension or clinically manifest atherosclerosis, have greater disability, psychomotor retardation, and cognitive impairment, and less psychomotor agitation, guilt, and insight as compared with younger-onset depressed patients without such medical comorbidity. However, they did not include additional comparison groups that might have clarified the meaning of their findings (e.g., later-onset depressives without CVRFs, and younger onset depressives with CVRFs).

Limitations of Empirical Support Thus Far

Having summarized the evidence in favor of the cerebrovascular model of depression, we now must consider the important limitations in available empirical support for the model. First, the generalizability of current findings to broader patient populations is unclear. Most of the depressed patients in the studies cited earlier were recruited from psychiatric treatment settings, in many cases (especially the neuroimaging reports) psychiatric inpatients with severe forms of depression, including melancholia. Older people are particularly unlikely to seek specialty care for depressive symptoms, although most do see their primary care physicians (Katon & Schulberg, 1992). Therefore, increasing interest has been paid to the public health importance of detection and treatment of depression in primary care settings. The relevance of the cerebrovascular model of depression to these efforts is at present largely unknown.

A related generalizability issue is that of so-called "subsyndromal" depressions. In nonpsychiatric settings, the majority of people suffering clinically significant depressive symptoms (in terms of distress and associated functional disability) do not meet criteria for major depressive disorder (Romanoski et al., 1992). A few meet criteria for dysthymic disorder, and some meet criteria for various definitions of minor depressive disorder, but many have symptoms that are too intermittent or otherwise variable to be diagnosable with any criteria-based disorder (Lyness et al., 1998d). If depression occurs along a biological gradient, then one would expect the cerebrovascular model of depression to apply to less severe forms, such as subsyndromal depressions. Alternatively, diagnosable depressive disorders may represent qualitatively (and therefore etiologically) distinct entities. Studies of the cerebrovascular model of depression might shed light on this question, but again rarely have included subsyndromally depressed subjects. In a recent report, Kumar et al. (1997) examined older patients with minor depression and found that the minor depressives had smaller prefrontal lobe volumes but did not differ in hyperintensity volumes from nondepressed subjects. Thus, there is little evidence to date specifically supporting the applicability of the cerebrovascular disease model to minor depression.

Aside from generalizability concerns, there are other limitations in support worth considering. In the cerebrovascular disease model, small vessel disease contributes to depression by functionally altering brain systems or sites that modulate mood state. Indeed, evidence is accumulating for the role of neurotransmitter alterations and regional anatomic specificity in depression associated with stroke (Robinson, 1997). However, to date, the neuroimaging abnormalities found in otherwise idiopathic later-life depression have been regionally nonspecific, involving relatively diffuse findings such as generalized cortical or central atrophy, periventricular white matter changes, or widespread subcortical hyperintensities (see reviews by Alexopoulos et al., 1997a; Krishnan, 1991). One recent study did note that depression was associated with left deep frontal white matter and left putaminal hyperintensities (Greenwald et al., 1998); these findings are consistent with what is known about the functional neuroanatomy of depression, but replication data are needed.

Other methodological issues affect interpretation of published studies. For example, some studies purporting to show that "vascular depression" differs phenomenologically from "nonvascular depression" included subjects with known strokes (i.e., subjects who suffered clinically manifest brain vascular disease). Inclusion of such subjects likely increased the likelihood of finding differences between the groups, since the contribution of stroke to depression is already well known. The question is whether the cerebrovascular model applies to patients with otherwise occult cerebrovascular disease (i.e., patients who would be diagnosed with idiopathic depression). To answer this question, studies must include only such patients, and must therefore exclude subjects with known strokes.

Finally, CVRFs may contribute to depression via psychological or psychosocial routes rather than through (or in addition to) direct effects on the brain, as discussed earlier. It is possible that the MRI hyperintensities associated with depression are in effect epiphenomena, that is, markers of the presence of CVRFs (which contribute to depression by nonneurological mechanisms) rather than representative of brain lesions causally contributory to depression. In other words, the presence of hyperintensities does not necessarily imply a pathophysiological role in depression: Association need not mean causation.

Conclusions and Future Directions

In summary, available evidence suggests the possibility that otherwise occult cerebrovascular disease may contribute to the pathogenesis of depression in some older patients, particularly those with more severe forms of major depression. However, it is premature to conclude that "vascular depression" due to otherwise occult structural disease exists as a pathophysiologically distinct entity, or that such an entity might apply to the majority of older people suffering clinically significant depressive symptoms and syndromes. To better determine the applicability of this etiological model, future research should incorporate the following themes and principles. To elucidate generalizability, studies must include patients from a variety of settings, including community, primary care, and medical specialty outpatients, as well as psychiatric treatment settings, and should include patients suffering a range of depressive symptom severity, including "subsyndromal" symptoms as well as major depression. To delineate anatomical mechanisms of depressive pathogenesis, neuroimaging research should move beyond description of MRI global hyperintensity (or other abnormality) burden and begin to determine if there is regional specificity of lesions that is consistent with models of the neuropathoanatomy of depression. Similarly, functional neuroimaging should be used in correlative fashion to determine if MRI hyperintensities in fact reflect altered brain function, and again to determine the regional specificity of such pathophysiology. At the same time, risk-factor studies still must play an important role in future research. Longitudinal studies are needed that examine the attributable of depression associated with the full range of CVRFs assessed individually and cumulatively. Optimally, such work would require large sample sizes followed for many years, from middle adulthood into later life. Finally, studies of CVRFs and of neuroimaging findings should simultaneously assess the possible moderating or mediating role of disability, personality, pain, social support, or other psychosocial factors.

ATHEROSCLEROSIS AND CYTOKINE-MEDIATED PATHOGENESIS OF DEPRESSION

A Proposed Model

One might postulate other pathobiological mechanisms by which CVRFs might contribute to depression. Building on data and theoretical models proposed by others (Katz 1996), we propose that the following model warrants empirical testing. We describe the model, then discuss the existing support for each of its components.

						↓ monoamine function		
Atherosclerosis (e.g., cardiovascular disorders)	→	↑ IL-1β	→	↑ CNS monoamine turnover	→	monoamine autotoxicity	→	depression

Atherosclerosis leads to elevated levels of interleukin-1-ß (IL-1ß) in plasma and in brain parenchyma. IL-1ß promotes central monoamine turnover, which in turn leads to long-term depletion of catecholamine systems due to the autotoxicity of monoamine metabolites. By this route (possibly along with other mechanisms), IL-1ß contributes to the pathogenesis of depression in later life and, therefore, so does atherosclerotic disease.

Model Components

Atherosclerosis. A common disease that rises in prevalence with age, atherosclerosis is the result of multiple complex factors that interact with each other to produce lesions in the walls of arteries and arterioles, eventually leading to complications related to partial or complete lumen occlusion (e.g., ischemia), vessel wall weakening (e.g., atherosclerotic aneurysms), or embolization of atherosclerotic plaque (producing end-organ ischemia or infarcts). These processes underlie most common cardiovascular and cerebrovascular diseases; as well, other CVRFs (e.g., hypertension, diabetes mellitus, cigarette smoking) are themselves contributors to the pathogenesis of atherosclerosis. Other known factors include lipid metabolism, hemodynamic stress, blood coagulation elements, behavioral factors (see later for discussion of depression as a risk factor for coronary artery disease), and, most relevant to the model being considered, inflammatory and immune mechanisms (Lopes-Virella & Virella, 1990; Massy & Keane, 1996; Ross, 1995).

It is only relatively recently that the importance of these latter mechanisms has been recognized and begun to be studied. Among the first steps in the formation of a fatty streak (the early atherosclerotic lesion) is the adherence of monocytes to intact endothelium. Macrophages and *T* lymphocytes accumulate subendothelially, some of the former transforming into foam cells. A variety of cytokines are secreted into the plaque and the bloodstream as part of this process; some cytokines may play a role in attracting further leukocytes. IL-1ß is one of the cytokines so secreted by a number of cells (macrophages and *T* lymphocytes, as well as smooth muscle and endothelial cells); while its full range of actions relevant to atherosclerosis is not known, it may alter endothelial cell function (promoting adhesion and coagulation) and stimulate smooth muscle cell proliferation while also affecting lymphocyte activity level.

Interleukin-1-ß. IL-1ß is a polypeptide secreted in response to inflammation (Dinarello, 1996; Dinarello & Wolff, 1993; Luster, 1998). It has a wide range of biological activity, including the stimulation of host defense processes and of the release of pituitary hormones (see later discussion of corticotropin-releasing factor). It also may induce fever and hypotension, and behavioral disturbances, including increased sleep (Krueger, Walter, Dinarello, Wolff, & Chedid, 1984), anorexia (Plata-Salaman, 1988), decreased locomotion (Crestani, Suguy, & Dantzer, 1991), decreased social exploration (Plata-Salaman, 1988), and "behavioral despair" (del Cerro & Borrell, 1990). Maes and colleagues (Maes, Bosmans, Meltzer, Scharpe, & Suy, 1993; Maes et al., 1995) have summarized a variety of findings consistent with the notion that depressive illness is associated with a systemic inflammatory and immune response. They (Maes et al., plain 1993) found that IL-1ß levels in the supernatant of mitogen-stimulated peripheral blood mononuclear cells were elevated in inpatients with major or minor depression as compared with normal controls, and that IL-1ß levels correlated with cortisol levels after a dexamethasone suppression test (a test of hypothalamo–pituitary–adrenal or HPA, axis function). They concluded that constituents of the immune response such as IL-ß1 may contribute to the HPA axis hyperactivity associated with depression, presumably at least partly due to the effects of IL-1ß in stimulating corticotropin-releasing factor (CRF) release (Sapolsky, Rivier, Yamamoto, Plotsky, & Vale, 1987).

While a number of other cytokines (and other substances) are also secreted as part of atherogenesis, the proposed model focuses on IL-1ß because it is the best studied of the cytokines, and in addition to the behavioral effects noted earlier, it also induces central norepinephrine, serotonin, and, possibly, dopamine release or turnover after central administration (see Katz, 1996, for review). As noted by Katz, while one would not expect peptides such as IL-1ß to be able to cross the blood–brain barrier, other routes may explain the central effects of peripherally administered cytokines, including leakage from circumventricular organs, actions at the level of brain capillary endothelial cells that trigger intracerebral changes, or release from activated lymphocytes that enter the brain parenchyma. Supporting this notion, Zalcman et al. (1994) found cytokine-specific alterations in murine central norepinephrine, serotonin, and dopamine function after peripheral administration of IL-1ß and other cytokines; the alterations in norepinephrine function did not correlate with IL-1ß-induced elevation of corticosterone levels. Therefore it is plausible, albeit unproved, that peripheral elevations in IL-1ß due to atherosclerosis may indeed have central effects such as stimulation of monoamine turnover.

Monoamine Autotoxicity. Given the above, some attention has been paid to cytokines as putative mediators of acute depressive symptoms accompanying acute illnesses ("sickness reactions") (Katz, 1996; Maes et al., 1993). However, our proposed model specifically focuses on the longer-term effects of IL-1ß elevation on the brain due to a *chronic* illness such as atherosclerosis. There is evidence that oxidative metabolites of monoamines are toxic to the monoaminergic neurons. While the greatest attention has been paid to oxidative destruction of dopaminergic systems because of the potential direct relevance to Parkinson's disease (Castano et al., 1997; Felten, Felten, Steece-Collier, Date, & Clemens, 1992; Schapira, 1997), similar processes likely affect other monoamine systems including serotonin (Gudelsky, 1996). Thus, greater cumulative catecholamine activity may lead to autotoxicity that produces cell death. This model has been hypothesized to underlie age-related decline in the dopaminergic nigrostriatal system. In our proposed model, chronic increased turnover of central monoamines (as stimulated by IL-1ß leads to autotoxicity, neuronal death, and ultimately, to decreased function of these systems.

Abnormalities in monoamine function associated with depressive conditions are complex, and while individual findings have been replicated, a comprehensive model of the functional neurobiology of depression has not yet been validated. However, the postulate that depletion of monoamine systems is a mechanism by which IL-1ß contributes to depression pathogenesis remains consistent with available empirical data, however oversimplified.

Limitations in Empirical Support for the Model and Challenges for Further Research

The proposed cytokine-mediated model has sufficient support and potential explanatory power to invite further testing. However, there are major gaps in supporting data, as well as measurement limitations and confounds that will affect the design and interpretation of needed studies. The model involves several steps for which there are at most only modest evidence, especially in humans. For example, there are virtually no data published regarding serum IL-1ß levels in patients with coronary artery disease or other specific atherosclerotic illnesses. Moreover, the model depends on a combination of these individual steps, which as yet have not been studied directly. Is atherosclerosis associated with increased central levels of IL-1ß levels? Does neuronal death from autotoxicity lead to depression? And most globally, yet most fundamentally, is the association between atherosclerosis and depression mediated by IL-1ß levels?

Investigators studying the relationships among atherosclerosis, depression, and IL-1ß levels must confront several complex confounding factors. Many systemic disease processes may influence IL-1ß levels. As noted earlier, IL-1ß stimulates release of CRF, so cortisol levels may mediate the relationship between IL-1ß and depression. Other cytokines are released as part of atherogenesis and might influence brain function in a fashion so as to affect depression, albeit the precise nature of such influences is not yet known. And, of course, many other physiological as well as psychosocial variables might influence depression.

As well, the optimal observational paradigm should include a measure of the amount of atherosclerotic burden in each subject. Such a measure does not exist. Atherosclerosis is typically measured by the presence (and, more crudely, by the severity of) clinically manifest diseases such as angina pectoris, myocardial infarction, stroke, intermittent claudication, and so on. But these disease manifestations generally do not occur until substantial atherosclerosis has been present and progressive for long periods of time, and the usual clinical measures of disease severity do not correlate well with the extent of atherosclerosis per se. The amount of atherosclerosis, or at least luminal occlusion, in a given vessel can be measured by angiography (and by ophthalmological exam for retinal arterial disease), but angiography is an invasive procedure that poses risks to subjects that are difficult to justify for a correlative study. In any case such procedures can only tell us the degree of atherosclerotic disease present in the vessels studied, which would be a poor proxy for quantifying the overall severity of systemic atherosclerosis.

Recognizing these limitations, research is warranted to test further the proposed model. Subject groups will need to chosen for the presence of one (or more) of the clinical manifestations of atherosclerosis, with comparison groups lacking such disease. Nonatherosclerotic diseases that are thought likely to confound IL-1ß levels, including illnesses known to produce a rise in other acute phase reactants, must be excluded or treated as covariates. For the reasons outlined earlier in the discussion of the cerebrovascular model, subjects should possess a wide range of depressive symptoms and syndromes, thus requiring recruitment from a variety of clinical and community settings. Ultimately, investigations should include measures of central catecholamine and serotonin function, although it may be sufficient initially to determine the role of IL-1ß as a mediator of the association between atherosclerotic disease group status and depression. Finally, while an association of IL-1ß with depression and cardiovascular disease would be consistent with the model we propose, it also is consistent with models in which depression contributes to cardiac illness (see following discussion).

DEPRESSION AS A RISK FACTOR FOR CARDIOVASCULAR DISEASE

Thus far, the discussion has focused on cardiovascular disease or other CVRFs as risk factors for depression. (We have maintained this focus because of our group's interest in elucidating pathogenic mechanisms of depression). Nonetheless, a growing body of evidence suggests that depression is a risk factor for the onset or course of cardiovascular disease. As reviewed recently (Glassman & Shapiro, 1998; Musselman, Evans, & Nemeroff, 1998), several studies have reported a high prevalence of major depression among patients hospitalized with coronary artery disease or myocardial infarction, albeit most have not found a correlation between cardiac disease severity and depression. Most longitudinal studies (Ford et al., 1998; Glassman & Shapiro, 1998; Musselman et al., 1998; Penninx et al., 1998) have found that depression is an independent risk factor for the subsequent development of myocardial infarction or cardiovascular-related death, even after controlling for other cardiovascular risk factors; one study found this to be true only for males in a population-based group (Hippisley-Cox, Fielding, & Pringle, 1998). Moreover, in an often-cited and well-designed study, major depression predicted poorer survival (due to cardiac death) at 6 months after myocardial infarction (Frasure-Smith, Lesperance, & Talajic, 1993). Both major depression and depressive symptom severity redicted 18-month survival (Frasure-Smith, Lesperance, & Talajic, 1995a), even after controlling for cardiac disease severity.

Since depression is a risk factor for cardiovascular disease, several lines of investigation have led to speculation about mechanisms by which depression might contribute to the pathogenesis of cardiac illness. "Mental stress," including negative emotions such as anger or anxiety, which are common concomitants of depression, are associated with cardiac risk factors (Fava, Abraham, Pava, Shuster, & Rosenbaum, 1996) and with recurrent cardiac events (Frasure-Smith, Lesperance, & Talojic, 1995b). Such emotions may acutely worsen myocardial ischemia (Gullette et al., 1997), presumably at least partly due to autonomically mediated increase in cardiac workload. Other potential physiological routes include the hypercortisolemia associated with depression (which may contribute to atherogenesis, hyperlipidemia, and hypertension), diminished beat-to-beat heart rate variability (Miyawaki & Salzman, 1991), autonomically mediated ventricular irritability and consequent proneness to arrhythmias (see Musselman et al., 1993, also Frasure-Smith et al., 1995b), and depression-associated alterations in platelet function that might contribute to thrombogenesis (Musselman et al., 1998).

It should be noted that there may be other, including nonphysiological, routes by which depression is a risk factor for cardiovascular disease. Other medical disorders comorbid with the depression may directly or indirectly worsen the course of the cardiac illness. Patients with depression may be less compliant with medications or with other recommendations regarding cardiac risk-factor reduction or cardiac disease treatment (Ziegelstein, Bush, & Fauerbach, 1998). And the role of psychosocial factors has been inadequately explored in studies that also rigorously examine cardiac disease severity and psychopathology, despite some evidence for the role of social support in cardiac patients (Oxman & Hull, 1997) and a voluminous older literature on personality attributes and cardiac illness, an area that needs reexamination using more recent personality and psychopathological assessment tools (Duberstein et al., 1999).

Finally, as noted earlier, it is worth viewing the cytokine model proposed in the previous section in light of these issues. If depression serves as a risk factor for cardiac disease (whether by direct physiological or indirect routes), then levels of IL-1ß may increase as a consequence of the worsened coronary (and other) atherosclerosis. Of course, this is not an either–or proposition; there may well be a vicious circle (or, to use 1990s phraseology, an endless loop) in which the IL-1ß in turn further exacerbates the depression. In other words, elevated IL-1ß in serum may at once *reflect* depression and *contribute* to it. Evidence that would more definitively support the proposed cytokine model would include demonstration of an interaction between cardiac disease and IL-1ß level in their association with depression, and longer-term longitudinal studies examining whether elevation in IL-1ß level (in association with atherosclerosis) can precede and predict the subsequent onset of depressive conditions.

SUMMARY AND FUTURE DIRECTIONS

In this chapter, we have reviewed the literature regarding CVRFs and depression in later life. We have critically examined two pathobiological models in which CVRFs contribute to depression. The theories considered were a structural model in which CVRFs lead to brain parenchymal damage via small vessel ischemic disease, and a cytokine model in which atherosclerosis leads to functional alterations in neurotransmitter systems underlying depressive pathogenesis. The former model has more direct empirical support regarding its applicability to older persons, yet both require considerable future investigative effort. As depression in later life is a heterogeneous condition, both theories (along with others) may be applicable; thus, research is needed to identify patient subgroups in which one or the other (or both) may be most relevant.

We also reviewed evidence for the role of depression in contributing to the pathogenesis of atherosclerotic heart disease. Many existing correlative data might, in fact, be used to support models in which the arrows point from depression toward cardiovascular disease, or from CVRFs toward depression. Investigators need to keep the potential for bidirectionality (or circular pathways) in mind as they interpret findings.

As well, we have noted the potential roles of psychological and psychosocial routes in the interface between CVRFs and depression. Investigators whose data only encompass the biomedical realm need to temper interpretation of their findings with explicit recognition that other variables may play crucial mediating or moderating roles. Better yet, future studies should include multidimensional assessments to test directly the importance of such routes.

Finally, it is our hope that the considerations in this chapter may foster similar rigorously critical attention to theory development and testing in other areas, including the relationships of other types of medical illnesses to later-life depression (and other forms of psychopathology). While it is humbling to recognize how much remains to be discovered, we find it encouraging that the frontiers of geriatric psychiatry have moved beyond mere correlative descriptions and evolved toward a process of model building and empirical validation. The existence of this volume is testimony to the strides that have been made and to the implications that such work has for clinical practice.

ACKNOWLEDGMENT: The preparation of this chapter was supported by NIMH Grant No. MH01113 to Jeffrey M. Lyness.

REFERENCES

Abrams, R. C., Rosendahl, E., Card, C., & Alexopoulos, G. S. (1994). Personality disorder correlates of late and early onset depression. *Journal of the American Geriatrics Society, 42,* 727–731.

Alexopoulos, G. S., Meyers, B. S., Young, R. C., Campbell, S., Silbersweig, D., & Charlson, M. (1997a). "Vascular depression" hypothesis. *Archives of General Psychiatry, 54,* 915–922.

Alexopoulos, G. S., Meyers, B. S., Young, R. C., Kakuma, T., Silbersweig, D., & Charlson, M. (1997). Clinically defined vascular depression. *American Journal of Psychiatry, 154,* 562-565.

American Heart Association. (1990). *Stroke risk factor prediction chart.* Dallas: Author.

Awad, I. A., Spetzler, R. F., Hodak, J. A., Awad, C. A., & Carey, R. (1986). Incidental subcortical lesions identified on magnetic resonance imaging in the elderly: I. Correlation with age and cerebrovascular risk factors. *Stroke, 17,* 1084–1089.

Baldwin, R. C., & Tomenson, B. (1995). Depression in later life: A comparison of symptoms and risk factors in early- and late-onset cases. *British Journal of Psychiatry, 167,* 649–652.

Broadhead, W. E., Blazer, D. G., George, T. K., & Tse, C. K. (1990). Depression, disability days, and days lost from work in a prospective epidemiological survey. *Journal of the American Medical Association, 264,* 2524–2528.

Bush, T. L., Miller, S. R., Criqui, M. H., & Barret-Connor, E. (1990). Risk factors for morbidity and mortality in older populations: An epidemiologic approach. In W. R. Hazzard, E. L. R. Andres, E. L. Bierman, & J. P. Blass (Eds.), *Principles of geriatric medicine and gerontology* (2nd ed.). New York: McGraw-Hill.

Caine, E. D., & Joynt, R. J. (1986). Neuropsychiatry . . . again. *Archives of Neurology, 43,* 325–327.

Caine, E. D., & Lyness, J. M. (2000). Cognitive disorders and secondary syndromes. In H. I. Kaplan & B. J. Sadock (Eds.), *Comprehensive textbook of psychiatry* (7th ed.). Baltimore: Williams, Lippincott & Wilkins.

Caine, E. D., Lyness, J. M., & King, D. A. (1993). Reconsidering depression in the elderly. *American Journal of Geriatric Psychiatry, 1,* 4–20.

Castano, A., Ayala, A., Rodriguez-Gomez, J. A., Herrera, A. J.,. Cano, J., & Machado, A. (1997). Low selenium diet increases the dopamine turnover in prefrontal cortex of the rat. *Neurochemistry International, 30,* 549–555.

Cohen-Mansfield, J., & Marx, M. S. (1993). Pain and depression in the nursing home: Corroborating results. *Journal of Gerontology, 48,* P96–P97.

Costa, P. T., Jr., & McCrae, R. R. (1987). Neuroticism, somatic complaints, and disease: Is the bark worse than the bite? *Journal of Personality, 55,* 299–316.

Crestani, F., Suguy, F., & Dantzer, R. (1991). Behavioral effects of peripherally injected interleukin-1: Role of prostaglandins. Brain Research, 542, 330–335.

del Cerro, S., & Borrell, J. (1990). Interleukin-1 affects the behavioral despair responses in rats by an indirect mechanism which requires endogenous CRF. *Brain Research 528,* 162–164.

Dinarello, C. A. (1996). Biologic basis for interleukin-1 in disease. *Blood, 87,* 2095–2147.

Dinarello, C. A., & Wolff, S. M. (1993). The role of interleukin-1 in disease. *New England Journal of Medicine, 328,* 106–113.

Duberstein, P. R., Seidlitz, L., Lyness, J. M., & Conwell, Y. (1999). Dimensional measures and the five factor model: Clinical implications and research directions. In E. Rosowsky, R. C. Abrams, & R. A. Zweig (Eds.), *Personality disorders in older adults: Emerging issues in diagnosis and treatment.* Hillsdale, NJ: Erlbaum.

Dyken, M. L., Wolf, P. A., Barnett, H. J. M., Bergan, J. J., Hass, W. K., Kannel, W. B., Kuller, L., Kurtzke, J. F., & Sundt, T. M. (Eds.). (1985). Risk factors for stroke. In *A statement for physicians by the Subcommit-*

tee on Risk Factors and Stroke of the Stroke Council. (Reprint No. 71-022-A). Dallas: American Heart Association.

Evans, A. S. (1978). Causation and disease: A chronological journey. *American Journal of Epidemiology, 108,* 249–258.

Fava, M., Abraham, M., Pava, J., Shuster, J., & Rosenbaum, J. (1996). Cardiovascular risk factors in depression. The role of anxiety and anger. *Psychosomatics, 37,* 31–37.

Felten, D. L, Felten, S. Y., Steece-Collier, K., Date, I., & Clemens, J. A. (1992). Age-related decline in the dopaminergic nigrostriatal system: The oxidative hypothesis and protective strategies. *Annals of Neurology, 32,* S133–S136.

Ford, D. E., Mead, L. A., Chang, P. P., Cooper-Patrick, L., Wang, N.-Y., & Klag, M. J. (1998). Depression is a risk factor for coronary artery disease in men: The Precursors study. *Archives of Internal Medicine, 158,* 1422–1426.

Frasure-Smith, N., Lesperance, F., & Talajic, M. (1993). Depression following myocardial infarction. Impact on 6-month survival. *Journal of the American Medical Association, 270,* 1819–1825.

Frasure-Smith, N., Lesperance, F., & Talajic, M. (1995a). Depression and 18-month prognosis after myocardial infarction. *Circulation, 91,* 999–1005.

Frasure-Smith, N., Lesperance, F., & Talajic, M. (1995b). The impact of negative emotions on prognosis following myocardial infarction: Is it more than depression? *Health Psychology, 14,* 388–398.

Glassman, A. H., & Shapiro, P. A. (1998). Depression and the course of coronary artery disease. *American Journal of Psychiatry, 155,* 4–11.

Greenwald, B. S., Kramer-Ginsberg, E., Krishnan, K. R., Ashtari, M., Auerbach, C., & Patel, M. (1998). Neuroanatomic localization of magnetic resonance imaging signal hyperintensities in geriatric depression. *Stroke, 29,* 613–617.

Greenwald, B. S., Kramer-Ginsberg, E., Krishnan, K .R. R., Ashtari, M., Aupperle, P. M., & Patel, M. (1996). MRI signal hyperintensities in geriatric depression. *American Journal of Psychiatry, 153,* 1212–1215.

Gudelsky, G. A. (1996). Effect of ascorbate and cysteine on the 3,4-methylenedioxymethamphetamine-induced depletion of brain serotonin. *Journal of Neural Transmission, 103,* 1397–1404.

Gullette, E. C. D., Blumenthal, J. A., Babyak, M., Jiang, W., Waugh, R. A., Frid, D. J., O'Connor, C. M., Morris, J. J., & Krantz, D. S. (1997). Effects of mental stress on myocardial ischemia during daily life. *Journal of the American Medical Association, 277,* 1521–1526.

Guralnik, J. M., & Simonsick, E. M. (1993). Physical disability in older Americans. *Journal of Gerontology, 48* (Special Issue), 3–10.

Hayward, C. (1995). Psychiatric illness and cardiovascular disease risk. *Epidemiology Review, 17,* 129–138.

Heckhausen, J., & Schulz, R. (1995). A life-span theory of control. *Psychological Review, 102,* 284–304.

Hippisley-Cox, J., Fielding, K., & Pringle, M. (1998). Depression as a risk factor for ischaemic heart disease in men: Population based case-control study. *British Medical Journal, 316,* 1714–1719.

Katon, W., & Schulberg, H. (1992). Epidemiology of depression in primary care. *General Hospital Psychiatry, 14,* 237–247.

Katz, I. R. (1996). On the inseparability of mental and physical health in aged persons: Lessons from depression and medical comorbidity. *American Journal of Geriatric Psychiatry, 4,* 1–16.

Kendler, K. S., Neale, M. C., MacLean, C. J., Heath, A. C., Eaves, L. J., & Kessler, R. C. (1993). Smoking and major depression. A causal analysis. *Archives of General Psychiatry, 50,* 36–43.

Kennedy, G. J., Kelman, H. R., & Thomas, C. (1990). The emergence of depressive symptoms in late life: The importance of declining health and increasing disability. *Journal of Community Health, 15,* 93–104.

King, D. A.,. Caine, E. D., Conwell, Y., & Cox, C. (1991). Predicting severity of depression in the elderly at six-month follow-up: A neuropsychological study. *Journal of Neuropsychiatry and Clinical Neurosciences, 3,* 64–66.

King, D. A., Cox, C., Lyness, J. M., & Caine, E. D. (1995). Neuropsychological effects of depression and age in an elderly sample: A confirmatory study. *Neuropsychology, 9,* 399–408.

King, D. A., Cox, C., Lyness, J. M., Conwell, Y. C., & Caine, E. D. (1998). Quantitative and qualitative differences in the verbal learning performance of elderly depressives and healthy controls. *Journal of the International Neuropsychological Society, 4,* 115–126.

Krishnan, K. R. (1991). Organic bases of depression in the elderly. *Annual Review of Medicine, 42,* 261–266.

Krishnan, K. R., Hays, J. C., & Blazer, D. G. (1997). MRI-defined vascular depression. *American Journal of Psychiatry, 154,* 497–501.

Krueger, J. M., Walter, J., Dinarello, C. A., Wolff, S. M., &.Chedid, L. (1984). Sleep-promoting effects of endogenous pyrogen (interleukin-1). *American Journal of Physiology, 246,* R994–R999.

Kumar, A., Miller, D., Ewbank, D., Yousem, D., Newberg, A., Samuels, S., Cowell, P., & Gottlieb, G. (1997). Quantitative anatomic measures and comorbid medical illness in late-life major depression. *American Journal of Geriatric Psychiatry, 5,* 15–25.

Kumar, A., Schweizer, E., Zhisong, J., Miller, D., Bilker, W., Swan, L. L., & Gottlieb, G. (1997). Neuroanatomical substrates of late-life minor depression. A quantitative magnetic resonance imaging study. *Archives of Neurology, 54,* 613–617.

Littlefield, C. H., Rodin, G. M., Murray, M. A., & Craven, J. L. (1990). Influence of functional impairment and social support on depressive symptoms in persons with diabetes. *Health Psychology, 9,* 737–749.

Lopes-Virella, M. F., & Virella, G. (1990). Immune mechanisms in the pathogenesis of atherosclerosis. In C. L. Malmendier, P. Alaupovic, H. Brewer (Eds.), *Hypercholesterolemia, hypocholesterolemia, hypertriglyceridemia* (pp. 383–392). New York: Plenum Press.

Luster, A. D. (1998). Chemokines–chemotactic cytokines that mediate inflammation. *New England Journal of Medicine, 338,* 436–445.

Lustman, P. J., Griffith, L. S., Gavard, J. A., & Clouse, R. E. (1992). Depression in adults with diabetes. *Diabetes Care, 15,* 1631–1639.

Lyness, J. M., & Caine, E. D. (1997). Secondary mood disorders: Terminology and determination of causality. In J. M. Lyness (Guest Ed.), *Affective Disorders Arising from Medical Conditions, Seminars in Clinical Neuropsychiatry, 2,* 228-231.

Lyness, J. M., Caine, E. D., Cox, C., King, D. A., Conwell, Y., & Olivares, T. (1998). Cerebrovascular risk factors and later life major depression: Testing a small vessel brain disease model of pathogenesis. *American Journal of Geriatric Psychiatry, 6,* 5–13.

Lyness, J. M., Caine, E. D., King, D. A., Conwell, Y., Cox, C., & Duberstein, P. R. (1999). Cerebrovascular risk factors and depression in older primary care patients. *American Journal of Geriatric Psychiatry, 7,* 252–258.

Lyness, J. M., Duberstein, P. R., King, D. A., Cox, C., & Caine, E. D. (1998c). Medical illness burden, trait neuroticism, and depression in older primary care patients. *American Journal of Psychiatry, 155,* 969–971.

Lyness, J. M., King, D. A., Cox, C., Yoediono, Z., & Caine, E. D. (1999). The importance of subsyndromal depression in older primary care patients. *Journal of the American Geriatric Society, 47,* 647–652.

Maes, M., Bosmans, E., Meltzer, H. Y., Scharpe, S., & Suy, E. (1993). Interleukin-1ß: A putative mediator of HPA axis hyperactivity in major depression? *American Journal of Psychiatry, 150,* 1189–1193.

Maes, M., Vandoolaeghe, E., Ranjan, R., Bosmans, E., Bergmans, R., & Desnyder, R. (1995). Increased serum interleukin-1 receptor-antagonist concentrations in major depression. *Journal of Affective Disorders, 36,* 29–36.

Massy, Z. A., & Keane, W. F. (1996). Pathogenesis of atherosclerosis. *Seminars in Nephrology, 16,* 12–20.

Meyerson, R. A., Richard, I. H., & Schiffer, R. B. (1997). Mood disorders secondary to demyelinating and movement disorders. In J. M. Lyness (Guest Ed.), *Affective Disorders Arising from Medical Conditions, Seminars in Clinical Neuropsychiatry, 2,* 252–264.

Miyawaki, E., & Salzman, C. (1991). Autonomic nervous system tests in psychiatry: Implications and potential uses of heart rate variability. *Integrated Psychiatry, 7,* 21–28.

Moritz, D. J., & Ostfeld, A. M. (1990). The epidemiology and demographics of aging. In W. R. Hazzard, E. L. R. Andres, E. L. Bierman, & J. P. Blass (Eds.), *Principles of geriatric medicine and gerontology* (2nd ed.). New York: McGraw Hill.

Musselman, D. L., Evans, D. L., & Nemeroff, C. B. (1998). The relationship of depression to cardiovascular disease. Epidemiology, biology, and treatment. *Archives of General Psychiatry, 55,* 580–592.

National Center for Health Sciences. (1992). *Vital statistics of the United States, 1989: Vol. II. Mortality, Part B* (DHHS Pub. No. (PHS) 92-1002). Washington DC: Public Health Service.

NIH Consensus Development Panel on Depression in Late Life. (1992). Diagnosis and treatment of depression in late life. *Journal of the American Medical Association, 268,* 1018–1024.

Oxman, T. E., & Hull, J. G. (1997). Social support, depression, and activities of daily living in older hearts surgery patients. *Journal of Gerontology: Psychological Sciences,* 52B, P1–P14.

Pantoni, L. & Garcia, J. H. (1997). Pathogenesis of leukoaraiosis: A review. *Stroke, 28,* 652–659.

Parmelee, P. A., Katz, I. R., & Lawton, M. P. (1991). The relation of pain to depression among institutionalized aged. *Journal of Gerontology, 46,* P15–P21.

Parmelee, P. A., Katz, I. R., & Lawton, M. P. (1992). Depression and mortality among institutionalized ages. *Journal of Gerontology, 47,* P3–P10.

Penninx, B. W. J. H., Guralnik, J. M., Mendes de Leon, C. F., Pahor, M., Visser, M. Corti, M.-C., & Wallace, R. B. (198). Cardiovascular events and mortality in newly and chronically depressed persons > 70 years of age. *American Journal of Cardiology, 81,* 988–994.

Plata-Salaman, C. (1988). Food intake suppression by immunomodulators. *Neuroscience Research, 3,* 159–165.

Popkin, M. K., & Andrews, J. E. (1997). Mood disorders secondary to systemic medical conditions. In J. M. Lyness (Guest Ed.), *Affective Disorders Arising from Medical Conditions, Seminars in Clinical Neuropsychiatry, 2,* 296–306.

Porsteinsson, A. P., Tariot, P. N., & Schneider, L. S. (1997). Mood disturbances in Alzheimer's disease. In J. M. Lyness (Guest Ed.), *Affective Disorders Arising from Medical Conditions, Seminars in Clinical Neuropsychiatry, 2,* 265–275.

Rabins, P. V., Pearlson, G. D., Aylward, E., Kumar, A. J., & Dowell, K. (1991). Cortical magnetic resonance imaging changes in elderly inpatents with major depression. *American Journal of Psychiatry, 148,* 617–620.

Robinson, R. G. (1997). Mood disorders secondary to stroke. In J. M. Lyness (Guest ed.), *Affective Disorders Arising from Medical Conditions, Seminars in Clinical Neuropsychiatry, 2,* 244–251.

Romanoski, A. J., Folstein, M. F., Nestadt, G., Chahal, R., Merchant, A., Brown, C. H., Greenberg, E. M., & McHugh, P. R. (1992). The epidemiology of psychiatrist-ascertained depression and DSM-III depressive disorders: Results from the Eastern Baltimore Mental Health Survey Clinical Reappraisal. *Psychological Medicine, 22,* 629–655.

Ross, R. (1995). Arteriosclerosis, an overview. In E. Haber (Ed.), *Molecular cardiovascular medicine* (pp. 11–30). New York: Scientific American.

Sapolsky, R., Rivier, C., Yamamoto, G., Plotsky, P. & Vale, W. (1987). Interleukin-1 stimulates the secretion of hypothalamic corticotropin-releasing factor. *Science, 238,* 522–524.

Schapira, A. H. (1997). Pathogenesis of Parkinson's disease. *Baillieres Clinical Neurology, 6,* 15–36.

Schleifer, S. J., Macari-Hinson, M. M., Coyle, D. A., Slater, W. R., Kahn, M., Gorlin, R., & Zucker, H. D. (1989). The nature and course of depression following myocardial infarction. *Archives of Internal Medicine, 149,* 1785–1789.

Schneider, L. S., Zemansky, M. F., Bender, M., & Sloane, R. B. (1992). Personality in recovered depressed elderly. *International Psychogeriatrics, 4,* 177–185.

Schulz, R., Heckhausen, J., & O'Brien, A. (1994). Control and the disablement process in the elderly. *Journal of Social Behavior and Personality, 9,* 139–152.

Sinclair, P., Lyness, J. M., Cox, C., Conwell, Y., King, D. A., & Caine, E. D. (submitted). Depression and examiner-rated versus self-reported functional disability in older primary care patients.

Smith, D. M., & Atkinson, R. M. (1997). Mood disorders secondary to drugs and pharmacologic agents. In J. M. Lyness (Guest Ed.), *Affective Disorders Arising from Medical Conditions, Seminars in Clinical Neuropsychiatry, 2,* 285–295.

Upadhyaya, A. K., Abou-Saleh, M. T., Wilson, K., Grime, S. J., & Critchley, M. (1990). A study of depression in old age using a single-photon emission computerized tomography. *British Journal of Psychiatry, 157*(Suppl. 9), 76–81.

Wells, K. B., Rogers, W. Burman, M. A., & Camp, P. (1993). Course of depression in patients with hypertension, myocardial infarction, or insulin-dependent diabetes. *American Journal of Psychiatry, 150,* 633–638.

Wells, K. B., Stewart, A., Hays, R. D., Burman, M. A., Rogers, W., Daniels, M., Barry, S., Greenfield, S., & Ware, J. (1989). The functioning and well-being of depressed patients: Results from the Medical Outcomes Study. *Journal of the American Medical Association, 262,* 914–919.

Williams, M. E. (1990). Why screen for functional disability in elderly persons? *Annals of Internal Medicine, 112,* 639–640.

Williamson, G. M., & Schulz, R. (1992). Pain, activity restriction, and symptoms of depression among community-residing elderly adults. *Journal of Gerontology, 47,* P367–P372.

Williamson, G. M., & Schulz, R. (1995). Activity restriction mediates the association between pain and depressed affect: A study of younger and older adult cancer patients. *Psychology and Aging, 10,* 369–378.

Zalcman, S., Green-Jonson, J. M., Murray, L. Nance, D. M., Dyck, D., Anisman, H., & Greenberg, A. H.

(1994). Cytokine-specific central monoamine attraction induced by interleukin -1, -2, and -6. *Brain Research, 643,* 40–49.

Ziegelstein, R. C., Bush, D. E., & Fauerbach, J. A. (1998). Depression, adherence behavior, and coronary disease outcomes. *Archives of Internal Medicine, 158,* 808–809.

Zubenko, G. S., Sullivan, P., Nelson, J. P., Belle, S. H., Huff, F. S., & Wolf, G. L. (1990). Brain imaging abnormalities in mental disorders of late life. *Archives of Neurology, 47,* 1107–1111.

4

Pain, Functional Disability, and Depressed Affect

GAIL M. WILLIAMSON

In the early 1990s, my colleagues and I became interested in teasing apart the association between pain, functional disability, and symptoms of depression. Although previous research had demonstrated an association between chronic pain and depression (e.g., Lindsay & Wyckoff, 1981; Moss, Lawton, & Glicksman, 1991; Parmelee, Katz, & Lawton, 1991; Von Knorring, Perris, Eisemann, Eriksson, & Perris, 1983; Waddell, 1987), these efforts had produced conflicting results; that is, even though the relation between pain and depression had been clearly established, the association typically was not a large one, and the direction of causality was open to question. For instance, is the intuitive approach correct, that is, that pain causes people to be depressed (e.g., Moss et al., 1991; Romano & Turner, 1985; Roy, Thomas, & Matas, 1984)? Or as some studies seemed to indicate (e.g., Crum, Cooper-Patrick, & Ford, 1994; Lefebvre, 1981; Leventhal, Hansell, Diefenbach, Leventhal, & Glass, 1996; Mathew, Weinman, & Mirabi, 1981), do depressed people simply report experiencing more symptoms, including pain, than nondepressed people? Or, are other factors responsible for both increased levels of experienced pain and more depressed affect?

Our interest in this issue was sparked by a paper published in 1991 by Parmelee et al. on the association between pain and depression in a sample of institutionalized elderly persons. Their results appeared to support the hypothesis that depressed individuals are more likely than nondepressed individuals to report experiencing pain; that is, pain was linearly related to level of depression. Institutionalized elderly adults classified as possible major depressives reported more pain than did those classified as minor depressives, and the latter group reported more pain than did those classified as nondepressed. Moreover, this effect was especially strong when physicians had diagnosed an illness that could justify pain.

GAIL M. WILLIAMSON • Department of Psychology, University of Georgia, Athens, Georgia 30602.

Physical Illness and Depression in Older Adults: A Handbook of Theory, Research, and Practice, edited by Gail M. Williamson, David R. Shaffer, and Patricia A. Parmelee. Kluwer Academic/Plenum Publishers, New York, 2000.

Of particular interest was the finding that although depression was correlated with functional disability, functional disability did not account for (i.e., did not mediate) the relation between pain and depression. Given that illness and dysfunction represent potential confounds in diagnosing depression in elderly patients (e.g., Blazer & Williams, 1980), these results were important because they suggested that, among institutionalized older adults, the relation between pain and depression may be independent of functional impairment. However, rather than simply interpreting their data as evidence that depressed affect leads to increased reports of pain, Parmelee et al. (1991) offered a process-oriented explanation; that is, institutionalized elderly persons may, over time, adopt the "sick role," with concomitant expectations of experiencing both depression and pain as a result of their illness (Waddell, 1987).

Parmelee et al. (1991) went a step further by hypothesizing that their findings for institutionalized older adults would generalize to community-residing elderly persons as well. If this were the case, like their counterparts who reside in an institutional setting, functional disability should *not* account for the relation between pain and depression among elders who are not institutionalized. However, we predicted a different pattern of results. Specifically, we expected that, in a sample of community-residing elderly adults, the degree to which pain restricted normal activities would account for a substantial portion of the association between pain and depression. Our reasoning was straightforward. First, elderly people residing in the community are more likely to be responsible for conducting routine daily activities (e.g., shopping, preparing meals, doing household chores, visiting others) than are those living in an institutional setting. Because research had strongly and consistently related pain to activity restriction (e.g., Moss et al., 1991), we expected that the degree to which pain causes functional impairment plays an important role in determining how difficult it is for older adults to carry on daily living. In turn, increased difficulty in conducting routine activities should be an important contributor to depressive symptomatology. Second, because community-residing individuals tend to be less physically impaired than those in institutions, they should be less likely to have experienced the process through which depressed affect may come to "sustain the psychological experience of pain" (Parmelee et al., 1991, p. 20) and to adopt the sick role (also see Baltes, Kinderman, Reisenzein, & Schmid, 1987). This rationale led to the first study in an ongoing program of research.

A STUDY OF GERIATRIC OUTPATIENTS

As reported in Williamson and Schulz (1992), the sample in this study consisted of 228 people, recruited from outpatient geriatric clinics, who had a wide variety of illness conditions. Patients were interviewed to obtain self-reports of pain (a 3-item measure assessing general pain, pain during the last week, and discomfort associated with illness) and symptoms of depression based on the Center for Epidemiologic Studies—Depression Scale (CES-D; Radloff, 1977). To measure functional disability, respondents indicated the extent to which nine areas of activity (self-care, care of others, eating habits, doing household chores, going shopping, visiting friends, working on hobbies, and maintaining friendships) were restricted by their illness or disability. We called this instrument the Activity Restriction Scale (ARS) and demonstrated that it had adequate internal reliability (i.e., Cronbach's alpha = .85; Williamson & Schulz, 1992)

Consistent with previous research, more pain was hypothesized to be bivariately related to higher levels of both activity restriction and depressed affect. More activity restriction was expected to be both bivariately and multivariately associated with more symptoms of depression. First, results confirmed bivariate hypotheses; that is, pain was positively correlated with activity restriction and depressed affect, and activity restriction was positively related to depressed affect (all *r*s > .30, $p < .001$). Second, as shown in Figure 4.1, results of path analyses confirmed our multivariate hypotheses. Pain had a small direct effect on depression and directly affected activity restriction, and the strongest direct effect on depression was exerted by activity restriction.

More importantly, additional analyses revealed that activity restriction partially mediated the association between pain and depressed affect (see Darlington, 1990, for the criteria for partial mediation). Thus, unlike Parmelee et al.'s (1991) results, the relation between pain and depression in community-residing older adults was partially attributable to functional impairment. Among noninstitutionalized elders, pain was a source of activity restriction which, in turn, contributed to depressed affect.

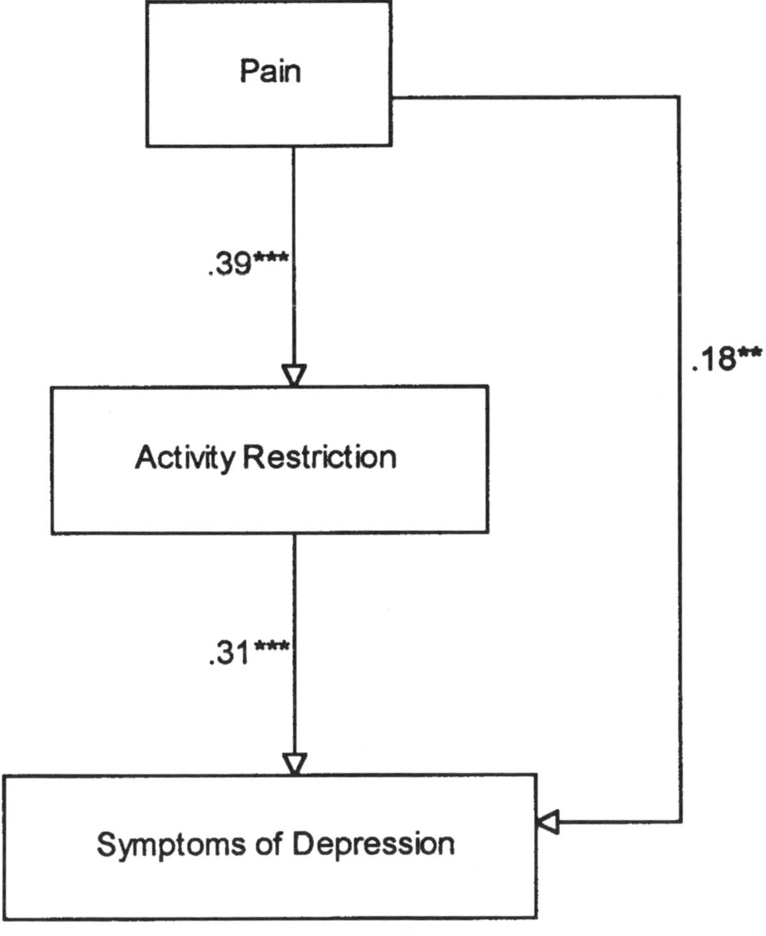

Figure 4.1. Path analysis results for geriatric outpatients: Pain, activity restriction, and depressed affect (from data reported in Williamson & Schulz, 1992).

However, we were not able to replicate the mediating effect of activity restriction with the full measure of activities of daily living (ADL; Center for the Study of Aging and Human Development, Duke University, 1978), also included in this study. On the other hand, a subset of five instrumental ADL items (i.e., transporting self, doing laundry, preparing meals, shopping, doing heavy work such as scrubbing floors and shoveling snow) produced the same pattern of results as those obtained with the ARS. In a later section, I consider how the ARS differs from traditional measures of ADL and possible meanings of these differences (also see Stump, Clark, Johnson, & Wolinsky, 1997, for a discussion of potential problems with ADL measures).

Results of our initial study were both encouraging and interesting, but they were from but one study. In addition, our findings had been based on cross-sectional data and, consequently, did not address issues related to change over time. From an intervention standpoint, it is important to determine whether changes in predictor variables (e.g., pain and activity restriction) are related to changes in outcomes (e.g., depression). We have subsequently found the mediating effect of activity restriction in several other populations. In the following sections, studies replicating the mediational role of activity restriction (in one case, employing longitudinal data) in the association between pain and depressive symptomatology are reviewed.

REPLICATING THE ACTIVITY RESTRICTION EFFECT

Recurrent Cancer

Patients with recurrent cancer are at risk for experiencing pain, restriction of normal activities, and depressive symptomatology. As reported in Williamson and Schulz (1995), longitudinal interview data were collected at two points in time, 8 months apart, from 268 recurrent cancer outpatients (*M* age = 62.4 years) who were undergoing palliative radiation therapy. They reported the pain they experienced as a result of their illness (type, intensity, severity, and presence in the last month), activities restricted by their illness (ARS), and symptoms of depression (CES-D).

Consistent with previous results, we predicted that activity restriction would mediate the relation between pain and symptoms of depression in analyses of cross-sectional data. In addition, we expected to find a similar pattern in longitudinal results; that is, increases in pain should predict increases in activity restriction, which in turn should predict increases in symptoms of depression. And this is precisely what we found. Cross-sectional analyses replicated our previous findings by showing that the effects of pain on symptoms of depression were partially mediated by activity restriction. Longitudinal analyses revealed that as pain increased over time, so did activity restriction, which in turn predicted increases in depressed affect.

Having obtained the same pattern of results both cross-sectionally and longitudinally, we became more confident about the strength and consistency of the activity restriction effect. At least, we were confident that the effect is consistent among older adults with various illness and disability conditions and when activity restriction is measured by the ARS. However, we could not say with certainty that these results would generalize to populations other than older adults. Nor could we be certain that the results were attributable to the conceptual construct of activity restriction rather than simply to the measure we had employed in these studies, an issue that concerned us, because results obtained with the ARS were difficult to replicate with traditional ADL measures (Williamson & Schulz, 1992). In addition, we could not rule out a negative reporting bias, since our data were derived exclusively from patient self-reports. We addressed these potential confounds in studies summarized in the following sections.

Chronic Pain in Children

To confront the age issue, we chose individuals from the other end of the life span (Walters & Williamson, 1999); that is, interview data were collected from 73 children (ranging in age from 5 to 18 years, mean age = 11.4) who were receiving outpatient therapy for recurrent pain associated with a chronic disorder (e.g., cancer, sickle-cell anemia, juvenile rheumatoid arthritis). This study also was designed to show that previous results could be obtained with a measure of activity restriction other than the ARS and could not be attributed solely to self-reporting bias. First, we employed a different set of instruments; children reported their level of pain using the Pediatric Pain Questionnaire (PPQ; Varni, Thompson, & Hanson, 1987) and their symptoms of depression using the Children's Depression Inventory (CDI; Kovacs, 1985). Second, rather than asking children about their own activity restriction, in separate interview sessions, primary caregivers (usually mothers) rated, on a different instrument, the extent to which illness caused the child's normal activities (e.g., attending school, playing with friends) to be restricted. Like our ARS, this measure had high internal reliability (Cronbach's alpha = .89).

Even with these modifications to our protocol and a much younger sample, results were the same as (and perhaps, stronger than) those in previous studies. Specifically, when caregiver-reported activity restriction was entered into the equation, the association between child-reported pain and depression became nonsignificant, indicating total mediation (see Darlington, 1990, for the criteria for total mediation). Moreover, the results could not be explained by negative self-reporting bias, since data came from both patients and caregivers. In summary, this study showed that the activity restriction effect generalizes to include a population other than older adults and a situation in which the construct of activity restriction is assessed with a measure conceptually similar to, but distinctly different from, the ARS. The results of an additional study (described later) further demonstrated that the effect cannot be attributed to the ARS measure per se and also showed that the mediating effect of activity restriction (1) is not simply a function of the interview format used in previous studies, and (2) applies to a sample of adults of all ages as well as to children and elderly persons.

Breast Cancer

Participants in this study had been diagnosed with Stage 1 ($n = 36$), Stage 2 ($n = 49$), or Stage 3 ($n = 10$) breast cancer (Williamson, in press). These women ranged in age from 26 to 75 years (mean age = 49.2) and were not interviewed face-to-face but rather, completed mail surveys, most often anonymously. They reported depressive symptoms (CES-D) and general bodily pain (one item from the Short-Form Health Survey SF-36; Ware & Sherbourne, 1992). In this study, the measure of activity restriction consisted of 16 items adapted from the SF-36 that assessed the extent to which vigorous and moderate daily activities were limited by health, whether problems with work or other regular activities had occurred during the last 4 weeks as a result of health status, and the extent and frequency to which health had interfered with social activities in the last 4 weeks (Cronbach's alpha = .92). In support of the validity of the construct of activity restriction, results indicated that activity restriction (as assessed by a measure other than the ARS) *totally* mediated the association between pain and depressed affect. Pain predicted depression to the extent that pain restricted normal activities.

Summary

At this point, we were satisfied that activity restriction mediates (either partially or totally) psychological adjustment to a variety of illness conditions and that the effect is widespread, encompassing individuals across the life span, from children to elderly adults. Moreover, our results did not appear to be attributable to reliance on specific types of methodology; that is, we found the same pattern in longitudinal as well as cross-sectional research (Williamson & Schulz, 1995) using various indicators of pain and different instruments to measure activity restriction and depression (e.g., Walters & Williamson, 1999; Williamson, in press). Furthermore, results were essentially the same when data were collected from dual sources rather than solely from patients (Walters & Williamson, 1999) and when different methods were used to gather data (i.e., face-to-face interviews vs. mail surveys). We then raised the issue of whether other factors play a role in the pain–activity restriction–depressive symptoms pathway.

DIFFERENTIAL EFFECTS OF AGE AND MULTIPLE ILLNESS CONDITIONS

Recall that our first study (Williamson & Schulz, 1992) did not replicate the findings of Parmelee et al. (1991) with respect to functional disability (activity restriction) as a mediating factor in the association between pain and depression. A major difference between the two studies was that Parmelee et al. studied institutionalized older adults, whereas we studied older adults residing in the community. Differences in results could be expected based solely on differences in samples (i.e., more responsibility for conducting daily activities and less physical impairment in those who are not institutionalized). Another difference is that those residing in institutions are likely to be, on average, older than those who are not institutionalized, and age appears to influence the association between pain and depressive symptomatology. For example, similar levels of disease-related chronic pain seem to be tolerated better by older than younger adults (Cassileth et al., 1984; Deal et al., 1985; Foley, 1985; Idler & Angel, 1990; Prohaska, Leventhal, Leventhal, & Keller, 1985). Most explanations for these findings revolve around a "process of normalization" (Idler & Angel, 1990) through which older individuals habituate to pain or cope with it better because they experience pain and disability more often and in greater degrees (Demlow, Liang, & Eaton, 1986; Ferrell, 1991; Ferrell, Ferrell, & Osterweil, 1990; Harkins, Kwentus, & Price, 1984).

To address these issues, in our study of patients with recurrent cancer (Williamson

& Schulz, 1995), we looked at the effects of age and one of the factors that goes along with age—the presence of multiple illness conditions. In addition to completing measures of depression, pain, and activity restriction, patients listed their other chronic health problems. The average cancer patient in this sample was coping with one or two other chronic ailments, most often arthritis, hypertension, diabetes, or heart disease.

The sample was split into a "younger" group (< 65 years old) and an "older" group (age 65 and older). Separate tests for mediation were calculated for each age group, and as predicted, the effects of pain on symptoms of depression were mediated *entirely* by activity restriction in the younger group but only *partially* in the older group. One reason for the reduced effect of activity restriction in older adults might be that older people—who are more likely to have chronic ailments—have had more opportunity to habituate to illness and associated declines in functional capacity. Some support for this idea would be obtained if the presence of chronic illnesses other than cancer could be shown to moderate the amount of activity restriction individuals of differing ages attributed to having cancer. Results confirmed this proposition; that is, in multiple regression analyses, the interaction between age and other chronic conditions was significant after controlling for the main effects of age, pain, and other chronic conditions. As shown in Figure 4.2, the pattern of this interaction was such that younger cancer patients with no chronic conditions other than cancer were most likely to perceive their routine activities as restricted by their cancer. In contrast, older patients reported relatively low levels of activity restriction due to cancer, and this was the case whether other chronic conditions were reported or not. Younger patients with chronic illness other than cancer reported activity restriction due to cancer at levels similar to

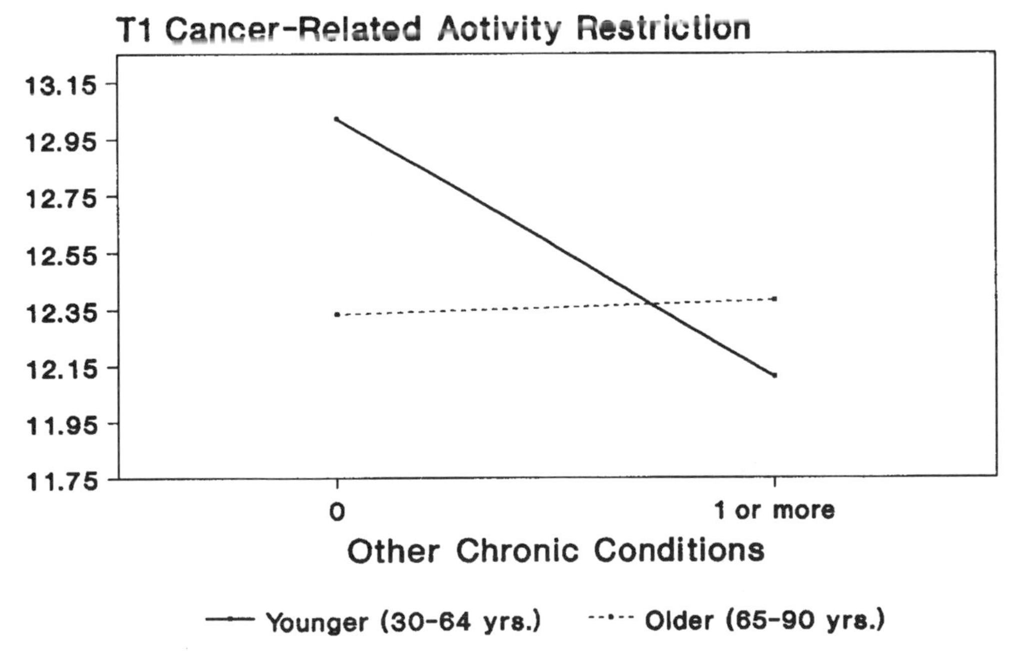

Figure 4.2. Recurrent cancer: Activity restriction as a function of age and other chronic conditions (Source: Williamson & Schulz, 1995; reprinted by permission).

those of older patients. Thus, it appears that dealing with chronic illnesses may cause younger people to respond in ways similar to their older counterparts.

Interestingly, however, younger and older adult recurrent cancer patients did not differ in reported pain, activity restriction, or depression (all *F*s < 1.06). Taken together with findings indicating that the activity restriction effect can be found in medically compromised individuals of all ages, these results suggest that it is not so much levels of pain and activity restriction or increasing age per se that matter, but rather the *process* through which activity restriction comes to mediate the pain–depression association. Specifically, it appears that experience is more critical than chronological age, an allegation consistent with results of our study of pediatric patients (Walters & Williamson, 1999); that is, we found the same patterns of results (1) in preadolescent children as in younger recurrent cancer patients, and (2) in postadolescent children as in older recurrent cancer patients. Among younger children (5–12 years of age) and younger adults (< 65 years of age), activity restriction totally mediated the association between pain and depressed affect. Thus, the younger groups in both studies were depressed by pain to the extent that pain interfered with normal activities. Among older children (13–18 years of age), activity restriction did not occupy an analogous mediating role, because pain was not associated with symptoms of depression in this group. Instead, pain exerted direct effects on activity restriction, and activity restriction exerted direct effects on depression. Similarly, among older adults (> 64 years of age), activity restriction only partially mediated the impact of pain on depression. Later in this chapter, I return to why results with younger and older children were comparable to those of younger and older adult cancer patients and note some possible meanings of these findings.

IMPLICATIONS FOR RESEARCH AND INTERVENTION

At this point, the data not only indicated that activity restriction is an important factor in the frequently observed association between pain and depression, but also that many questions remained unanswered. In the following sections, a few of these questions are considered in some detail, along with directions for future research and intervention strategies. In a subsequent chapter, several other issues are raised, along with an account of the evolution of a model specific to the relevance of activity restriction in the association between health-related stressors and symptoms of depression (Williamson & Shaffer, Chapter 9, this volume).

How Does Activity Restriction Differ from Activities of Daily Living?

At the onset of this program of research, we were interested in how functional disability might qualify the association between pain and depression. However, it soon became clear that functional disability can be operationalized in more than one way (for similar conclusions, see Rodgers & Miller, 1997; Stump et al., 1997). In gerontological research, it most often is conceptualized (e.g., Lawton & Brody, 1969; Center for the Study of Aging and Human Development, Duke University, 1978) as the amount of help one needs and/or receives with a variety of normal activities, ranging from basic ADLs (e.g., eating, personal grooming) to more instrumental ADLs (IADLs; e.g., transporting self, preparing meals, shopping). Basic ADL needs are most likely to be found in the more advanced stages of disease and disability, while IADL needs are apparent in earlier stages (e.g., Rodgers & Miller, 1997). A classic example is the case of Alzheimer's disease (AD), in which IADL needs begin early, and basic ADL deficits increase with illness severity (Haley & Pardo, 1989). Recall that in our initial activity restriction study (Williamson & Schulz, 1992), we were not able to replicate the mediating effect of activity restriction with a traditional measure of ADL (Center for the Study of Aging and Human Development, Duke University, 1978). However, a subset of five IADLs (i.e., transporting self, doing laundry, preparing meals, shopping, doing heavy work such as scrubbing floors and shoveling snow) produced the same pattern of results as those obtained with the ARS. In addition, we were able to replicate the activity restriction effect in subsequent research with measures that are conceptually similar to the ARS (Walters & Williamson, 1999; Williamson, in press). Taken together, the data suggest that it is the relatively early signs of loss of mobility and independence (as assessed by the ARS and similar instruments, and by a subset of IADL items) that contribute to the association between pain and depression. Institutionalized older adults are more likely than their community-residing counterparts to be beyond this stage. Thus, another explanation for differences between our results and those of Parmelee et al. (1991) is that by the time elderly people are placed in an institution, they have transitioned from requiring help with IADLs to needing assistance with basic ADLs. In other words, these individuals may already have passed through the most distressing stages of the first signs of functional impairment such that they have come to accept their limitations and needs for care. Those whose impairment has not progressed beyond needing assistance with IADLs may perceive their current condition as forecasting things to come and, consequently, be especially focused on continuing to perform tasks more or less as usual in order to forestall being placed in an institution. We should then expect that restriction of normal instrumental activities would be more important (i.e., have more impact on the relation between pain and depression) among older adults who are still residing in the community.

To be sure our results were not specific to the ARS, in subsequent studies (Walters & Williamson, 1999; Williamson, in press), we employed measures of activity restriction that differed from our ARS. And we found the same pattern of results, lending increased confidence in the conceptual construct of activity restriction. Still, regardless of the activity restriction measure used, we have always asked participants to report subjectively perceptions of ability to conduct activities as usual. In contrast, traditional ADL instruments more objectively measure how much help people need (and/or receive) with normal activities. The subjective versus objective distinction is an important one (for similar opinions, see Bernard et al., 1997; Gurland, Katz, & Pine, Chapter 15, this volume). First, perceptions constitute a more psychological assessment of functional disability and thus should be more closely related than objective measures to psychological outcomes. Second, and perhaps more critical, actual needs may matter less than whether those needs are being adequately met.

What do these findings suggest in terms of intervention and future research? It seems clear that when the goal is to assess needs for assistance, ADL measures pinpoint the areas in which older adults can benefit from additional help (Rodgers & Miller, 1997). On the other hand, when the goal is to determine psychological adjustment, a measure of perceived activity restriction may be more useful. Such measures identify not only areas in which help is needed but also how people react to pain and disability, both of which are potential points of intervention. For example, interventions should be most effective when they ameliorate the restricted activities about which people are most distressed. Like others (e.g., Satariano, 1997), we also advocate taking a close look at aspects of the physical environment for clues about interventions that can promote independence and mobility.

Partial or Total Mediation?

In two of the studies described earlier (geriatric outpatients and older adult cancer patients), activity restriction *partially* mediated the association between pain and depression, indicating that pain has direct effects as well as indirect effects on depression through its impact on activity restriction. However, in three studies (younger adult cancer patients, younger children in chronic pain, and breast cancer patients), the mediating effect was *total*, indicating that pain only leads to depression to the extent that pain influences ability to conduct normal activities. In a subsample in one study (older children in chronic pain), pain directly affected activity restriction, activity restriction directly affected depression, and there was no mediating effect of activity restriction at all, because child-reported pain and depression were not correlated.

Is the distinction between levels of mediation important? I think it may be. Off the cuff, one might suppose that the differences between partial and total mediation could be explained by the instruments used to measure activity restriction; that is, partial mediating effects were found when the ARS was used (i.e., in geriatric outpatients and recurrent cancer patients), and total mediating effects were found when other measures of activity restriction were used (i.e., in children with chronic pain and breast cancer patients). Other results, however, suggest a more complex interpretation. First, partial mediation by ARS was found only in the older adult cancer patients; in younger adult cancer patients, mediation by ARS was total. Second, total mediation by caregiver-reported activity restriction, using a measure different from the ARS, was observed only in preadolescent children with chronic pain; no mediating effect was apparent in postadolescent children.

This pattern of findings suggests that it is not the specific operationalization of activity restriction that matters but, rather, individuals' prior experience with pain. The potentially critical intersection between age and experience is explored in more detail in the next section.

Is It Age or Is It Experience?

Our study of cancer patients (Williamson & Schulz, 1995) indicated that activity restriction mediates the pain–depression pathway to a lesser extent among older than younger adults. These results are consistent with other research showing that pain appears to be tolerated better by older than younger adults (Cassileth et al., 1984; Deal et al., 1985; Foley, 1985; Idler & Angel, 1990; Prohaska et al., 1985; Williamson, in press). It is thought that such findings reflect a "process of normalization" (e.g., Idler & Angel, 1990) through which older individuals habituate to pain and functional disability because they experience them more often and in greater degrees as a result of age-related decrements in health. A related possibility is that older adults are less distressed by pain and functional impairment because, as one ages, declining health is viewed as normative (Ferrell, 1991; Ferrell et al., 1990; Harkins et al., 1984). Similarly, older adults may be less distressed by restriction of normal activities because they have different expectations about functional ability compared to their younger counterparts (e.g., Clark & Maddox, 1992; Demlow et al., 1986; Heckhausen & Schulz, 1995; Schulz & Heckhausen, 1996).

An expanded interpretation seems viable at this juncture. Results obtained from our study of adult cancer patients support the notion that older adults handle pain and functional disability better than do younger adults (Williamson & Schulz, 1995). But is this really a function of old age? If so, how can these results be reconciled with our findings that the reactions to chronic pain of preadolescent children looked very much like those of our younger (< 65 years of age) cancer patients, while those of postadolescent children were similar to our group of cancer patients over age 65 (Walters & Williamson, 1999; Williamson & Schulz, 1995)? Perhaps, chronological age matters less than (albeit, age-related) experience. Put simply, the postadolescent group had been dealing with their chronically painful conditions longer than had the younger children and, like the younger adults with other chronic conditions (and the older adults regardless of whether they had other conditions) in our study of cancer patients, seemed to have accepted pain as a routine part of life and so were less distressed by it. Further research could test this hypothesis by studying age peers who have suffered chronic pain for varying lengths of time.

This experience hypothesis is consistent with research indicating that, over time, most people do adapt even to the most severe forms of disability (e.g., Schulz & Decker, 1985), and our data suggest that this adjustment can occur at any age. However, adjustment may be more difficult among those who have not yet reached an age where functional limitation is normal or expected and those whose life experience has not previously involved chronic pain and disability. What does this imply for intervention? Our data concur with observations by Bruce, Seeman, Merrill, and Blazer (1994) and Cassileth et al. (1984) that individuals who have had little or no previous experience with illness- and disability-related activity restriction are likely to become depressed when their first debilitating illness episode strikes. Assuming that intervention resources are limited, it would seem prudent to intervene with this group first. In particular, chronological age and illness experience appear to interact such that those who are younger *and* who have not previously experienced illness-related activity restriction are likely to become depressed and, perhaps, will benefit most from programs designed to reduce activity restriction. However, as work by Erdal and Zautra (1995) indicates, the transition from acute to chronic illness may warrant psychological intervention as well. We should not assume that "an individual knows how to adjust to an illness downturn because she or he has experienced it before" (Erdal & Zautra, 1995, p. 576).

CONCLUDING COMMENTS

In terms of research initiatives, controlled experimental studies are needed in order to show that changing one aspect of a person's life will, in fact, produce changes in well-being. Optimum research would involve random assignment intervention studies, for instance, to evaluate (1) the utility of strategies aimed at increasing ability to conduct routine activities in the presence of pain, (2) the effects of decreased activity restriction on depressive symptomatology, and (3) the extent to which patients are willing to make trade-offs between discomfort and activity restriction when given the choice. It may be that some patients will tolerate higher levels of pain in order to maintain desired activities. Identifying differences between these individuals and those who will not tolerate similar levels of discomfort is another important topic for future research.

However, in the absence of such controlled studies, longitudinal research can help determine whether changes in predictor variables are related to changes in outcomes. Our study of patients with recurrent cancer (Williamson & Schulz, 1995) represents a step in this direction by showing that as pain increased over time, so did activity restriction, and increased activity restriction in turn predicted increased depression. A more positively valenced way to interpret these results is that interventions aimed at decreasing pain and activity restriction should promote better psychological adjustment.

ACKNOWLEDGMENTS: Research summarized in this chapter was supported by Grant No. MH41887 from the National Institute of Mental Health and Grant No. CA48635 from the National Cancer Institute (R. Schulz, principal investigator), and by fellowships from the Institute for Behavioral Research at the University of Georgia and the Southeastern Center for Cognitive Aging (G. M. Williamson, principal investigator). Manuscript preparation was facilitated by Grant No. AG15321 from the National Institute on Aging (G.M. Williamson, principal investigator).

REFERENCES

Baltes, M. M., Kinderman, T., Reisenzein, R., & Schmid, U. (1987). Further observational data on the behavioral and social world of the institutions for the aged. *Psychology and Aging, 2,* 390–403.

Bernard, S. L., Kincade, J. E., Konrad, T. R., Arcury, T. A., Rabiner, D. J., Woomert, A., DeFriese, G. H., & Ory, M. G. (1997). Predicting mortality from community surveys of older adults: The importance of self-rated functional ability. *Journal of Gerontology, 52,*155–163.

Blazer, D., & Williams, C. D. (1980). Epidemiology of depression and dysphoria in an elderly population. *American Journal of Psychiatry, 137,* 439–443.

Bruce, M. L., Seeman, T. E., Merrill, S. S., & Blazer, D. G. (1994). The impact of depressive symptomatology on physical disability: MacArthur Studies of Successful Aging. *American Journal of Public Health, 84,* 1796–1799.

Cassileth, B. R., Lusk, E. J., Strouse, T. B., Miller, D. S, Brown, L. L., Cross, P. A., & Tenaglia, A. N. (1984). Psychosocial status in chronic illness: A comparative analysis of six diagnostic groups. *New England Journal of Medicine, 311,* 506–511.

Clark, D. O., & Maddox, G. L. (1992). Racial and social correlates of age-related changes in functioning. *Journal of Gerontology, 47,* 222–232.

Crum, R. M., Cooper-Patrick, L., & Ford, D. E. (1994). Depressive symptoms among general medical patients: Prevalence of a one year outcome. *Psychosomatic Medicine, 56,* 109–117.

Darlington, R. B. (1990). *Regression and linear models.* New York: McGraw-Hill.

Deal, C. L., Meenan, R. F., Goldenberg, D. L., Anderson, J. J., Sack, B., Pastan, R. S., & Cohen, A. S. (1985). The clinical features of elderly-onset rheumatoid arthritis. *Arthritis and Rheumatism, 28,* 987–994.

Demlow, L. L., Liang, M. H., & Eaton, H. M. (1986). Impact of chronic arthritis in the elderly. *Clinics in Rheumatic Diseases, 12,* 329–335.

Duke University, Center for the Study of Aging and Human Development. (1978). *Multidimensional functional assessment, the OARS methodology: A manual* (2nd ed.). Durham, NC: Duke University.

Erdal, K. J., & Zautra, A. J. (1995). Psychological impact of illness downturns: A comparison of new and chronic conditions. *Psychology and Aging, 10,* 570–577.
Ferrell, B. A. (1991). Pain management in elderly people. *Journal of the American Geriatric Society, 39,* 64–73.
Ferrell, B. A., Ferrell, B. R., & Osterweil, D. (1990). Pain in the nursing home. *Journal of the American Geriatric Society, 38,* 409–414.
Foley, K. M. (1985). The treatment of cancer pain. *New England Journal of Medicine, 313,* 84–95.
Haley, W. E., & Pardo, K. M. (1989). Relationship of severity of dementia to caregiving stressors. *Psychology and Aging, 4,* 389–392.
Harkins, S. W., Kwentus, J., & Price, D. D. (1984). Pain and the elderly. In C. Benedetti, C. R. Chapman, & G. Moricca (Eds.), *Advances in pain research and therapy* (Vol. 7, pp. 103–121). New York: Raven Press.
Heckhausen, J., & Schulz, R. (1995). A life-span theory of control. *Psychological Review, 102,* 284–304.
Idler, E. L., & Angel, R. J. (1990). Age, chronic pain, and subjective assessments of health. *Advances in Medical Sociology, 1,* 131–152.
Kovacs, M. (1985). The Children's Depression Inventory. *Psychopharmacology Bulletin, 21,* 995–998.
Lawton, M. P., & Brody, E. M. (1969). Assessment of older people: Self-maintaining and instrumental activities of daily living. *Gerontologist, 9,* 179–188.
Lefebvre, M. F. (1981). Cognitive distortion and cognitive errors in depressed psychiatric and low back pain patients. *Journal of Consulting and Clinical Psychology, 49,* 517–525.
Leventhal, E. A., Hansell, S., Diefenbach, M., Leventhal, H., & Glass, D. C. (1996). Negative affect and self-report of physical symptoms: Two longitudinal studies of older adults. *Health Psychology, 15,* 193–199.
Lindsay, P., & Wyckoff, M. (1981). The depression–pain syndrome and its response to antidepressants. *Psychosomatics, 22,* 571–577.
Mathew, R., Weinman, M., & Mirabi, M. (1981). Physical symptoms of depression. *British Journal of Psychiatry, 139,* 293–296.
Moss, M. S., Lawton, M.P., & Glicksman, A. (1991). The role of pain in the last year of life of older persons. *Journal of Gerontology, 46,* 51–57.
Parmelee, P. A., Katz, I. R., & Lawton, M. P. (1991). The relation of pain to depression among institutionalized aged. *Journal of Gerontology, 46,* 15–21.
Prohaska, T. R., Leventhal, E. A., Leventhal, H., & Keller, M. L. (1985). Health practices and illness cognition in young, middle aged, and elderly adults. *Journal of Gerontology, 40,* 569–578.
Radloff, L. (1977). The CES-D Scale: A self-report depression scale for research in the general population. *Applied Psychological Measurement, 1,* 385–401.
Rodgers, W., & Miller, B. (1997). A comparative analysis of ADL questions in surveys of older people. *Journal of Gerontology, 52,* 21–36.
Romano, J. M., & Turner, J. A. (1985). Chronic pain and depression: Does the evidence support a relationship? *Psychological Bulletin, 97,* 18–34.
Roy, R., Thomas, M., & Matas, M. (1984). Chronic pain and depression: A review. *Comprehensive Psychiatry, 25,* 96–105.
Satariano, W. A. (1997). Editorial: The disabilities of aging—looking to the physical environment. *American Journal of Public Health, 87,* 331–332.
Schulz, R., & Decker, S. (1985). Long-term adjustment to physical disability: The role of social support, perceived control, and self-blame. *Journal of Personality and Social Psychology, 48,* 1162–1172.
Schulz, R., & Heckhausen, J. (1996). A life span model of successful aging. *American Psychologist, 51,* 702–714.
Stump, T. E., Clark, D. O., Johnson, R. J., & Wolinsky, F. D. (1997). The structure of health status among Hispanic, African American, and white older adults. *Journal of Gerontology, 52,* 49–60.
Varni, J. W., Thompson, K. L., & Hanson, V. (1987). The Varni/Thompson Pediatric Questionnaire. *Pain, 28,* 27–38.
Von Knorring, L., Perris, C., Eisemann, M., Eriksson, U., & Perris, H. (1983). Pain as a symptom in depressive disorders: I. Relationship to diagnostic subgroup and depressive symptomatology. *Pain, 15,* 19–26.
Waddell, G. (1987). A new clinical model for the treatment of low-back pain. *Spine, 12,* 632–644.
Walters, A. S., & Williamson, G. M. (1999). The role of activity restriction in the association between pain and depressed affect: A study of pediatric patients with chronic pain. *Children's Health Care, 28,* 33–50.

Ware, J. E., Jr., & Sherbourne, C. D. (1992). The MOS 36-item Short-Form Health Survey (SF-36). *Medical Care, 30,* 473–483.

Williamson, G. M. (in press). Extending the Activity Restriction Model of Depressed Affect: Evidence from a sample of breast cancer patients. *Health Psychology.*

Williamson, G. M., & Schulz, R. (1992). Pain, activity restriction, and symptoms of depression among community-residing elderly. *Journal of Gerontology, 47,* 367–372.

Williamson, G. M., & Schulz, R. (1995). Activity restriction mediates the association between pain and depressed affect: A study of younger and older adult cancer patients. *Psychology and Aging, 10,* 369–378.

5

The Role of Everyday Events in Depressive Symptoms for Older Adults

ALEX J. ZAUTRA, AMY S. SCHULTZ, and JOHN W. REICH

Research efforts to date show there is no simple explanation for depressive symptoms among the elderly (Gatz & Hurwicz, 1990). A review of the current literature on the relationship between aging and depression consistently implicates a third variable, functional disability; persons with more disability tend to be more depressed, and functional disability tends to increase with age (Lewinsohn, Rohde, Fischer, & Seeley, 1991; Zeiss, Lewinsohn, Rohde, & Seeley, 1996). In fact, when the level of physical disability is controlled, illness per se does not predict degree of depression (Williamson & Schulz, 1992; also see Williamson & Shaffer, Chapter 9, in this volume). Any understanding of the relationship between aging and depression, therefore, should incorporate disability and functional limitation variables. Given these parameters, the question then becomes how to determine what other factors are responsible for depression in older adults.

We present in this chapter one approach to identifying those other factors related to depression among the elderly. The approach taken here is "event-based" in that it systematically assesses the everyday lives of older adults. In a very broad sense, physically disabling illnesses and injuries may be considered a major life event, provoking a lasting change in functional capacity. If event assessments of everyday life are comprehensive, as they should be, the empirical yield from analyses of those assessments can be useful in a number of ways. Events can be found to be related to depression, or they can be unrelated, and both types of information are useful for setting the parameters of interventions for helping older adults cope with advancing age and possible in-

ALEX J. ZAUTRA, AMY S. SCHULTZ, and JOHN W. REICH • Department of Psychology, Arizona State University, Tempe, Arizona 85287,

Physical Illness and Depression in Older Adults: A Handbook of Theory, Research, and Practice, edited by Gail M. Williamson, David R. Shaffer, and Patricia A. Parmelee. Kluwer Academic/Plenum Publishers, New York, 2000.

creases in depression. To the extent that physical illness or other age-related stressors can be described in terms of everyday life difficulties that are modifiable, we open new doors for preventive interventions for older adults (Roberts, Kaplan, Shema, & Strawbridge, 1997).

The roots of this event-based approach were established by the life events scaling techniques of Holmes and Rahe (1967). A number of inventories for assessing events have been developed since the original Holmes and Rahe formulation. Recent conceptualizations of events have led to more refined analyses of the properties of events over and above their simple frequency of occurrence. Reliable relationships with various indices of mental health have been found (see review by Zautra, Affleck, & Tennen, 1994) to support the conceptual logic behind the development of event-based approaches. This chapter is in line with this research tradition.

We first discuss the development of our scale for assessing events, the Inventory of Small Life Events (ISLE; Zautra, Guarnaccia, & Dohrenwend, 1986). The complete set of items and their properties (see next section) are presented in the Appendix. We then describe an empirical study of the relationship between ISLE events and two major life events, major stressors of older adults, physical disability and bereavement, and present analyses of how interactions of these two classes of events relate to depression in our samples of older adults.

CONCEPTUALIZATIONS UNDERLYING THE INVENTORY OF SMALL LIFE EVENTS

The ISLE was developed as a small-event measure to complement the Psychiatric Epidemiology Research Interview (PERI) Major Life Events Scale (Dohrenwend, Krasnoff, Askenasy, & Dohrenwend, 1978). Zautra et al. (1986) established explicit criteria for ISLE item inclusion. In order to be included the event had to (1) be an observable change in a person's life; (2) have a discrete beginning, (3) be objectively classifiable as either desirable or undesirable; and (4) be rated as requiring an average of 250 or less life change units using the parameters of Dohrenwend et al. (1978).

These criteria were chosen to improve the psychometric properties of this scale and remove potential confounds. The requirement that events on the ISLE be observable changes minimized concerns that the measure of small events would be actually measuring depressive states or other psychiatric symptoms. Indeed, other event measures have been criticized for ignoring issues of confounding (Dohrenwend & Schrout, 1985). This requirement removed subjective feeling states from the pool of small-event items. Establishing the independence of events a priori from distress measurement is especially important if methodologically sound advances are to be made in stress and coping research.

The requirement that events on the list identify changes with a discrete beginning eliminated routine day-in and day-out occurrences from the ISLE. The discrete beginning ensured that the ISLE was measuring specific event transactions. Without a distinct boundary such as this, it would be difficult to know when one event stopped or another began. For example, some measures list events such as "troubled relationship with boss." This item describes an ongoing dispute but not an event, or even a series of events. This is not to say that ongoing difficulties will not be reflected in the ISLE measure through their effect on daily events. The ISLE does not assess ongoing problematic situations directly; rather, it detects their results when such situations erupt into recurrent undesirable events. With the inclusion of this criterion, the ISLE maintains the distinction of measuring event transactions apart from routine daily time budgeting and permits a quantitative assessment of daily and weekly fluctuations in life stressors; desirable events are also formally assessed with the same ISLE instrument.

For events still to be considered small, the upper magnitude score of items chosen for inclusion in the ISLE is 250 life change units (Holmes & Rahe, 1967). This cutoff point was also set to ensure that items on the ISLE did not overlap with the set of major events found on its companion instrument from the PERI and other inventories assessing major life experiences. ISLE items were written to be either clearly desirable or undesirable to facilitate research that investigates the differential effects of desirable and undesirable event occurrences on psychological outcome. In addition to the previous, single linear dimension of "magnitude-of-required-readjustment," the desirability information on ISLE items allows research to address event desirability as a variable of interest.

As originally constructed, the ISLE contains 178 items. Of these, 98 are desirable and 80 are undesirable (see below). The events on the ISLE are divided into 13 categories by topical area: school, recreation, religion, money and finance, transportation, children, household, relations with family, love and marriage, crime and legal, social life, health and illness, and work. Exemplary items include the following: "Did poorly on an important test" (undesirable), "Learned that child(ren) did particularly well in a school project" (desirable), "Neighborhood noise disrupted sleep" (undesirable), and "Praised by spouse/mate" (desirable). The Appendix contains a complete list of all items, grouped by topical area, along with items gleaned from write-in events provided by elderly subjects during telephone assessments of everyday events of 265 older adults.

INITIAL TEST OF ITEM PROPERTIES

We have subjected the scale to several tests to examine how well the items meet the criteria listed. Utilizing responses from samples of 39 undergraduate raters and 14 experts in the field of life event research, standard magnitude estimation procedures provided consensus ratings of readjustment, control, desirability, and personal causation of items on the original scale. The results of those tests are reported in detail (Zautra et al., 1986). After the ISLE was revised for use with an older adult population, the scale was further evaluated. Readjustment ratings were made by 10 experts in aging and life event research. Control, causation, and desirability ratings were obtained by 5 experts and 12 female older adult interviewers, who had conducted telephone assessments of older adults. We have summarized those ratings in the Appendix and review those findings briefly here.

There was good agreement overall between naive and expert raters on ISLE event characteristics. Most of the events had average readjustment scores of 250 or less. Seven items that had scores higher than 300 from both the college students and expert raters were not considered small events. For example, the event, "Stopped smoking," had an average readjustment rating of 319. Twenty-eight items (e.g., problem with insurance company or government benefits) were judged to be ambiguous with regard to size. We have recommended retaining those items for further study to determine whether they should be classified as major or minor stressors. The ratings of control and causality were less distinct. Many events appear to have arisen from transactions in which control and causality were shared between the participant and others in their social networks. For example, responsibility for the event "Argument with family member" is clearly shared, and the ratings reflect that property.

Also included in Appendix are the rates of occurrence and recurrence of these small events for an elderly sample to be described presently. Frequency of occurrence may be useful as a means of identifying stressful and also pleasurable events for elderly persons in different living environments. The frequencies may also be used to select subsets of events to trim the scale of items that occur infrequently or, alternatively, to investigate the impact of relatively rare events among a sample of elderly. The recurrence probabilities were computed using the first interview as the standard. The number of times an event was reported again across nine monthly reassessments was divided by nine for those subjects who reported the event occurring at the initial interview.

A total of 11 items showed less than 90% agreement in ratings of desirability: "Installed a security system in your home," for example, was rated as undesirable by 15% of the raters. Three of these ambiguous items were not expected to be uniformly desirable or undesirable because they referred to medical practices (e.g., "stopped taking a drug prescribed by a doctor").

These initial tests of the ISLE revealed not only that the scale appears to meet the criteria established for a measure of small events but also that such tests are important to conduct; one cannot assume the event list would meet criteria based on face-valid evidence. Some events clearly were not small; others failed to be rated as clearly desirable or undesirable. Some researchers have argued that what matters most is how the subjects themselves appraise these events. While such idiographic ratings are of great importance (Lazarus & DeLongis, 1983; Lazarus, DeLongis, Folkman, & Gruen, 1985), we do not think these person-specific measures should substitute for good event-specific measures. In our view, events ought to represent person–environment transactions in a way that is as free as possible from psychological reactions to events. Psychological appraisals should be measured as a separate facet of the life–stress process.

ASSESSING THE RELATIONSHIP BETWEEN SMALL DAILY EVENTS AND MAJOR EVENTS

Our next step was to determine the relationship between our ISLE small daily events measure and the two major events measured in our study of the stressors of older adults: recent physical disability and recent conjugal bereavement. These two major stressors set the context of daily living for many older adults. From our event-based approach, the question is the extent to which these two stressors lead to different patterns of daily living, the latter assessed in this study as event scores on the ISLE. In the following sections, we first examine differences in depression between these two groups. We then perform detailed analyses of relationships between these major stressors, the ISLE measure of everyday events, and depression.

THE LIFE EVENTS AND AGING PROJECT

The data for our analyses of major and small daily events are derived from our study of aging and life events, the Life Events and Aging Project (LEAP).

A total of 246 noninstitutionalized community residents between the ages of 60 and 80 were recruited for participation in the study. Of these, 62 persons had disabilities, suffering from illness, accident, or worsening of an existing physical condition, resulting in at least moderately limited activity within 3 months of initial screening. Sixty-one persons were conjugally bereaved within 6 months of the initial screening, and 123 were control participants matched (with either a bereaved or disabled subject) on sex, age (within 5 years), and income (or neighborhood residence if income was missing). Control participants were not conjugally bereaved within 2 years of initial screening, were not functionally disabled, and had not been disabled for the year prior to initial screening.

The average age of the participants was 70 years: 77% were female and 1.6% were of minority status. Of those with disabilities, 62.5% were high school graduates compared with 75.6% of the controls and 82.5% of those who were recently bereaved ($ps < .05$). Median annual income among bereaved persons was \$15,000 compared with \$14,000 among controls and \$11,600 among disabled persons ($p < .05$). Disabled subjects reported an average of 13.6 functional limitations on the Gurland Activity Limitations Scale (Teresi, Golden, Gurland, Wilder, & Bennet, 1983). Participants from other groups were screened out if they had high numbers of activity limitations. The bereaved group and controls had an average of 1.29 and 1.32 limitations, respectively.

In order to recruit participants, a broad base of contacts in the community was established. A total of 13 organizations assisted in locating participants. Permission was obtained to identify prospective bereaved subjects from vital statistics records. Once identified, these residents were contacted first by mail and then by a telephone call from a project staff member. Control groups were constituted through agency-assisted screening, by neighborhood canvassing, and through participant and interviewer referrals. In targeting areas for canvassing, neighborhoods with a high proportion of older adults, including bereaved and disabled residents, were chosen. These included approximately 20 mobile home parks and two census tracts in which more than 20% of the residents were over 60 years of age.

Initial contacts were followed up in telephone screening, by trained interviewers. A total of 710 residents was contacted. Of these, 160 (23%) refused to participate and 281 (40%) agreed to participate but did not meet screening criteria. After being accepted into the study, participants were assigned to a trained, female older adult interviewer. Only data from the first monthly interview were used in the present investigation. Participants received a stipend in return for their participation. A total of 206 participants had no missing data and constitute the sample in this study. Those subjects excluded from the study because of missing data were compared to those remaining on all measures. One significant difference was found; subjects with missing data reported fewer undesirable events at the second interview than subjects without missing data ($t(64) = 3.75$, $p < .01$). No other differences were observed, leading us to surmise that the requirement that all subjects have no missing data did not bias the sample significantly.

GROUP DIFFERENCES IN FREQUENCY OF EVERYDAY EVENTS

Because of the comprehensive nature of the ISLE, was obtained an unusually complete picture of the occurrence of small events for each individual's life. Therefore, findings were interpreted in terms of both significant differences between groups of people undergoing various major life events and similarities among those same groups, regardless of major life events. In this way, we could present a fuller understanding of the relations between major life events and small life events.

The first question we addressed was the extent to which the two major stressors—physical disability and conjugal bereavement—led to both different and similar patterns is subjects' daily living compared to matched controls. (Both the matched control group for those with disabilities and the matched control group for those recently bereaved were combined into one group for ease of comparison). To answer this question, mean frequencies of the 13 small events were calculated within the three groups and compared with a one-way three-level analysis of variance (ANOVA). Further t-tests were then computed where applicable to determine which of the three groups differed from each other.

The majority of significant differences fell between the bereaved group and either the disabled and/or control groups. Bereaved subjects reported fewer undesirable events in the financial area compared to disabled subjects, ($t(111) = 2.55$, $p < .05$), and controls ($t(164) = 2.36$, $p < .05$). As a direct consequence of the death of their partner, they reported fewer love desirable events (i.e., started a love affair, had long conversation with spouse/mate) compared to disabled subjects, ($t(111) = 5.43$, $p <.01$, and controls, ($t(164) = 5.61$, $p <.01$, and fewer love undesirable events as compared to disabled subjects, ($t(111) = 2.90$, $p < .01$, and controls, ($t(164) = 3.59$, $p < .01$). As a group, they also reported fewer desirable events in legal matters compared to disabled subjects ($t(111) = 2.04$, $p < .05$). Not surprisingly, disabled individuals reported significantly more desirable health (promotion) events compared to controls ($t(165) = 2.42$, $p < .05$), and more undesirable health events compared to bereaved subjects, ($t(111) = 5.15$, $p < .01$), and controls, ($t(165) = 5.73$, $p > .01$). The three groups reported similar numbers for school, recreation, religion, desirable financial, transportation, children, household, family, crime, social, and employment events. These numerous areas of equality may be referenced as sections of life not necessarily affected by the two major events of disability and bereavement. Table 5.1 provides a summary of group differences and similarities among disabled, bereaved, and control groups in the occurrence of the 13 small desirable and 13 small undesirable events. Overall, the findings show expected effects of major stressors on everyday life and little evidence for the spread of stressful events to areas outside the domain of the major stressor.

THE RELATION BETWEEN EVERYDAY EVENTS AND DEPRESSION

The next question we addressed was how everyday events are related to depression. We were interested in the impact of small desirable, small undesirable, health desirable, and health undesirable events, and whether that impact would differ depending on the type of major life stress participants were experiencing. The 12 categories of small desirable events and 12 matching categories of small undesirable events (not

Table 5.1. Group Differences in Mean Numbers of Small Events

Type of event	Disabled	Bereaved	Controls
School desirable	.14	.25	.15
School undesirable	.14	0	.01
Recreation desirable	1.51	2.13	1.83
Recreation undesirable	.28	.11	.15
Religion desirable	.53	.64	.55
Religion undesirable	.04	.07	.06
Money desirable	1.21	1.39	1.21
Money undesirable	.86[a]	.48[b]	.81[a]
Transportation desirable	.35	.21	.27
Transportation undesirable	.19	.14	.23
Children desirable	.98	1.27	1.07
Children undesirable	.25	.27	.15
Household desirable	.33	.45	.42
Household undesirable	1.14	.80	1.0
Family desirable	1.14	1.03	.93
Family undesirable	.16	.07	.09
Love desirable	1.32[a]	0[b]	1.27[a]
Love undesirable	.42[a]	0[b]	.53[a]
Crime desirable	.07[a]	0[b]	.02
Crime undesirable	.25	.11	.05
Social desirable	3.21	3.91	3.57
Social undesirable	.77	.63	.57
Employment desirable	.28	.57	.49
Employment undesirable	.21	.45	.20
Health desirable (promotion)	1.09[a]	.87	.67[b]
Health undesirable	6.51[a]	3.59[b]	3.59[b]

Note. All means in rows with different superscripts are significantly different from each other.

including health) were combined to produce overall scores on small desirable and small undesirable events, respectively. Health events were not included in this general group because they formed a special class of life experiences. We measured depression with 10 items from the Veit and Ware (1983) Mental Health Inventory, which formed a depression factor in our confirmatory factor analyses (Zautra, Guarnaccia, & Reich, 1988).

Pearson correlations were calculated among small desirable, small undesirable, health desirable, health undesirable, and depression events by group. For disabled persons, small desirable events were associated with less depression, and health undesirable events were associated with more depression. For bereaved persons, health undesirable events were associated with more depression. For the control group, small undesirable events, health undesirable events, and health desirable events were all associated with more depression; these results are presented in parentheses in Table 5.2.

Because desirable and undesirable events typically show positive correlations with each other (Zautra et al., 1988) yet correlate in opposite directions with some measures of mental health, we expected that suppressor effects would be operating such that the correlation between undesirable event scores and depression might change after "correcting" for the effects of small desirable events. Likewise, the correlation between small desirable events and depression might also be influenced by the pres-

Table 5.2. Correlation between Small Event Indices and Depression for Older Adult Disabled, Bereaved, and Control Groups

	Disabled n = 51–54		Bereaved n =28–51		Controls n = 101–104	
Small desirable	–.48*	(–.30*)	–.11	(–.08)	–.27*	(–.11)
Small undesirable	.42*	(.13)	.10	(.08)	.32*	(.21*)
Health undesirable	.24	(.24)	.49*	(.48*)	.33*	(.39*)
Health desirable	–.06	(–.08)	–.09	(.02)	.15	(.26*)

Note. Partial correlations for desirable (controlling for undesirable) and undesirable (controlling for desirable) are presented first. Raw correlations reported in parentheses.
$^{*}p < .05$, $p < .08$

ence of small undesirable events that are correlated with desirable events. Therefore, we calculated partial correlations between (1) small desirable events and depression, controlling for small undesirable events; (2) small undesirable events and depression, controlling for small desirable events; (3) health desirable events and depression, controlling for health undesirable events; and (4) health desirable events and depression, controlling for health undesirable events.

As expected, many of the relations between the events and depression changed when controlling for each counterpart measure as a function of group membership (see Table 5.2). For disabled persons, small desirable events showed a stronger relation to decreased depression, small undesirable events showed a newly significant positive relation to increased depression, and health undesirable events showed a tendency toward increased depression. In contrast, no relations for those recently bereaved subjects had significant changes when controlling for counterpart measures. For the control group, small desirable events showed a stronger relation to decreased depression, and small undesirable events showed a stronger relation to increased depression. Conversely, health undesirable events showed a reduced relation to increased depression, and the relation between health desirable events and depression lost significance. To determine if groups significantly differed on correlations between events and depression, a t test for significantly different partial correlations was computed. It showed a significant difference in disabled and bereaved participants in the relation between small desirable events and depression, ($t(104) = 2.05$, $p < .05$); people with disabilities displayed a stronger association between reported small desirable events and decreased depression. We explored these findings further in the next series of analyses.

GROUP DIFFERENCES IN CONTRIBUTION OF SMALL EVENTS TO DEPRESSION

The results thus far suggested that the major comparison for how small events relate to depression is between subjects recently bereaved compared to both those with disability and those in the control group. The issue of how events relate to depression differently within these groups was then further explored. In these tests, we were interested in the extent to which everyday events mediated the relation between the major stressors of disability and bereavement and depressive symptoms, and whether desirable events in everyday life served to protect elders from depression.

To answer the first question, we performed two separate regressions, each contrasting one major life event group (disabled or bereaved) with all of the other groups combined. We dummy coded a variable to score 1 for a single group and 0 for all other groups. We then performed a 5-step regression of dummy-coded group (Step 1), small desirable and small undesirable events (Step 2), group by small events interactions (Step 3), desirable and undesirable health events (Step 4), and group by health events interactions (Step 5) on depression. The interactions provided tests of differences between groups in the role of everyday events.

The results (see Table 5.3) showed that with the inclusion of health events, having a disability lost its significance for predicting depression. Conversely, both small desirable events and the interaction between group and small desirable events remained significant predictors throughout the regression sequence, including the addition of health events. This shows the importance of health events in relation to depression for disabled individuals and adds support to the comparatively greater importance of desirable events in reducing depression for those with disability, both individually and compared to the other groups (as was reflected in the previous test of differences between correlations).

In contrast to those with disability, the relationship between subjects with bereavement and depression grew stronger when small desirable and undesirable events were included. Table 5. 4 shows these relationships. Both small desirable events and undesirable health events also predicted depression throughout the regression series but did not account for the higher depression scores for those recently bereaved. These results suggest that neither small desirable events nor small undesirable events played a key role in the depression of bereaved persons.

SUMMARY AND CONCLUSIONS

In this chapter we have explored the role of everyday events in the lives of older adults. In order to pursue this question, we have developed a new inventory designed to assess thoroughly the everyday events of older adults. Since the stresses of life, particu-

Table 5.3. Regression of Disabled Group (vs. Others) and Events on Depression

	Steps							
	1		2		3		4	
	b	SE	b	SE	b	SE	b	SE
Disabled group (vs. others)	.22**	.13	.18**	.13	.15*	.13	.02	.14
Small desirable events			–.28**	.01	–.18*	.01	–.18*	.01
Small undesirable events			.23**	.02	.17	.02	.02	.02
Disabled group*small desirable events					–.22*	.03	–.18*	.02
Disabled group*small undesirable events					.15	.04	.15	.04
Desirable health events							.12	.06
Undesirable health events							.33**	.02
R^2	.04		.11		.13		.23	

Note. $N = 224$
$^*p < .05$, $^{**}p < .01$

Table 5.4. Regression of Bereaved Group (vs. Others) and Events on Depression

	Steps							
	1		2		3		4	
	b	SE	b	SE	b	SE	b	SE
Bereaved group (vs. others)	.13	.14	.17*	.13	.15*	.14	.18**	.13
Small desirable events			–.32**	.01	–.37**	.01	–.30**	.01
Small undesirable events			.29**	.02	.34**	.02	.15	.02
Bereaved group*small desirable events					–.09	.03	.06	.03
Bereaved group*small undesirable events					–.10	.05	–.05	.05
Desirable health events							.08	.06
Undesirable health events							.38**	.02
R^2	.02		.10		.11		.25	

Note. $N = 224$
$^*p < .05$, $^{**}p < .01$

larly bereavement and disabling illness, play such a prominent role in defining the adaptation challenges of older adults, we focused attention on those two experiences. We contrasted the quality of everyday life between those elders with recent spousal bereavement and controls, and between elders with a disabling illness and controls. As expected, these high-stress groups reported more depression than those without either recent spousal bereavement or a disabling illness. Of interest, though, were the effects of major stressors on minor ones. In general, there were few surprises in these findings; in most cases, differences in small events between groups could be readily predicted as a consequence of the major stressor. Rarely did we observe effects of these major stressors spreading to other areas of life. These findings do not preclude differences in perceptions among elders concerning changes in quality of their everyday lives following spousal bereavement or reduction in functional abilities. Indeed, many elders feel and report a sense of loss that is much more pervasive than what we find when counting the events in their everyday lives.

What we are able to measure are key areas of everyday life in which the actual number of desirable and/or undesirable transactions with the social and physical environment differs between groups. In doing so, our methods provide a means for focusing theory and interventions on the variables of interest. Thus, for most older adults who profess widespread changes in the quality of their lives following major calamities, we may reinterpret their affliction as one involving only a few that that (over)generalize the impact of those major stressors. Likewise, we may revise our own social theories about the impact of these major life events among older adults, holding in check the temptation to predict widespread changes in everyday life for those experiencing loss of spouse and those coping with disabling illnesses. We would suggest that this result potentially has direct relevance for therapeutic practice with such populations.

Our methods provide a means of evaluating the role of everyday stressors as important correlates of depression and everyday desirable events as potential sources of recovery from depressive symptoms for different groups. Our results show that the relationships between everyday life events and depressive symptoms differ substantially depending upon the source of the depression. For those in our study who were depressed as a consequence of a recent death of the spouse, everyday events were inconsequential. The exception to these findings were daily health symptoms that, when present, led to more depression for all groups. Even when accounting for the effects of those health events, we were unable to account for the elevated levels of depression for the recently bereaved in comparison to controls by measures of daily life events. We infer from these data that theories of depression among bereaved persons need to look to mechanisms other than stresses of everyday life to understand grieving. We would encourage more attention to processes associated with loss of meaning and purpose in life as central to the depression following death of a spouse. Likewise, interventions designed to speed recovery from loss-related depression may find little value in a focus on everyday life events, whether it is the reduction of stress or promotion of desirable social life events.

Our findings for spousal bereavement are not likely to hold for younger age cohorts. During child rearing and working to sustain economic well-being, spousal bereavement can have a dramatic effect on everyday life. Recently, Pillow, Zautra, and Sandler (1996) found that many everyday stressors, measured with a modified version of our scale, were associated with spousal bereavement in their study of families with young children. Unlike the current study, we also found clear evidence that small stressors did account for a significant portion of the psychological distress of the surviving spouse.

The depression that results from functional disability is very different from depression as a consequence of bereavement for the elderly subjects in our study. Everyday life events appear to play a central role in maintaining depressive symptoms. Differences between those with disability and other groups relative to depression could be fully accounted for by differences in the quality of everyday life events between groups. Indeed, interventions oriented toward coping with everyday life stressors, including social difficulties and daily health complaints, would appear to have promise given the findings we uncovered in our studies (see chapter by Williamson & Shaffer, Chapter 9, this volume).

Of particular interest in our investigations is the strong role for everyday desirable events in lowering depression among disabled persons. The impact of these events was stronger for those with disability than the other groups. This finding supports the use of behavioral interventions that focus on improvements in the quality of everyday life experiences as a means of lifting depressive symptoms for disabled elderly. We think there are important implications for theories of depression as well. First, it is apparent that the stresses of everyday life are not the sole contributor to depressive symptoms for those with disability. The relative absence of desirable events plays a key part in maintaining depression even after accounting for various forms of undesirable and otherwise stressful events. Needles and Abramson (1990) have recently developed a new model of recovery from depression in which desirable events and attributions about those events are the key factors. Our findings provide partial support for that model and urge further theoretical work along these lines, such as that provided in Williamson and Shaffer's Activity Restriction Model (see Chapter 9, this volume).

In summary, what we offer with our focus on small events in the lives of the elderly is attention to details that may inform us of influential aspects of everyday life that have been neglected in theories of depression. The measurement of everyday life events also provides a means of testing our model about how life has changed for elderly adults, identifying parts of it that may incorrectly ascribe important roles for everyday life events. Our findings reveal how different sources of depression among the elderly require attention to different aspects of their adjustment to life's difficulties.

We offer use of the inventory itself, and the initial norms on frequencies of occurrence of the events found in the Appendix, for use in describing the quality of the everyday lives of the elderly in other contexts. Investigators may wish to focus on subsets of items depending upon the needs of their research. We have found it useful to study interpersonal events, for example, in the study of disease activity among arthritis patients (e.g., Zautra et al., 1997). Event measurement using our methods also provides a useful backdrop by which to differentiate the role of cognitive and affective responses to events from the events themselves, as a recent study of ours on desirable and undesirable event efficacy demonstrates (Zautra, Reich, & Newsom, 1995). Unless one measures events independently of reactions to those events, it is impossible to distinguish effects due to the events themselves versus effects due to our responses to those events. In the study of the etiology and course of depression, this distinction is especially important to retain.

Future studies may also benefit by comparing and contrasting the findings for perceptions of stress with reports of stressful events. The individual differences in reports of events are sizable; whereas some elderly adults have few events, both desirable and undesirable, others have many. Appraisals of the stressfulness of events are also highly variable; some elders have highly anxious profiles, while others appear serene when faced with stressors. Individuals also differ substantially in their physiological responses to stressful events. One person may register higher blood pressure and quicker recovery to resting levels following stress than another. Cortisol, prolactin, and other stress hormones also map out unique physiological response profiles (Calloway & Dolan, 1989; McEwen, 1998). Which responses to stressful events lead to better health? This is a key question in research in psychosomatic medicine. Indeed, there is evidence that the difficulties in adaptation for persons with depressive disorders are most evident in their responses to stressful situations. The role of desirable events in daily life has been neglected in most research on depression, and we urge that greater attention be paid to how affective qualities of daily life experience may contribute to mental health among older adults. During stressful or otherwise distressing events, desirable affective experience may play a key role in adaptation for older adults (Reich & Zautra, 1988). By attending to individual and group differences in desirable events, investigators and practitioners alike will gain insights into the full range of the life experiences for older adults and be better able to understand the contribution of positive states to mental and physical health.

REFERENCES

Calloway, P., & Dolan, R. (1989). Endocrine changes and clinical profiles in depression. In G. W. Brown & T. O. Harris (Eds.), *Life events and illness* (pp. 139–160). New York: Guilford Press.

Dohrenwend, B. S., Krasnoff, L., Askenasy, A. R., & Dohrenwend, B. P. (1978). Exemplification of a method for scaling life events: The PERI Life Events Scale. *Journal of Personality and Social Psychology, 19,* 205–229.

Dohrenwend, B. P., & Schrout, P. E. (1985). "Hassles" in the conceptualization and measurement of life stress variables. *American Psychologist, 40,* 780–785.

Gatz, M., & Hurwicz, M. L. (1990). Are older people more depressed? Cross-sectional data on Center for Epidemiological Studies Depression Scale factors. *Psychology and Aging, 5,* 284–290.

Holmes, T. H., & Rahe, R. H. (1967). The social readjustment scale. *Journal of Psychosomatic Research, 11,* 213–218.

Lazarus, R. S., & DeLongis, A. (1983). Psychological stress and coping in aging. *American Psychologist, 38,* 245–254.

Lazarus, R. S., DeLongis, A., Folkman, S.,& Gruen, R. (1985). Stress and adaptational outcomes: The problem of confounded measures. *American Psychologist, 40,* 770–779.

Lewinsohn, P. M., Rohde, P., Fischer, S. A., & Seeley, J. R. (1991). Age and depression: Unique and shared effects. *Psychology and Aging, 6,* 247–260.

McEwen, B. S. (1998). Protective and damaging effects of stress mediators. *Seminars in Medicine of the Beth Israel Deaconess Medical Center, 338*(3), 171–179.

Needles, D. J., & Abramson, L.Y. (1990). Positive life events, attributional style, and hopefulness: Testing a model of recovery from depression. *Journal of Abnormal Psychology, 99*(2), 156-165.

Pillow, D. R., Zautra, A. J., & Sandler, I. (1996). Major life events and minor stressors: Identifying mediational links in the stress process. *Journal of Personality and Social Psychology, 70*(2), 381-394.

Reich, J. W. & Zautra, A. J. (1988). Direct and stress-moderating effects of positive life experiences. In L. H. Cohen (Ed.), *Life events and psychological functioning: Theoretical and methodological issues.* (pp. 149–181). Newbury Park, CA: Sage.

Roberts, R. E., Kaplan, G. A., Shema, S. J., & Strawbridge, W. J. (1997). Prevalence and correlates of depression in an aging cohort: The Alameda County Study. *Journal of Gerontology: Social Sciences, 52B,* S252–S258.

Teresi, J. A., Golden, R. R., Gurland, B. J., Wilder, D. E., & Bennet, R. G. (1983). *Construct validity of indicator scales develop from the comprehensive assessment and interview schedule.* (Unpublished manuscript, available from the Center for Geriatrics, Columbia University, 100 Haven Ave., New York, NY 10032).

Veit, C. T., & Ware, L. E., Jr. (1983). The structure of psychological distress and well-being in general populations. *Journal of Consulting and Clinical Psychology, 51,* 730–742.

Williamson, G. M., & Schulz, R. (1992). Pain, activity restriction, and symptoms of depression among community-residing elderly adults. *Journals of Gerontology, 47*(6), 367–372.

Zautra, A. J., Affleck, G. & Tennen, H. (1994). Assessing life events among older adults. In M.P. Lawton & J. A . Teresi (Eds.), *Annual review of gerontology and geriatrics* (pp. 324–352). New York: Springer.

Zautra, A. J., Guarnaccia, C. A., & Dohrenwend, B. P. (1986). The measurement of small life events. *American Journal of Community Psychology, 14,* 629–655.

Zautra, A. J., Guarnaccia, C. A., & Reich, J. W. (1988). Factor structure of mental health measures for older adults. *Journal of Consulting and Clinical Psychology, 56*(4), 514–519.

Zautra, A. J., Hoffman, J., Potter, P. T., Matt, K. S., Yocum, D., & Castro, L. (1997). Examination of changes in interpersonal stress as a factor in disease exacerbations among women with rheumatoid arthritis. *Annuals of Behavioral Medicine, 19*(3), 279–386.

Zautra, A. J., Reich, J. W., & Newsom, J. T. (1995). Autonomy and sense of control among older adults: An examination of their effects on mental health. In L. Bond, S. Cutler, & A. Grams (Eds.), *Promoting successful and productive aging.* (pp. 103–170). Newbury Park, CA: Sage.

Zeiss, A. M., Lewinsohn, P. M., Rohde, P., & Seeley, J. R. (1996). Relationship of physical functional impairment to depression in older people. *Psychology and Aging, 11,* 572–581.

APPENDIX: RESEARCH INVENTORY OF MAJOR AND SMALL LIFE EVENTS FOR OLDER ADULTS*

Major and Small Event Readjustment, Control, and Cause Ratings for Older Adults Organized by Desirability with Monthly Probability of Occurrence & Reoccurrence

Category	Mean re adjustment	Control over event	Cause of event	Monthly probability of occurrence	Monthly probability of reoccurrence
School Section					
Undesirable school events					
Major magnitude event(s)					
Could not pay tuition when due	289.3	M	M	.002	.000
Small magnitude event(s)					
Did poorly on an important test	207	C	I	.001	.000
Excluded from participation in a valuable course	204	U	E	.002	.000
Homework assignments became extra heavy	190.5	U	E	.004	.023
Were late in registering for a class	125.5	C	I	.001	.000
Had to miss class(es) because of family or work demands	120	M	E	.004	.013
Desirable school events					
Major magnitude event(s)					
Started school or a training program after not going to school for a long time[1]	340	C	I	.016	.024
Graduated from school or training program	317.5	C	I	.008	.007
Taught a class[2]	288.5	C	I	–	–
Small magnitude event(s)					
Took a stimulating class/seminar	159	C	I	.018	.053
Got a good grade on a difficult test	154.8	C	I	.009	.037
Completed work on an interesting school project	154	C	I	.009	.071
Passed a course	138.2	C	I	.008	.030
Obtained convenient class schedule	116	M	I	.012	.044
Recreation Section					
Undesirable recreation events					
Major magnitude event(s)					
Pet was very sick and needed extra attention	260.5	U	E	.026	.118
Had to stop a hobby, sport or recreational activity	258.1	M	I	.075	.076

*Zautra & Guarnaccia (1988, November). Paper presented at the 41st Annual Meeting of the Gerontological Society of America, San Francisco, CA.

Category	Mean re adjustment	Control over event	Cause of event	Monthly probability of occurrence	Monthly probability of reoccurrence
Small magnitude event(s)					
Got minor injury from physical exercise	179.5	M	I	.036	.035
Called off planned (weekend or longer) vacation	170.5	C	I	.063	.055
Desirable recreation events					
Major magnitude event(s)					
Took up a hobby or other recreational activity[1]	281	C	I	.095	.094
Ambiguous magnitude event(s)					
Acquired a pet	231.6	C	I	.017	.048
Small magnitude event(s)					
Went on a vacation (for a weekend or longer)	209.5	U	E	.213	.187
Attended concert, play, lecture, etc.[2]	126.5	C	I	–	–
Went to a sporting event	123.4	C	I	.086	.139
Took a pleasurable trip (to the beach, a drive in the country, etc.)	115.5	C	I	.274	.260
Went to club or organized group meeting	108	C	I	.484	.505
Visited a gallery, exhibit or museum	104	C	I	.099	.086
Went swimming, biking, etc.[2]	106.6	C	I	–	–
Went to an activity at a senior center[2]	99.5	C	I	–	–
Went shopping for pleasure	95.5	C	I	.541	.523
Religion Section					
Undesirable religion events					
Major magnitude event(s)					
Had to attend a funeral service[1]	294	M	E	.083	.092
Small magnitude event(s)					
Broke an important rule or commandment of your religion	211	C	I	.025	.094
Priest/Rabbi/Minister could not see you when you asked	173.5	U	E	.007	.118
Desirable religion events					
Ambiguous magnitude event(s)					
Began to get involved with a church/ religious group	227	C	I	.082	.048
Small magnitude event(s)					
Did church volunteer work[2]	148.5	C	I	–	–
Had a call/visit from pastor/priest or church member[2]	123	M	E	–	–
Observed a religious holiday	119.3	C	I	.193	.139
Attended a particularly satisfying religious program or service	119	C	I	.295	.187
Made donations to a church or charity[2]	83.1	C	I	–	–
Watched or listened to TV or radio service[2]	74.5	C	I	–	–

Category	Mean re adjustment	Control over event	Cause of event	Monthly probability of occurrence	Monthly probability of reoccurrence
Money and Financial Section					
Undesirable money and financial events					
Major magnitude event(s)					
Repossession of a car, furniture or other items bought on installment plan	379.5	U	M	.002	.004
Pension/federal aid, etc. was cut or lost[2]	374	U	E	–	–
Had to take charity	364.5	U	E	.013	.078
Ran out of money and could not cover living expenses this month	354.5	M	M	.048	.166
Failed to qualify for funds to pay medical expenses	352	U	E	.011	.027
Went on welfare	321	U	M	.002	.000
Suffered a financial loss/loss of income/ loss of property not related to work	285	U	E	.021	.039
Ambiguous magnitude event(s)					
Problems with insurance company or government benefits[2]	260	U	E	–	–
Small magnitude event(s)					
Ran out of money and could not help child(ren) with finances	256.5	M	M	.023	.058
Received threatening news from a creditor (by phone or mail)	256	U	M	.027	.084
Turned down for credit (for example charge card, loan)	244	U	M	.010	.025
Rent or mortgage payment was increased	233	U	E	.098	.074
Had an unexpected expense over $50.00 but under $500.00	228	U	E	.305	.254
Did not get pension, social security (or other government) check on time	216.5	U	E	.021	.051
Did not get unemployment compensation on time	192	C	I	.001	.000
Found a large unfavorable error in your check book balance	192	C	I	.033	.058
Questionable desirability money and financial events					
Major magnitude event(s)					
Started buying a car, furniture or other large items bought on installment plan[1]	264	C	I	.018	.023
Took out a mortgage	260.5	C	I	.007	.004
Desirable money and financial events					
Major magnitude event(s)					
Went off welfare	278	M	M	.002	.016
Small magnitude event(s)					
Qualified for public assistance program[2]	206	U	M	–	–
Paid off debt	200.5	C	I	.098	.116
Increase in income[2]	182.5	C	I	–	–

Category	Mean re adjustment	Control over event	Cause of event	Monthly probability of occurrence	Monthly probability of reoccurrence
Rent or mortgage payment was reduced	158	U	E	.021	.014
Developed a budget that would save money	140	C	I	.052	.029
Put money in savings	139.5	C	I	.363	.330
Had government benefits extended (unemployment, medicare)	133	U	E	.081	.059
Received money from insurance, government or interest payment[2]	128.7	C	M	–	–
Received money as a refund	121	M	M	.139	.163
Had financial improvement not related to work	119	C	M	.000	–
Got a stock dividend	111	M	E	.276	.176
Won a small amount of money (under $50.00)	90.3	M	M	.121	.079
Received money (under $50) as a gift	84.5	U	E	.114	.085
Small magnitude event(s)					
Loaned money[2]	139	C	I	–	–
Transportation Section					
Undesirable transportation events					
Ambiguous magnitude event(s)					
Involved in a traffic accident in which there were no injuries	260.5	M	M	.011	.012
Small magnitude event(s)					
Car broke down[1]	227	U	E .0		
Got a traffic ticket for a moving violation (for example, for speeding, or going through a red light)	180	C	I	.010	.001
Public transportation used broke down or stopped running	156	U	E	.004	.000
Locked keys in car[2]	138.5	C	I	–	–
Were a passenger in a car/bus with a poor driver	135.5	M	M	.047	.077
Got a parking ticket	91.5	C	I	.012	.117
Questionable desirability transportation events					
Small magnitude event(s)					
Used Dial-A-Ride or bus[2]	87.5	C	I	–	–
Desirable transportation events					
Small magnitude event(s)					
Bought a vehicle[2]	221.5	C	I	–	–
Took a trip other than a vacation/ recreation trip	162.5	C	I	.046	.065
Got your car fixed	112	C	I	.183	.170
There was an improvement in public transportation used	96	U	E	.015	.044
Found a new convenient parking place for car	63.6	M	M	.018	.126

Category	Mean re adjustment	Control over event	Cause of event	Monthly probability of occurrence	Monthly probability of reoccurrence
Household Section					
Undesirable household events					
Major magnitude event(s)					
Lost a home through fire, flood or other disaster	890	U	E	.000	–
Moved to a worse residence or neighborhood	478	M	M	.002	.052
Water or storm damage to your home[2]	280.5	U	E	–	–
Ambiguous magnitude event(s)					
Amount of living space in the house was reduced	260	M	M	.022	.024
Small magnitude event(s)					
Elevator broke down	217.5	U	E	.021	.191
Plumbing etc. broke down	216.5	U	E	.086	.078
Home had too much heat for a day or more	204.5	U	E	.058	.084
Home had too little heat for a day or more	195	U	E	.025	.059
Household appliance broke down stopped running well	165.5	U	E	.110	.083
Had to do home repairs or extra work around the house[2]	155.5	M	M	–	–
Neighbor noise disrupted sleep	143	U	E	.079	.116
Repair person or apt. super failed to fix something properly	138	U	E	.044	.085
Had to wait a long time for a repair person to arrive at your home	121.5	U	E	.044	.075
Were locked out of your home	113.6	M	I	.046	.058
Saw unwanted household pest (roaches, mouse, spider, etc.)	82.1	U	E	.232	.222
Household item broke (glass, dish, etc.)	65.5	M	M	.121	.124
Questionable desirability household events					
Major magnitude event(s)					
Moved to a residence or neighborhood no better or worse than the last one	412.5	C	I	.003	.046
Desirable household events					
Major magnitude event(s)					
Remodel a home or had a home built[1]	548	C	I	.001	.000
Moved to a better residence or neighborhood	420	C	I	.013	.007
Addition to home	291	C	I	.000	–
Small magnitude event(s)					
Amount of living space in the home was increased	197	C	I	.014	.036
Obtained household help	170	C	I		
Built or repaired something	140	C	I	.118	.145

Category	Mean re adjustment	Control over event	Cause of event	Monthly probability of occurrence	Monthly probability of reoccurrence
Finished big cleaning job in the house	134	C	I	.165	.165
Bought needed household appliance/item	116	C	I	.094	.065
Helped others with household work[2]	99	C	I	–	–
Child and Family Section[3]					
Undesirable child and family events					
Major magnitude event(s)					
Child died[1]	1036	U	E	.001	.000
Hit or abused by child/grandchild(ren)	591.8	U	E	.001	.000
Family member other than spouse/mate or child died	472.2	U	E	.009	.000
Child moved[2]	375	U	E	–	–
Ambiguous magnitude event(s)					
Child(ren) became sick and needed your attention	289.2	U	E	.049	.079
Argued with son-in-law or daughter-in-law	278	M	M	.014	.017
Had a serious argument with family member (not spouse/mate or child)[1]	262	M	M	.029	.050
Small magnitude event(s)					
Criticized by child(ren)	244.8	U	E	.047	.106
Child(ren) had major financial trouble[2]	238	U	E	–	–
Discovered that child(ren) has problems with his/her spouse	235	U	E	.041	.130
Fought with child(ren)	232.3	M	M	.021	.058
Child(ren) broke a major rule of the family	222	U	E	.011	.019
Had an argument with a family member (not spouse/mate or child)	209	M	M	.031	.055
Asked to baby-sit by children when you did not want to	205	U	E	.023	.108
Criticized or blamed for something by family member (not spouse or child)	199.5	U	E	.045	.049
Child(ren) wanted to visit when you could not or did not want to see them	191	U	E	.021	.036
Usual visit with your child(ren) was canceled or postponed	189	U	E	.030	.042
Discovered that child(ren) has problem(s) at work	186.5	U	E	.041	.093
Forced to visit with family member (not spouse/mate or child) when you did not want to	168.2	U	E	.019	.019
Questionable desirability child and family events					
Major magnitude event(s)					
New person moved into the household	315.4	M	M	.016	.008
Small magnitude event(s)					
Person moved out of the household	252.7	M	M	.013	.024

Category	Mean re adjustment	Control over event	Cause of event	Monthly probability of occurrence	Monthly probability of reoccurrence
Desirable child and family events					
Major magnitude event(s)					
Birth of a grandchild[2]	298	U	E	–	–
Small magnitude event(s)					
Child(ren)/grandchild(ren) visited[2]	173.5	M	M	–	–
Helped member of family (besides spouse/mate or child) with a personal problem	173.1	C	I	.103	.133
Helped child(ren) with a personal problem	169.3	C	I	.115	.180
Made an extra visit to child(ren) and/or grandchild(ren	161.5	C	I	.136	.149
Relieved of watching grandchild(ren) by friend, family member or spouse	161.5	U	E	.029	.050
Child(ren) had major job change[2]	153.5	U	E	–	–
Visited child(ren)/grandchild(ren)[2]	152.5	C	I	–	–
Child(ren) did something especially nice for you	145	U	E	.405	.405
Took grandchild(ren) to entertaining event	130	C	I	.079	.117
Praised by family member (not spouse/ mate or child)	126	U	E	.176	.185
Talked with family member (besides spouse/mate or child) you had not seen in a long time	123.7	C	M	.548	.515
Taught child(ren) or grandchild(ren) something new	115.6	C	I	.104	.233
Heard child(ren) or grandchild(ren) praise you to others	113	U	E	.137	161
Learned that child(ren) did particularly well at work	106.5	U	E	193	.196
Received a gift from a family member (not spouse/mate or child)	88	U	E	.186	.156
Talked on the phone with child(ren)/ grandchild(ren)[2]	76.5	M	I	–	–
Love and Marriage Section					
Undesirable love and marriage events					
Major magnitude event(s)					
Spouse/mate died	1225	U	E	.002	.000
Divorce	820	M	M	.001	.000
Married couple separated	606.5	M	M	.000	–
Spouse/mate engaged in marital infidelity	550	U	E	.000	–
Were hit by spouse or mate	495	U	M	.001	.030
Engaged in marital infidelity	481	C	I	.001	.000
Ended a love affair	447.5	C	M	.003	.004
Engagement was broken	435.5	M	M	.000	
Hit spouse/mate	432	C	I	.005	.000
Relations with spouse/mate changed for the worse without separation or divorce	394	M	M	.004	.030
Had a sexual problem with spouse/mate	392	M	M	.001	.059

Category	Mean re adjustment	Control over event	Cause of event	Monthly probability of occurrence	Monthly probability of reoccurrence
Small magnitude event(s)					
Saw spouse/mate flirt with another person	237	U	E	.007	.007
Spouse/mate was away from home overnight unexpectedly	203.3	U	E	.010	.054
Disagreed with spouse/mate about child(ren)	196.5	C	M	.031	.110
Argued with spouse/mate about something other than child(ren)	192	M	M	.071	.181
Spouse/mate stopped being affectionate for a day or more	188	M	M	.024	.090
Criticized by spouse/mate	174.1	U	M	.088	.258
Were critical of spouse/mate	171	C	I	.079	.265
Desirable love and marriage events					
Major magnitude event(s)					
Married[1]	500	C	I	.001	.000
Married couple got together again after separation	414	M	M	.000	–
Became engaged	359.5	C	I	.001	.000
Started a love affair	342	C	I	.004	.026
Small magnitude event(s)					
Relations with spouse/mate changed for the better	263	C	M	.022	.085
Celebrated special occasion with spouse/mate	179	C	I	.097	.201
Received a special gift from spouse/mate	165	U	E	.077	.163
Expressed love to spouse/mate	143	C	I	.248	.560
Had long conversation with spouse/mate	132	C	M	.175	.407
Crime and Law Section					
Undesirable crime and law events					
Major magnitude event(s)					
Went to jail	695	M	M	.000	–
Convicted or found guilty of a crime	622.8	M	M	.000	–
Physically assaulted or attacked	495.5	U	E	.000	–
Arrested	430.8	M	M	.000	–
Robbed	401	U	E	.006	.002
Didn't get out of jail when expected to	364	U	E	.000	–
Involved in a traffic accident in which there were injuries	362.6	U	M	.003	.000
Accused of something for which a person could be sent to jail	346.8	M	M	.001	.000
Burglarized	336.8	U	E	.005	.000
Physically harassed	321.5	U	E	.009	.092
Involved in a lawsuit	291.8	M	M	.009	.033
Ambiguous magnitude event(s)					
Suspected of doing something illegal by the authorities	275	U	E	.001	.000

Category	Mean re adjustment	Control over event	Cause of event	Monthly probability of occurrence	Monthly probability of reoccurrence
Got involved in a court case	248	U	E	.005	.040
Your personal property was damaged	244.8	U	E	.009	.038
An unsuccessful attempt was made to steal your property	241.8	U	E	.004	.018
Small magnitude event(s)					
Received harassing phone calls	220.5	U	E	.000	–
Broke a minor law (misdemeanor or lessor crime)	217.3	C	I	.005	.002
Discovered a friend or neighbor was a victim of a crime	189.8	U	E	.032	.036
Installed a new security system to protect your possessions	160.3	C	I	.011	.035
Cheated or short-charged in a store	141	U	E	.030	.021
Questionable desirability crime and law events					
Small magnitude event(s)					
Changed your normal route around town to avoid unsafe neighborhood	148.3	C	I	.003	.000
Changed locks in your home or auto	117.5	C	I	.016	.108
Desirable crime and law events					
Major magnitude event(s)					
Acquitted or found innocent of a crime[1]	468	M	E	.000	–
Released from jail	450.8	U	M	.000	–
Small magnitude event(s)					
Recovered property stolen from you	191.5	M	M	.004	.009
Began carrying a device to protect yourself from possible assailants	167.3	C	I	.012	.054
Social Life Section					
Undesirable social life events					
Major magnitude event(s)					
Close friend or relative died	525.5	U	E	.077	.058
Close friend or relative had to move away (institutionalized, nursing home, extended hospitalization)	348	U	E	.025	.037
Broke up with a friend[1]	328	C	M	.009	.052
Pet died/was lost	325	U	E	.005	.000
Close friend(s) left the neighborhood	272.5	U	E	.089	.088
Small magnitude event(s)					
Had problems with neighbors[2]	215	M	M	–	–
Not invited to a party given by friends	154	U	E	.015	.036
Criticized by friend/acquaintance	151.8	U	M	.060	.070
Argued with friend/acquaintance	142.5	M	M	.034	.042
Spouse/mate had argument with friend/neighbor	138.8	U	E	.009	.080

Category	Mean re adjustment	Control over event	Cause of event	Monthly probability of occurrence	Monthly probability of reoccurrence
Friend/acquaintance fails to show up for scheduled meeting	133.5	U	E	.028	.032
Overslept and was either late or missed an appointment	124.5	C	I	.031	.054
Friend/acquaintance did not return your call	100.5	U	E	.033	.036
Met an unfriendly or rude person	91.5	U	E	.146	.122
Desirable social life events					
Small magnitude event(s)					
Made a new friend/acquaintance	167	C	M	.388	.355
Spent a holiday with friends/family[2]	166	C	I	–	–
Had a party or other social gathering	157.5	C	I	.359	.330
Kissed and/or had other pleasing physical contact with friend/acquaintance	110.5	C	M	.426	.397
Went out with friend(s) (party dance, movie, night club, etc.)	106.5	C	I	.418	.408
Visited with family members	100.2	C	I		
Invited out by friend/acquaintance unexpectedly	97	U	E	.449	.412
Celebrated a birthday for a family member[2]	94.5	C	I	–	–
Received a gift or gave gift to friend/family member[2]	93	C	M	–	–
Visited with friends[2]	88.5	C	I	–	–
Received a compliment from a friend/ acquaintance	71.8	U	E	.539	.509
Played a sport, game, or cards with friend(s)	74.5	C	I	.423	.439
Found extra time for privacy	68	C	I	.333	.324
Received letter from a friend or family member[2]	64.5	U	E	–	–
Talked on phone with friends[2]	60	C	I	–	–
Employment Section					
Undesirable employment events					
Major magnitude event(s)					
Fired	577.5	U	M	.001	.000
Laid off	469.5	U	E	.002	.000
Suffered a business loss or failure	454	M	M	.002	.000
Demoted at work	426	U	E	.001	.000
Changed jobs for a worse one	384	C	M	.002	.000
Turned down for a job	348	U	E	.005	.002
A supervisor threatened to fire you	315.5	U	E	.002	.000
Took a cut in wage or salary without a demotion	313.5	U	E	.003	.040
Found out that was *not* going to be promoted at work	297.5	U	M	.001	.000
Heard rumors of layoffs that would affect your position	295.5	U	E	.010	.046
Your authority to make decisions at work was reduced	281	U	E	.004	.068

Category	Mean re adjustment	Control over event	Cause of event	Monthly probability of occurrence	Monthly probability of reoccurrence
Did not get an expected wage or salary increase	279.5	U	E	.001	.000
Got a negative job performance review	273	M	M	.002	.000
Ambiguous magnitude event(s)					
Had added pressure to work harder/faster	271	U	E	.018	.037
People under your supervision failed to get work done on time	257	U	E	.008	.018
Had trouble with boss	249.2	M	M	.004	.032
Received less pay than expected	235.7	U	E	.000	–
Small magnitude event(s)					
Had to move to a worse desk/office/work station	226.7	U	E	.004	.007
Criticized by superior at work	215.5	M	M	.004	.000
Had to work overtime when you did not want to	210.5	U	E	.015	.029
Disagreement with others about your job assignment	210.5	M	M	.007	.047
There was not enough work to keep busy	190	U	E	.008	.048
The office ran out of supplies you needed to do your job	164.5	U	E	.011	.011
Questionable desirability employment events					
Major magnitude event(s)					
Stopped working, *not retirement*, for an extended period	388.5	C	I	.003	.000
Quit job	385	C	I	.003	.000
Small magnitude event(s)					
Sharply reduce work load	203.5	M	M	.006	.044
Desirable employment events					
Major magnitude event(s)					
Retired	565	C	I	.002	.000
Started a business or profession	457	C	I	.001	.000
Started employment for the first time	419	C	I	.002	.000
Returning to work after not working for a long time	393.5	C	I	.005	.007
Promoted[1]	374	M	I	.000	–
Expanded business or professional practice	323.5	C	I	.002	.000
Changed jobs for a better one	312.4	C	I	.002	.000
Ambiguous magnitude event(s)					
Given more authority to make decisions on your job	255.5	M	M	.012	.037
Got a substantial increase in wage or salary without a promotion	246.5	M	M	.003	.052
Had significant or important success in work	238.2	C	I	.009	.032

Category	Mean re adjustment	Control over event	Cause of event	Monthly probability of occurrence	Monthly probability of reoccurrence
Small magnitude event(s)					
Received an award or special praise for your work	233.3	M	I	.013	.041
Given a merit raise	233.5	M	M	.005	.016
Pressure on you to work harder and/or faster was reduced	210	U	E	.004	.000
Started an interesting project at work	203	M	M	.037	.048
Solved a complicated problem at work	202.1	C	I	.016	.034
Did volunteer work	201.5	C	I	.091	.143
Completed work on a major task or project	199.5	C	I	.022	.060
Praised at work by superior	185.5	M	M	.033	.124
Moved to a better desk/office/work station	183.5	M	M	.005	.000
Helped by a fellow employee on a task	126	M	M	.025	.076
Health Section					
Undesirable health events					
Major magnitude event(s)					
Cancers, tumors, or leukemia (except skin cancers)	1063	U	E	.004	.023
Stroke	1001.5	U	E	.000	—
Vision disorder such as glaucoma, blindness, or other severe eye problems	813.3	U	E	.036	.098
Neurological disorders (such as epilepsy, Parkinson's Disease, meningitis, cerebral palsy, convulsions, scizures, MS, muscular dystrophy, etc.)	829	U	E	.004	.080
Unable to get treatment for a serious illness or injury	764.5	U	E	.029	.059
Heart disease, angina, or heart attack	764	U	E	.022	.067
Serious injury occurred or got worse	694.7	U	E	.021	.022
Spinal injury or disease	668.5	U	E	.010	.033
Serious lung troubles such as asthma, emphysema, chronic bronchitis, or tuberculosis	661	U	E	.028	.087
Serious physical illness started or go worse	642.4	U	E	.092	.012
Kidney or bladder disease	622.1	U	E	.012	.023
Skin disorders such as pressure sores, leg ulcers, severe burns, skin cancers, or other dermatological problems	598	U	E	.018	.033
Liver disease	589.9	U	E	.000	—
Surgery or hospitalizations[2]	583	U	E	—	—
Diabetes, sugar diabetes or hypoglycemia	577	U	E	.014	.156
Broken or injured bones	572	U	E	.009	.020
Hearing impairment	512	U	E	.014	.032
Chronic back pain	510	U	E	.037	.075
Arthritis or rheumatism	502.2	U	E	.070	.114
Effects of polio	497.5	U	E	.000	—
Ulcers or other stomach or intestinal disorders	459.4	U	E	.022	.073
Anemia	446.9	U	E	.006	.065
Speech impediment or impairment	436	U	E	.002	.229

Category	Mean re adjustment	Control over event	Cause of event	Monthly probability of occurrence	Monthly probability of reoccurrence
Circulation trouble in arms or legs	424.4	U	E	.024	.032
Hardening of the arteries	419.5	U	E	.001	.049
High blood pressure or hypertension	405.7	U	E	.021	.049
Thyroid or other glandular disorders	388	U	E	.001	.000
Tried to stop smoking but were not successful[2]	315.6	C	I	–	–
Ambiguous magnitude event(s)					
Your allergy flared up	302.7	U	E	.205	.282
Suffered a minor physical injury (minor sprain, pulled muscle, cut or bruise)	293.9	U	E	.150	.162
Increased your use of medication	283.6	M	M	.127	.113
Tried to improve your diet but were not successful	278.5	C	I	.101	.105
Small magnitude event(s)					
Contracted cold or flu	261.1	U	E	.140	.114
Had trouble sleeping on one or more nights	247	U	E	.506	.527
Got sick to your stomach from something you ate	230.1	M	M	.156	.200
Air pollution caused you discomfort (difficulty with eyes, nose, breathing etc.)	211	U	E	.265	.303
Began a day with physical pain or discomfort	206	U	E	.312	.341
Got very tired in a short time	204	U	E	.420	.465
Questionable desirability health events					
Major magnitude event(s)					
Stayed in a nursing home/convalescent center[2]	596.5	U	E	–	–
Saw someone for advice and help with emotional or nervous problem	402.5	C	I	.048	.115
Small magnitude event(s)					
Started taking a drug not prescribed by a doctor	232.5	C	I	.026	.026
Desirable health events					
Major magnitude event(s)					
Physical health improved[1]	562	M	M	.184	.182
Purchased medical device such as a hearing aid, cane, dentures, etc.[2]	375.7	C	I	–	–
Participated in a self-help group	354.5	C	I	.064	.140
Stopped smoking	307.5	C	I	.013	.091
Had a physical exam, and were found in better health than expected	282.5	M	I	.137	.106
Cut down on smoking successfully	281	C	I	.053	.232

Category	Mean re adjustment	Control over event	Cause of event	Monthly probability of occurrence	Monthly probability of reoccurrence
Ambiguous magnitude event(s)					
Decreased your use of alcohol	279.7	C	I	.038	.109
Changed to a more healthy diet	273.5	C	I	.188	.184
Began exercise routine to improve health	271.5	C	I	.179	.182
Decreased your use of medication	244.1	C	I	.107	.135
Saw a doctor/obtained medical care[2]	240.5	C	I	–	–
Health symptoms					
Major magnitude event(s)					
Pain in the heart or tightness or heaviness in the chest	442.2	U	E	.181	.260
Had pains in the back or spine	402.7	U	E	.391	.443
Had frequent headaches	357.1	U	E	.197	.234
Had frequent cramps in the legs	345.4	U	E	.302	.349
Had repeated pains in the stomach	329.9	U	E	.127	.133
Ambiguous magnitude event(s)					
Had swollen ankles	307.1	U	E	.269	.379
Had stiffness, swelling or aching in any joint or muscle	304.4	U	E	.463	.469
Had trouble breathing or shortness of breath	304.1	U	E	.217	.297
Experienced physical pain later in the day	281.4	U	E	.168	.182

Note: Readjustment ratings were made by 10 experts in life event research. Control, cause, and desirability ratings were made by five experts in life events research and 12 female older adult interviewers from the Life Events and Aging Project. Procedures are similar to those described in Zautra, Guarnaccia, and Dohrenwend (1986) except nonambiguous desirability ratings were dichotomized. Control ratings are categorized as (C) Controllable, (U) Uncontrollable, or (M) Mixed Controllability. Cause ratings are categorized as (I) Internally Caused, (E) Externally Caused, or (M) Mixed Causation.

[1]These events had the readjustment rating shown supplied to event raters to provide a benchmark within each category.

[2]Events marked have no monthly probability or occurrence or reoccurrence, as they were write-ins rather than contained on the original inventory.

[3]The Child and Family Section was originally two separate sections, one for children and grandchildren, and the other for other family members; thus, there are two benchmark items within this section.

6

Caregiving and Detrimental Mental and Physical Health Outcomes

JAMILA BOOKWALA, JENNIFER L. YEE, and RICHARD SCHULZ

Providing care to an ill family member is a stressful experience, one that is consistently associated with psychiatric morbidity and, occasionally, adverse physical health outcomes. In the present chapter, we begin with a summary of the prevalence estimates of family caregiving reported in recent national surveys. Next, we describe findings on the links among caregiving and psychiatric morbidity and/or detrimental physical health outcomes that have been published since the earlier reviews by Schulz and his colleagues (Schulz, O'Brien, Bookwala, & Fleissner, 1995; Schulz, Visintainer, & Williamson, 1990). Studies that treat mental health (e.g., depressive symptoms, depression, anxiety) and/or physical health indicators (e.g., self-rated health, physical symptomatology and illness, health-related behaviors, health service utilization, immune and physiological responses) as outcome variables are reviewed. A section is devoted to a discussion of the specific association (or lack thereof) between caregivers' psychiatric morbidity and indicators of physical health, and finally, the correlates of the mental and physical health effects of caregiving are discussed.

NATIONAL ESTIMATES OF FAMILY CAREGIVING

Several national surveys have been conducted to assess the prevalence of caregiving to the elderly in the United States and to characterize those who provide care to older adults. These surveys include the Channeling Study of Informal Caregivers, conducted

JAMILA BOOKWALA, • Department of Psychology, Abington College, Pennsylvania State University, Abington, Pennsylvania 19001. **JENNIFER L. YEE** • Center for Social and Urban Research, University of Pittsburgh, Pittsburgh, Pennsylvania 15213. **RICHARD SCHULZ** • Department of Psychiatry and Center for Social and Urban Research, University of Pittsburg, Pittsburg, Pennsylvania 15213.

Physical Illness and Depression in Older Adults: A Handbook of Theory, Research, and Practice, edited by Gail M. Williamson, David R. Shaffer, and Patricia A. Parmelee. Kluwer Academic/Plenum Publishers, New York, 2000.

in 1980 (Stephens & Christianson, 1986); the National Informal Caregivers Survey, a component of the National Long-Term Care Survey, conducted in 1982 and 1989 by the Bureau of the Census for the Department of Health and Human Services, for which results have been published for the 1982 data (Stone, Cafferata, & Sangl, 1987); and the National Alliance for Caregiving Informal Care Survey, administered in 1996 (National Alliance for Caregiving and the American Association of Retired Persons, 1997). Table 6.1 displays the definitions of caregiving employed in each study, prevalence estimates, and characteristics of caregivers based on these surveys.

Definitions of Caregiving

As seen in Table 6.1, estimates of family caregiving obtained by these surveys vary considerably due to the adoption of different definitions of caregiving. The National Alliance of Caregiving Survey and the Administration on Aging report adopted broad definitions of caregiving. The National Alliance of Caregiving Survey included individuals providing any kind of unpaid assistance or support in the past year to someone over age 50. Thus, this definition includes those who provided care for those with acute health problems, as well as those caring for someone with a chronic condition. The Administration on Aging included individuals providing unpaid assistance to an impaired older person in need of ongoing assistance. In contrast, the National Informal Caregivers Survey and the Channeling Study of Informal Caregivers adopted more restrictive definitions of caregiving. For the National Informal Caregivers survey, caregivers were defined as those providing assistance with one or more basic activities of daily living (ADL). The Channeling Study of Informal Caregivers, which utilized the most restrictive definition of caregiving, interviewed caregivers of elders who had at least two moderate ADL limitations and three severe instrumental activities of daily living (IADL) limitations.

Prevalence of Caregiving

The estimates of the prevalence of caregiving from the National Informal Caregivers Survey are lower than those of the National Alliance Survey and the Administration on Aging report. The Stone et al. (1987) report based on the 1982 National Informal Caregivers Survey estimates that approximately 2.2 million people in the United States provide informal care to 1.6 million disabled elders. In contrast, the Administration on Aging (1997) states that approximately 22 million people provide care to approximately 5 million elders in the United States. The National Alliance for Caregiving Survey also estimates that there are 22,411,200 English-speaking caregiving households in the United States. Because the National Alliance for Caregiving Survey probably included individuals providing acute as well as chronic care, the estimate from this survey is higher than those from the National Informal Caregivers Survey and the Channeling Study. Taken together, these surveys estimate that a sizable proportion of people in the United States are currently providing informal care to impaired elders. Data from the United States Bureau of the Census (1995) project that even more people will be providing informal care to impaired elders in the future because of the rapid expansion of the population of older adults, and because of restrictions in the availability of formal health care services. These statistics show that the population for those over 85

Table 6.1. Prevalence Estimates and Caregiving Characteristics Based on National Caregiving Surveys

Survey	Administration on Aging Report (1997)		Channeling Study (Stephens & Christianson, 1986)		National Alliance for Caregiving (National Alliance for Caregiving and the American Association of Retired Persons, 1997)		National Informal Caregiver Survey (Stone, Cafferata, & Sangl, 1987)	
Sample	Gender	75% women	N = 1,940		N = 1,509		N = 1,924	
	Employed	50%	Gender	73.2% women	Gender	72.5% women	Gender	72% women
			Race	72.4% white	Race	86.1% of caregiving households estimated to be white	Race	79.5% white
			Kinship ties	31.1% adult daughters 22.7% spouses	Kinship ties	40% care for mother/mother in law 5% care for spouse	Kinship ties	37.5% adult children 35.5% spouses
			Married	62.1%	Married	66%	Married	69.5%
			Employed	33%	Employed	64%	Employed	30.9%
			Recipient Gender	71.6% women	Recipient Gender	mostly women	Recipient Gender	60% women
			Recipient Race	72.8% white				
			Lives with recipient	56.5%	Lives with recipient	20%	Lives with recipient	73.9%
	Age		Age		Age		Age	
	M = 46		40 or under	12.2%	M = 46.15		M = 57.5	
	12% are 65 or older		40 or under	12.2%	Under 35	22.3%	14–44	21.6%
			41–54	23.2%	35–49	39.4%	45–64	41.4%
			55–64	24.8%	50–64	26.0%	65–74	25.4%
			65–75	25%	65+	12.4%	75+	10.1%
			Over 75	14.9%	Recipient M = 77		Recipient M = 77.7	
Sampling plan	Details not provided		Interviews were conducted in 1982 with informal caregivers identified by elderly individuals included in the Channeling Demonstration Project. These elderly persons were referrals to one of 10 Channeling Projects, which were located in Baltimore, MD; Houston, TX;		Two samples were used to conduct the survey in 1996. The first was a fully-replicated, stratified, single-stage Random Digit Dialing sample of U.S. telephone households (N = 754) generated by ICR Survey research group. The second sample (oversample of minorities) was extracted		The Informal Caregivers Survey was a component of the National Long-Term Care Survey. The Sample for the NTCS was drawn from Medicare Health Insurance Skeleton Eligibility Write-Off Files. A random sample of 36,000 was screened for the NTCS	

(Continued)

Table 6.1. (*Continued*)

Survey	Administration on Aging Report (1997)	Channeling Study (Stephens & Christianson, 1986)	National Alliance for Caregiving (National Alliance for Caregiving and the American Association of Retired Persons, 1997)	National Informal Caregiver Survey (Stone, Cafferata, & Sangl, 1987)
		Middlesex County, NJ; Eastern KY; Southern ME; Miami, FL; Renssalear County, NY; Cuyahoga County, OH; and Philadelphia, PA.	from ICR's Excel Omnibus Service, and included individuals who had previously identified themselves as black, Hispanic, or Asian and spoke English.	and a sample of 6,393 elders who had long-term problems with atleast one ADL or one IADL were interviewed. The Informal Caregivers Survey, was conducted by the National Opinion Research Center; 1,924 interviews were completed with helpers identi fied by respondents in the NTCS who were assisting with at least one ADL.
Caregiving definition	Unpaid care provided to a person age 65 or older who has an impairment that limits that person's independence and requires some level of ongoing assistance.	Individuals who were identified by elders at risk for institutionalization (elders who had two moderate ADL limitations and three severe IADL limitations) as the person who helped them the most.	Individuals age 18 and older who provided unpaid care to a relative or friend who is age 50 or older to help them take care of themselves. Caregiving may include help with personal needs or household chores. It might be taking care of a person's finances, arranging for outside services, or visiting regularly to see how they are doing. This person need not live with you.	Caregivers were defined as those providing assistance with at least one ADL.
Prevalence	22 million people caring for approximately 5 million elders	Not provided	22,411,200 English speaking caregiving households in the United States	2.2 million people in the Unites States provide informal care to 1.6 million disabled elders.

Caregiving tasks	Not provided	IADLs		IADLs		IADLs	
		Shopping/ transportation	51.6%–7.7%	Shopping/ transportation	77.3%–79.3%	Shopping/transportation	86.2%
						Household Chores	80.6%
		Household chores	77.9%	Household chores	60%–73.6%	Finances	49.2%
		Finances/organizing services	59.3%–4.5%	Finances/organizing services	53.9%–55.6%	Medications	53.1%
						ADLs	
		Medications	58.9%	Medications	37.3%	Hygiene	67.2%
		ADLs		ADLs		Mobility	45.7%
		Dressing/getting out of bed/chair	45.2%–2.8%	Dressing/getting out of bed/chair	31.4%–36.8%	Hours spent caregiving	5.7 per day
		Bathing/toileting	39.8%–46.2%	Bathing/toileting	26.2%–26.6%		
		Feeding/continence	39.0%–39.8%	Feeding/continence	13.6%–19.2%		
		Hours spent caregiving	4 per day	Hours spent caregiving	17.9 per week		
Interference with work (of those employed)	Not provided	Worked fewer hours	32.5%	Worked fewer hours	7.3%	Worked fewer hours	21.0%
		Gave up working	9%	Rearranged work schedules	49.4%	Rearranged work schedules	29.4%
				Took a leave of absence	10.9%	Took a leave of absence	18.6%
				Gave up working	6.4%	Gave up working	8.9%
Caregiving strain	Not provided	Emotional strain	49.7%	Emotional strain	25%		
		Physical strain	38.2%	Physical strain	12%		
		Financial strain	18.5%	Financial strain	7%	Not provided	

years of age (those who are most likely to report disability) has experienced the most dramatic increase among the population of elders (274% increase between 1960 to 1994). The Bureau of the Census projects that this trend will continue in future decades. As a result, more family members in their 50s and 60s will be needed in future decades to care for those who live beyond age 85. The parent–support ratio, which equals the number of persons aged 65 and older per every 100 persons ages 50 to 64, illustrates the increasing need for more informal caregivers in the future. This ratio tripled from 3 to 10 between 1950 and 1993. During the next six decades, the parent–support ratio is expected to triple again to 29.

Profiles of Care Recipients

Although the definitions of caregiving vary, all involve providing unpaid assistance or support to an older person. As seen in Table 6.1, these national surveys characterize the "typical" elderly care recipient as white, female, with an average age of approximately 77–78 years (National Alliance of Caregiving and the American Association of Retired Persons, 1997; Stephens & Christianson, 1986). The National Alliance of Caregiving reported that only a minority of caregivers indicated that they were caring for someone with a dementia-related illness (22.4%); approximately 60% reported caring for someone with a physical health condition.

Profiles of Caregivers

In general, as shown in Table 6.1, the majority of informal caregivers are women, white, and married. In addition, most caregivers are middle-aged daughters. However, a sizable minority of caregivers (between 12% and 40%) are elders themselves. The American Psychological Association (1997) reports that an older spouse is most likely to be the caregiver for an impaired, community-dwelling older adult, and that the proportion of older adults needing personal assistance with everyday activities increases with age. For example, 9% of those ages 65–69 and up to 50% of those age 85 years or older report needing assistance with day-to-day living. In terms of kinship ties, national surveys estimate that spouses comprise a sizable minority of caregivers and, predictably, are more prevalent among caregivers over age 65. With regard to living arrangements, the National Alliance Survey, which utilized a broad definition of caregiving, reported that the majority of caregivers did not live with the care recipient. The Channeling Study and the National Informal Caregivers Survey, which used more restrictive definitions, on the other hand, reported that most caregivers lived with the recipient.

Time Spent Caregiving and Tasks Performed

In general, caregivers report spending several hours per day on caregiving tasks and almost all caregivers (90 to 98%) in these national surveys mention assisting with at least one IADL. Among IADLs, caregivers report helping the most with shopping and transportation, followed by household chores, finances and organizing services, and administering treatments or medications. For ADLs, the national surveys estimate that 51%–71% of caregivers help with at least one ADL. The most frequently reported ADL assistance includes helping the care recipient with dressing and getting in and out of a bed or chair, followed by bathing and toileting, and feeding and continence. Across all surveys, women and spouse caregivers reported that they were more likely to perform personal care tasks for the care recipient and spend more time on caregiving tasks compared to men and nonspouse caregivers. Women, nonspouse caregivers were the most likely to mention assisting the care recipient with household tasks.

Effects of Caregiving on Work and Other Responsibilities

The National Alliance Survey reports that the median household income for caregivers was $35,000 (in 1996) and that 35.3% graduated from high school, 22.5% had some college, and 20.1% were college graduates. With regard to employment status, the national surveys report that one- to two-thirds of caregivers are employed. Data from these surveys showed that caregiving can interfere with caregivers' work responsibilities. As seen in Table 6.1, results from these surveys revealed that some employed caregivers worked fewer hours, rearranged their work schedules, took a leave of absence, or gave up working entirely because of their caregiving responsibilities. In addition, the National Alliance Survey showed that caregivers reported that their caregiving responsibilities interfered with other aspects of their lives, such as leisure time (55%) and time with other family members (43%).

Caregiver Stress and Health Status

The Channeling Study and the National Alliance for Caregiving Survey provided data concerning the emotional, physical, and financial strain of caregiving. Results from both surveys demonstrated that a sizable proportion of caregivers experienced emotional, physical and/or financial strain as a result of caregiving. Also, results from the National Alliance of Caregiving Survey revealed that 15% of caregivers experienced physical, or mental health problems. Furthermore, Stone et al.'s report on the National Informal Caregivers survey and the Channeling Study showed that over one-third of caregivers rated their health as only fair or poor.

To summarize, national surveys estimate that as many as 22 million people are providing informal care to impaired elders. Although the "typical" caregiver is a white, middle-aged woman with at least a high school education, who is caring for a parent, elderly spouses comprise a significant proportion of the caregivers to older adults. The results of the national surveys indicate that caregiving has a significant impact on caregivers' work-, family-, and leisure-related activities, and can be stressful, with detrimental mental and physical health effects for many caregivers. In the sections that follow, we detail recent findings on the mental and physical health effects of caregiving and their interrelationships.

CAREGIVING AND PSYCHIATRIC MORBIDITY

By the middle of this decade, the stresses of caregiving became undeniably linked to detrimental mental health outcomes (Schulz et al., 1995; Schulz et al., 1990; Wright, Clipp, & George, 1993). In the most recent of these reviews, Schulz et al. (1995) summarized findings related to the psychiatric morbidity effects of dementia caregiving from studies published during the first half of this decade. They observed that, in terms of psychiatric morbidity, in virtually all the reviewed studies caregivers reported elevated depressive symptomatology and, in studies that used diagnostic measures, caregivers were characterized by higher rates of affective and anxiety disorders.

Caregiving studies published since 1995 corroborate the strong association between caregiving and psychiatric morbidity noted by Schulz et al. (1990, 1995). Table 6.2 presents a detailed summary of these findings. Like their predecessors, more recent caregiving studies continue to rely predominantly on self-report measures of depressed affect. The Center for Epidemiological Studies—Depression scale (CES-D; Radloff, 1977) is perhaps the most widely used among the self-report measures. Scores on the 20-item CES-D can range from 0–60, and scores of 16 or higher indicate that an individual is at risk for clinical depression. Typical population scores on the CES-D range from 7.4 to 9.4 (Blazer, 1994). Another frequently used measure is the Beck Depression Inventory (BDI; Beck, Ward, Mendelson, Mock, & Erbaugh, 1961), on which total scores of 5, 8, and 16 represent cutoffs for mild, moderate, and severe depression, respectively. Other, less commonly used self-report measures of psychological distress include the Symptom Checklist (SCL-90; Derogatis, 1983), the Brief Symptom Inventory (BSI), which is a short version of the SCL-90, the General Health Questionnaire (GHQ; Goldberg & Hillier, 1979), and the Zung Self-Rated Depression Scale (Zung, 1965, 1975). A subgroup of studies published during the last 4 years have used a diagnostic interview to determine the prevalence of mood and/or anxiety disorders among dementia caregivers. Common diagnostic tools include the Structured Clinical Interview for DSM-III-R (SCID; Spitzer, Williams, Gibbon, & First, 1992), the Diagnostic Interview Schedule (DIS-III-R; Di Nardo, O'Brien, Barlow, Waddell, & Blanchard, 1983), and the Hamilton Rating Scale for Depression, which provides an interviewer-rated continuous index of depressive symptoms (HRSD; Hamilton, 1967).

Caregivers of Patients with Alzheimer's Disease and Related Dementias

Research indicates that simply living with a cognitively impaired spouse can have a significant impact on an elderly individual's mental health (Moritz, Kasl, & Berkman, 1989), especially in marriages marked by close spousal relationships (Tower, Kasl, & Moritz, 1997). Approximately 1.9 million Americans ages 65 or older suffered from some form of Alzheimer's disease (mild, moderate, or severe) in 1995, and slightly more than 1 million people with Alzheimer's disease ages 65 or older required personal care. It is projected that in the year 2015, 2.9 million Americans over age 65 will have some form of Alzheimer's disease, and 1.7 million will require personal assistance (General Accounting Office, 1998). Because of the progressive overall deterioration characteristic of dementia, caring for these individuals can pose some of the most extreme challenges faced by caregivers; consequently, compared to caregivers of individuals diagnosed with other health conditions, dementia caregivers may experience the most detrimental mental health outcomes. Indeed, in a recent study that directly compared dementia spouse caregivers with spouse caregivers of Parkinson's disease patients, the former were observed to have significantly more depressive symptoms and anxiety than the latter (Hooker, Monahan, Bowman, Frazier, & Shifren, 1998).

Consistent with the pattern of findings described by Schulz et al. (1995), the most recent studies on dementia caregivers continue to find substantial evidence of psychiatric morbidity. As Table 6.2 indicates, the vast majority of studies have found elevated self-reports of depressive symptomatology among caregivers compared to control

Table 6.2. Psychiatric and Physical Morbidity Effects in Caregiving Studies Published Since Schulz et al. (1995)

Authors	Sample	Patient impairment	Health measures	Psychiatric morbidity	Physical morbidity
Bass, Noelker, & Rechlin (1996)	*n* = 401, recruited from nonprofit case management agencies; *M* age = 57.6 yrs.; 75.6% female; 41% White; 18.7% spouses, 44.9% adult children, 23.7% other relative, 12.7% friend.	Elder persons who receive case management service	Psychiatric: CES-D. Physical: Physical health deterioration (e.g., more aches and pains, having pains, having less energy) as a result of caregiving	*M* = 10.37, with about 23% identified as at risk for clinical depression. Lower CES-D scores associated with use of health care service when patient more disabled and household service when patient exhibit more behavior problems.	CGs reported considerable negative physical health deterioration (*M* = 5.7 out of a possible score of 15, with high scores indicating greater physical health deterioration).
Beach et al. (1998)[a]	*n* = 299 spouse CGs, recruited from a population-based sample surveying elderly individuals; *M* age at entry in survey = 71.5 yrs.; 49.9% female; 90.4% white; age- and sex-matched nonCGs, *n* = 381.	Elders who had difficulty with ≥ 1 IADL or ADL due to physical or health problems, or problems of confusion	Psychiatric: DIS-III-R for anxiety and depression diagnosis. Physical: functional limitations, single-item self-rated health.	CGs experienced significantly more depression and anxiety symptoms (*M* = .31 and .95, respectively) than nonCGs (*M* = .09 and .34 for depression and anxiety, respectively), as assessed by the DIS-III-R. CGs' poorer self-rated health was associated with more symptoms of anxiety and depression at T1 and T2, and poorer self-reported functioning was associated with more anxiety symptoms at T1 and T2. Greater psychiatric morbidity also associated with more CG strain or burden; more anxiety symptoms related to more negative life events.	CGs perceived their health to be significantly poorer (*M* = 3.2) than did nonCGs (*M* = 3.6). Poorer self-rated health was related to greater stress.
Bookwala & Schulz (1998)[a]	*n* = 378 spouse CGs, recruited from a population-based sample surveying elderly individuals; *M* age = 72.4 yrs.; 49.9% female; 89.3% White.	Elders who had difficulty with ≥ 1 IADL or ADL due to physical or health problems, or problems of confusion	Psychiatric: 10-item CES-D. Physical: single-item self-rated health.	*M* = 5.85 (approx. equivalent to *M* = 11.7 on the 20-item scale). Higher depressive symptoms in CGs who experienced more strain, scored higher on neuroticism and/or lower on mastery, rated their health more poorly, and whose spouses exhibited more behavior problems.	30% rated health as "fair" or "poor."

(Continued)

Table 6.2. (*Continued*)

Authors	Sample	Patient impairment	Health measures	Psychiatric morbidity	Physical morbidity
Burton et al. (1997)[a]	n = 434 spouses, recruited from a population-based sample s urveying elderly individuals; categorized as high-level CGs (spouse had ≥ 1 ADL impairment) vs. low-level CGs (spouse had ≥ 1 IADL impairment; 43.8% aged 65–74 and 56.2% aged 75+; 51.3% female; 90% white; age- and sex-matched n onCGs, n = 385.	Elders who had difficulty with ≥ 1 IADL or ADL due to physical or health problems, or problems of confusion	Psychiatric: None. Physical: Health-related behaviors (eating < 3 meals/day, not enough time to exercise, smoking, alcohol intake, not getting enough rest when healthy or sick, forgetting to take medications, delaying a physician visit when ill, missing a doctor's appointment, missing a flu shot, running out of medications.	N/A.	High-level CGs were more likely to not get enough rest, not have enough time to exercise, not have time to rest to recuperate from illness, and forget to take prescription medications than nonCGs. No differences on missing meals, doctor appointments, flu shots, and medication refills. Larger proportions of CGs with a strong sense of control had good preventive health behaviors (e.g., not missing meals, not missing flu shots) compared with CGs with a weak sense of control.
Canning, Dew, & Davidson (1996)	n = 83 at T1, recruited by interviewing heart transplant recipients; 39.8% 50 yrs or older; 85.5% female; 93.6% white; 8 6.6% spouse; n = 72 at T2, n = 65 at T3.	Heart transplant recipients	Psychiatric: SCL-90, SCID Physical: Checklist of chronic health problems	SCL-90: Compared to a normative sample of community residents (M = .25), CGs had significantly higher means 2 mos after the patient's transplant surgery (M = .40); no differences at 7 and 12 mos. post-surgery (CG M = .32 and community sample norm = .28). 25.9% CGs scored above the clinical cut-point (> 1 *S.D.* above the normative M) at 2-mos, 23.4% at 7-mos., and 15.5% at 12-mos. postsurgery. 12.1% CGs had clinically elevated levels of	20.5% CGs reported ≥ 3 current physical illnesses or problems soon after the patient's surgery.

				psychological distress at all 3 time points. SCID: 28.9% CGs diagnosed with a history of major depression or generalized anxiety disorder presurgery; however, a history of psychiatric illness not related to greater psychological distress postsurgery. More depressive symptoms related to a closer CG-patient relationship, less social withdrawal, more self-reported illnesses. CGs who reported more chronic illnesses soon after the patient's surgery were more likely to have higher psychological distress 7 and 12 mos. after the surgery.	
Cochrane, Goering, & Rogers (1997)	*n* = 1,512, recruited from a population based random sample. 5.8% spouse, 35.4% child or sibling, 7.8% parent, 20.8% friend, 30.3% other relative.	Persons with serious or chronic mental, physical, or developmental problems	Psychiatric: University of Michigan composite international diagnostic interview Physical: Disability, service utilization, checklist of health problems and symptoms.	CGs had a significantly higher prevalence of affective and anxiety disorders (6.3% and 17.5%, respectively) than nonCGs (4.2% and 10.9%, respectively). CGs also reported using outpatient mental health services on a more frequent basis than the nonCGs.	Higher prevalence of physical health problems in the previous year among caregivers (68.8%) than noncaregivers (62.7%). CGs and nonCGs did not differ in number of days of total disability, but significantly more CGs reported having "limited-activity days" (23.8%) and "days that required extreme effort" (21.7%) compared to nonCGs (18.2% and 13.9%, respectively).
Collins & Jones (1997)	*n* = 48 spouse CGs, recruited through referrals by a psychiatric service; *M* age = 74.6 yrs.; 50% female.	Dementia	Psychiatric: GHQ. Physical: None.	8% of CG husbands and 65% of CG 3 wives obtained a score of $\geq$ 12 on the GHQ, marking psychiatric "caseness."	N/A.

(*Continued*)

Table 6.2. (*Continued*)

Authors	Sample	Patient impairment	Health measures	Psychiatric morbidity	Physical morbidity
Draper et al. (1995)	*n* = 99, recruited through local community referrals; *M* age = 74 yrs.; 48% female; 90% spouses; median *CG* duration = 24 mos.	51% dementia and 48% stroke	Psychiatric: GHQ. Physical: Single item self-rated health.	None reported. CGs at greatest risk for psychological morbidity were female, reported a poorer quality of life, and provided care to more behaviorally disturbed patients.	None reported.
Fingerman et al. (1996)	*n* = 81, recruited through adult day care centers and caregiver respite programs; *M* age = 59.7 yrs.; 85% female; mostly White; 54% spouses; *M* CG duration = 56.2 mos.	Frail elderly requiring assistance with IADLs	Psychiatric: BDI. Physical: None.	*M* = 10.8 at T1 and 9.8 at T2. A decline in depressive symptoms over time associated with higher levels of internal resourcefulness. More depressive symptoms associated with greater use of avoidant coping.	N/A.
Fuller-Jonap & Haley (1995)	*n* = 52 husbands, recruited through various community sources; *M* = 74.5 yrs.; 100% white; comparison group *n* = 53.	Alzheimer's disease	Psychiatric: BSI, psychotropic drug use. Physical: symptom checklist, single-item self-rated health, no. of physician visits.	CGs did not differ significantly from nonCGs in their self-reported use of psychotropic medication for help with sleep, nerves, or depression.	CGs reported more respiratory problems than comparisons; no differences on self-reported rates of other health conditions such as cardiovascular, digestive, and musculoskeletal problems. CGs reported more difficulty with sleep, taking rest periods, and exercising regularly than nonCGs.
Gignac & Gottlieb (1996)	*n* = 87, recruited through respite/day support programs; *M* age 70.2 for spouses and 49.6 for intergenerational family CGs; 70% female; 60% spouses; *M* CG duration = 30.4 mos.	Dementia	Psychiatric: Hopkins Symptom Checklist for anxiety and depression. Physical: None.	Psychiatric: No means reported. CGs who appraised themselves as having no control over the CG situation or their emotions were more depressed over time. Anxiety *decreased* among CGs who believed that their coping behaviors had non efficacious outcomes.	N/A.

Glaser & Kiecolt-Glaser (1997)[b]	n = 71, recruited through various community sources; M age = 60.6 yrs.; 82% female; 92% white; 48% spouses; M CG duration = 7.7 yrs.; n = 58 nonCGs, similar on age and family income.	Dementia	Psychiatric: None. Physical: Immune response (antibody titers, neutralizing antibody titers, T-cell response) to latent herpes simplex virus type I; health-related behaviors (alcohol intake, smoking, weight change, body mass, physical exercise, caffeine intake, sleep, medication use).	N/A.	Poorer immune response in CGs than nonCGs to infection by a latent herpes simplex virus; re health behaviors, CGS only reported less sleep than nonCGs.
Haley et al. (1995)	n = 175, recruited through memory disorders clinic and special community outreach efforts to reach black CGs; M age = 74 yrs.; 70% whites and 78% blacks female total, 60% white; 51% of white CGs were spouses and 29% of black CGs; M CG duration = 40 mos. for white CGs and 36 mos. for black CGs; nonCGs matched as closely as possible on sociodemographic variables, n = 175.	Alzheimer's disease and related dementias	Psychiatric: CES-D; BSI. Physical: symptom checklist; single-item self-rated health; health care utilization (physician visits, % hospitalized); medication use.	Significantly higher psychological distress among white and black CGs on the CES-D (M = 16.44 and 12.16, espectively) and the BSI-Depression rscale (M = 58.26 and 53.17, espectively) than white and black rnoncaregivers (M = 11.21 and 12.43 on the CES-D and M = 53.75 and 51.69 on the BSI-Depression scale, respectively); white CGs were more depressed than their black counterparts, and manifested the highest prevalence of clinically significant elevation on both the CES-D (scores of $\geq$ 16) and the BSI (overall score or at least 2 subscale scores of $\geq$ 63): 47% on the CES-D and 55% vs. 34% on the BSI.	For both whites and blacks, no differences between CGs and nonCGs on self-rated health (M range = 1.94 to 2.27), physician visits (M range = 4.41 to 5.51), percent hospitalized (% range = 15.71 to 17.14), and medication use (M range = 1.42 to 1.80).

(Continued)

Table 6.2. (*Continued*)

Authors	Sample	Patient impairment	Health measures	Psychiatric morbidity	Physical morbidity
Hooker et al. (1998)	*n* = 88 spouses of Alzheimer's disease patients and n = 87 spouses of Parkinson's disease patients, recruited from a wide variety of sources; *M* age = 70.2 & 67.1 yrs, respectively; 59.1% and 63.2% female, respectively; 95.% and near-100% white, respectively; *M* CG duration = 4.4 yrs. and 7.6 yrs., respectively.	Elderly individuals diagnosed with diagnosed with Alzheimer's disease (AD) or Parkinson's disease (PD)	Psychiatric: CES-D, state and trait anxiety. Physical: Current health index, 5-item self-rated health, checklist of chronic conditions.	AD CGs reported more depressive symptoms (*M* = 15.1) and state anxiety (*M* = 40.8) than PD CGs (9.8 and 34.9 for depressive symptoms and state anxiety, respectively). Poorer mental health was related to more stress, higher neuroticism, lower optimism, and poorer self-rated health.	*M* self-rated health = 13.9 each for AD and PD CGs; *M* current health index = 31.3 for AD CGs and 31.6 for PD CGs; *M* chronic conditions = 2.4 for AD CGs and 2.3 for PD CGs. After controlling for other variables, AD CGs had significantly better physical health than PD CGs. Poorer physical health was related to higher neuroticism, lower optimism, and greater stress.
Irwin et al. (1997)	*n* = 100 spouse CGs, recruited ADRC or local community support groups; *M* age = 71; 57% female; 87% white; age- and sex-matched nonCGs, n = 33.	Alzheimer's disease	Psychiatric: BDI; HDRS Physical: symptom checklist; medical "caseness"; immune (natural killer cell activity), neuro-endocrine (adreno-corticotropic hormone [ACTH], β endorphin, cortisol, prolactin), and sympathetic (plasma concentra-tions of epinephrine, norepinephine, neuropeptide Y) measures.	BDI: *M* = 0.6 for CGs and 0.2 for nonCGs. HDRS: *M* = 4.2 for CGs and 2.6 for nonCGs. No relationship between depressed affect and immune, neuroendocrine, or sympathethic measures.	No significant difference between CGs and nonCGs on any health measure. Basal ACTH levels were higher among CGs who reported an extreme mismatch in the amount of care they provided and the amount of respite time they received compared to CGs who did not experience such an extreme discrepancy.

Jutras & Lavoie (1995)	n = 292, drawn from a population-based sample of eligible households; M age = 60.1 yrs.; 53.4% female; 67% spouses; 2 comparison groups matched on age and sex: those living with a nonimpaired elder (Group 1) and those not living with an elder (Group 2).	Patients had to meet at least one of the following: confined to bed or to a chair most of the day because of health; needs help of another person with personal care needs; needs help of another person looking after personal affairs; or has periods of confusion or frequent important memory losses	Psychiatric: Brief version of Hopkins Symptom Distress Checklist; presence of severe psychological problems. Physical: Global health index tapping mobility restriction, activity limitation, presence of chronic conditions, presence of health-related symptoms.	Significantly more CGs reported lower psychological well-being (38.3%) than those in both Group 1 (24.3%) and Group 2 (29%). More CGs also had mental health problems (15.1%) and psychiatric symptoms (15.2%) compared to Group 1 (7.5% and 12.3% for mental health problems and psychiatric symptoms, respectively).	Significantly more CGs reported disabilities and chronic conditions (65.7%) than Group 1 (53.8%). More CGs than those in Group 1 also reported suffering from diabetes (4.5% vs. 1%) and hay fever (5.8% vs. 3.1%). More CGs than those in Group 2 reported suffering from back problems (14.0% vs. 9.6%) but this trend was reversed for asthma (2.7% vs. 4.1%).
Kiecolt-Glaser et al. (1996)[b]	n = 32, spouse CGs, recruited through various community sources; M age = 73.1; 56 % female, 93% white; control group n = 32 matched on age, sex, and SES.	Progressive dementia	Psychiatric: BDI Physical: Plasma albumin levels, body mass, prevalent illnesses, immune response (antibody response, cytokine response) to influenza virus; health-related behaviors.	CGs reported more depression than nonCGs (M = 5.92 vs. 2.69). No significant relationship between caregivers' depressive symptoms and their immune response.	Caregivers showed a poorer immune response after exposure to an influenza virus vaccine than nonCGs; CGs reported significantly less physical activity and/or sleep than nonCGs. No differences between CGs and nonCGs on alcohol consumption, smoking, weight change, and body mass index.
King & Brassington (1997)	Study 1: n = 103 CGs, identified in a random population-based s urvey sample, M age = approx. 73 yrs.; 71% female; 88% white; M CG duration = 5 yrs. Compared to nonCG survey respondents.	Study 1: sick or frail elder requiring care on a daily or weekly basis with personal care needs	Psychiatric: Study 1: none; Study 2, BDI, Taylor Manifest Anxiety Scale (TMAS).	Study 1: N/A. Study 2, the exercise and control groups obtained significantly higher depression scores on the BDI (M = 11.7 and 10.4, respectively) and anxiety scores on the TMAS (M = 7.6	Study 1: no differences in the proportion of CGs who engaged in physical activity (6.7% women and 17.6% men) compared to nonCGs (9% women and 16.6% men).

(*Continued*)

Table 6.2. (*Continued*)

Authors	Sample	Patient impairment	Health measures	Psychiatric morbidity	Physical morbidity
	Study 2: 23 spouse CGs randomly assigned to either an exercise training program or a delayed exercise control program, recruited through citywise promotion; *M* age = 60.2 (exercise group) and 63.2 (control group) yrs.; 87% female.	Study 2: Individuals with Alzheimer's disease or related dementias	Physical: Study 1, physical activity patterns; Study 2, none (random assignment onto exercise and control groups).	and 7.2, respectively) than similarly aged nonCG populations.	Study 2: no difference in depressed affect or anxiety between the exercise and control groups.
Lawrence, Tennstedt, & Assman (1998)	*n* = 118, identified by members of a representative sample of older adults recruited for a longitudinal study; 69% female; 27% spouses, 73% adult children.	Elderly disabled individuals requiring assistance with ADLs	Psychiatric: 5-item CES-D. Physical: None.	*M* = 2.9 (11.6 according to the 20-item version). More depressive symptoms associated with more patient problem behaviors, greater role overload, a greater sense of role captivity associated with the caregiving role, and a less close relationship with the patient.	N/A.
Li, Seltzer, & Greenberg (1997)	*n* = 103 CG wives and 149 daughters, recruited as a subset from a larger probability sample; *M* age = 70.9 and 57.5, respectively; 99% and 96% white; *M* CG duration = 7.1 and 9.4 yrs., respectively.	Elderly individuals needing care for a variety of illnesses, including arthritis, asthma, blindness, cancer, dementia	Psychiatric: CES-D. Physical: Single-item self-rated health.	Wives reported significantly more depressive symptoms than daughters (*M* = 11.9 and 9.3, respectively). More depressive symptoms related to being younger, being the spouse, more patient problem behaviors, receiving less emotional support, lower levels of social participation, and poorer self-rated health.	Wives reported significantly poorer global health than daughters (*M* = 2.85 and 3.09, respectively).
Majerovitz (1995)	*n* = 54 spouse CGs, recruited through various community sources; *M* age = 70.5 yrs.; 69% female; 94% white.	Dementia	Psychiatric: CES-D. Physical: self-rated health compared to other people their age.	*M* = 13.69. More depressive symptoms related to more problem behaviors in the patient; CGs with less family adaptability who provided more hours of care reported more depressive symptoms.	81% reported their health to be "good" or "excellent" compared to others of their own age.

Martire, Stephens, & Atienza (1997)[c]	n = 118 employed CG daughters/daughters-in-law, recruited via various community sources; M age = 48.5 yrs.; 91% white; M CG duration = 5.7 yrs.	Ill or disabled elderly parent who needed assistance with shopping, personal care, perparing meals, or supervision due to memory problems	Psychiatric: CES-D. Physical: 3-item self- rated physical health.	M = 16.83; 44% scored ≥ 16. Satisfaction in the caregiving role was associated with less depressive symptoms; higher caregiving stress was related to more depressive symptoms for CGs reporting low or high work satisfaction, especially high and for CGs working a low number of hours.	M = 9.72 (range = 4–12). Satisfaction in the caregiving and work roles was associated with better self-rated physical health. High caregiving stress was related to worse self-rated health for CGs working a low number of hours.
Malone Beach & Zarit (1995)	n = 57 female CGs, recruited through a variety of community sources; M age = 58.3 yrs.; 36% wives, 63% daughters/daughters-in-law; 1% other; M CG duration = 4.9 yrs.	Dementia	Psychiatric: CES-D; HDRS. Physical: None.	M CES-D = 15.3 and HDRS = 8.0. Higher depression related to less instrumental support, but not emotional and informational support.	N/A.
Meshefedjian et al. (1998)	n = 321, identified by elderly participants of a population-based random sample survey; M age = 61.5 yrs.; 79.4% female; 30.8% spouses, 44.6% adult children, 24.6% other.	Dementia	Psychiatric: CES-D. Physical: None.	M = 8.8. More depressive symptoms were related to being the patient's spouse, being female, and caring for a patient with more functional impairment and behavior problems.	N/A.
Miller et al. (1995)	n = 215 spouses, recruited through referrals from various community agencies; M age = 74.7 yrs.; 64% female; 64% white.	Dementia	Psychiatric: CES-D. Physical: Single-item self-rated health.	M = 12.7 for entire sample, with 37% coring ≥16, indicating risk for clinical depression. White CGs scored higher on the CES-D scores than blacks (M = 14.9 vs. 8.8, respectively). More depressive symptoms were related to lower mastery and control, and poorer self-rated health. For CGs with lower mastery, greater distress with patient's behavior problems was related to more depressive symptoms.	M = 2.8 for entire sample. Blacks rated their health to be poorer (M = 2.5) than whites (M = 2.9).

(*Continued*)

Table 6.2. ***(Continued)***

Authors	Sample	Patient impairment	Health measures	Psychiatric morbidity	Physical morbidity
Mui (1995)	*n* = 437 spouse CGs of elderly national survey participants; 37% husbands & 18% wives 80 yrs. or older.	Frail elderly with functional and/or cognitive impairment	Psychiatric: None. Physical: Single-item self-rated health; IADLs and ADLs.	N/A.	63% of CG wives and 50.7% of CG husbands reported "fair" or "poor" health; *M* self-rated health = 2.81 and 2.50 and *M* functional limitations = 1.86 and 1.87, respectively. Poorer self-rated health was related to being female, greater stress, and poorer functional status. For CG wives, poorer self-rated health was associated with greater perceived unmet needs in care recipients and for CG husbands, poorer health was associated with longer caregiving duration.
Penning (1998)	*n* = 687 adult child caregivers to disabled parents, identified through systematic probability sampling techniques; *M* age = 47.9 yrs.; 76% female; 57% primary CGs.	A long-term (≥ 6 mos.) illness, physical disability, mental handicap, or mental health/behavioral problem	Psychiatric: None. Physical: Single-item self-rated health.	N/A.	15.9% rated their health as "fair" or "poor." Poorer self-rated health was associated with more parental functional impairment, more parental behavior problems, more hours of care provision, being older, less educated, coresiding with the parent, and not being a parent to a child living at home.
Pariente et al. (1997)	*n* = 18 mothers, recruited through local centers for disabled individuals; *M* age = 44.3 yrs.;	Individuals with mental retardation, autism, or	Psychiatric: BDI; State-Trait Anxiety Inventory; checklist	CGs reported significantly more psychiatric symptoms than nonCGs (*M* = 7.1 vs. 3.7), but no differences	CGs had lower levels of *T* cells, a higher percentage of *T* suppressor/cytotoxic

	age- and sex-matched nonCGs, n = 18.	quadriplegia	of psychiatric symptoms. Physical: immune response (leukocyte count, T cells, T helper cells, T suppressor/ cytotoxic cells.	on the BDI (M = 10.2 and 7.6, respectively). CGs also had significantly higher levels of trait anxiety than nonCGs (41.8 vs. 34.8, respectively) but did not differ on state anxiety. 33% CGs suffered from clinically significant depression.	cells, and a lower T helper: suppressor ratio compared to nonCGs. When grouped by age, older CGs had lower numbers of T cells and T helper cells and higher ntibody titers for cytomega-alovirus compared to matched nonCGS; no differences between younger CGs and their nonCG controls.
Pruchno, Burant, & Peters (1997)[d]	n = 252 daughters/daughters-in-law caring for an elderly parent/parent-in-law, recruited through community outreach techniques; M age = 48.3 yrs.; 93.9% white; comparison groups: husbands and coresident children.	Assistance needed with at least one ADL one ADL	Psychiatric: CES-D. Physical: None.	M CES-D for CGs = 10.66, for husbands = 6.74, for children = 12.05. Families characterized by high agreement in the assessment of the care recipient's behavior problems were significantly less depressed than those from families with moderate or low agreement.	N/A.
Pruchno, Peters, & Burant (1995)[d]	n = 140 daughters/daughters-in-law caring for an elderly parent/ parent-in-law, recruited through community outreach techniques; M age = 49.4 yrs.; 92.1% white; comparison groups: husbands and coresident children.	Assistance needed with at least one ADL	Psychiatric: CES-D. Physical: Single item self-rated health.	CGs' M CES-D = 12.37, higher than their husbands (M = 7.84). Poorer self-rated health and negative stress appraisals predicted higher CES-D.	CGs had poorer self-rated health (M = 3.17, range 1–4 where higher = better) than their coresident offspring (M = 3.39).
Pruchno, Patrick, & Burant (1996)	n = 487 aging CG mothers of child with developmental disability and n = 351 of child with schizophrenia, recruited through a variety of community sources; M age = 65.7 and 64.4, respectively; 87.7% and 89.2% white, respectively.	Adult children with developmental disability or schizophrenia	Psychiatric: CES-D. Physical: 4-item self-rated health	CGs of schizophrenics reported significantly more depressive symptoms than those caring for a developmentally disabled child (M = 10.4 vs. 8.1). More depressive symptoms were associated with patients exhibiting more problem behaviors, more CG burden, and poorer self-rated and more self-reported illnesses health; no relationship quality with patient.	No differences between CGs of schizophrenics (M = 9.30) and CGs of developmentally disabled (M = 9.52) on self-rated health.

(*Continued*)

Table 6.2. (*Continued*)

Authors	Sample	Patient impairment	Health measures	Psychiatric morbidity	Physical morbidity
Rauktis, Koeske, & Tereshko (1995)	n = 106, identified as primary caregiver by mentally ill individuals selected randomly from a larger group of 243 prospective participants; M age = 59 yrs.; 86% female; 98% white; 75% parents.	Various mental illnesses: 75% schizophrenic, 19% with bipolar affective disorder, 2% each unipolar depression, personality disorder, other diagnosis	Psychiatric: CES-D; psychological distress (composite of guilt, stigma, fear, and worries). Physical: None.	M = 0.68 (i.e., a mean total of 13.6); M distress = 2.3 (range = 1–4). Greater distress related to more negative interactions.	N/A.
Redinbaugh, MacCallum, & Kiecolt-Glaser (1995)[b]	n = 103, recruited from a variety of community resources; M age = 56 yrs.; 71% female; 93% white.	Alzheimer's disease	Psychiatric: SCID, HDRS. Physical: None.	SCID: 33% CGs diagnosed with episodic depression, 20% with chronic depression; 47% never depressed. 55% of the episodic group and 11% of the chronic group experienced a pre-CG depressive episode. HDRS: At all three time points of the study, the chronic depression group scored significantly higher (M range 12.2–13.7) than the never-depressed group (M range = 3.46–3.81) and the episodically depressed group (M range = 6.00–7.24). Less positive social support and more negative life events were related to chronically elevated depressive symptoms	N/A.
Rose-Rego et al. (1998)	n = 99 spouse CGs, recruited from the local ADRC; M age = 70 yrs for women and 72 yrs. for men; 62% female; 97% and 92% white, respectively; n = 113 nonCGs.	Alzheimer's disease	Psychiatric: CES-D. Physical: Single-item self-rated health.	CGs scored higher on the CES-D than nonCGs, M = 9.31 for CG males and 16.64 for CG females vs. 4.30 for nonCG males and 4.31 for nonCG females. Depressive symptoms were higher in female CGs than male CGs.	CGs rated their health to be poorer than nonCGs, M = 2.68 for CG males and 2.20 for CG females vs. 2.82 for nonCG males and 2.95 for nonCG females. Female CGs had poorer self-rated health than male CGs.

Schulz et al. 1996)[a]	n = 392 spouses, recruited from a population-based sample surveying elderly individuals; categorized as living with (but providing no care to) disabled spouse (n = 75), providing care to disabled spouse (n = 138), and providing care + reporting caregiving strain (n = 179); overall, M age = approx. 76 yrs.; approx. 49% female; approx. 89% White.	Elders who had difficulty with ≥ 1 IADL or ADL due to physical or health problems, or problems of confusion	Psychiatric: 10-item CES-D. Physical: Blood pressure, blood chemistries (cholesterol, glucose, albumin, creatinine), medication use, health-related behaviors (alcohol intake, smoking, not getting enough rest, no time to see doctor or missed doctor's appointment), single-item self-rated health.	After adjusting for known covariates, CES-D scores were significantly higher among CGs who reported experiencing caregiving strain (M = 7.5) compared to nonCGs (M = 4.0), spouses who lived with but provided no assistance to a disabled spouse (M = 4.3), and CGs who experienced no strain (M = 4.3).	After adjusting for known covariates, CGs who experienced strain perceived their health to be significantly poorer (M = 3.1) than did nonCGs and CGs who experienced no strain (M = 3.4 each). Significant differences in medication use between strained CGs and nonCGs disappeared after they adjusted for known covariates. Compared to nonCGs, CGs with strain were more likely to report that they did not get enough rest, had no time to exercise, and did not have enough time to rest when they were sick. Those who lived with but provided no care for the disabled spouse were more likely than nonCGs to report that they did not have enough time to rest when sick.
Seltzer, Greenberg, & Krauss (1995)	n = 105 aging mothers of mentally ill adults (MI) and 389 of mentally retarded adults (MR), recruited through state/county agency, media ads, other participants; M age = approx. 66 yrs.	Mentally ill or mentally retarded adult children	Psychiatric: CES-D. Physical: single-item self-rated health.	CGs of MI had significantly higher CES-D scores than those of MR (M = 11.35 vs. 9.58). For CGs of MR, more depressive symptoms related to poorer self-rated health, more patient behavior problems, use of less problem-focused coping and more emotion-focused coping; for CGs of MI, more depressive symptoms related to use of more emotion-focused coping.	M for CGs of MI = 2.99 and CGs of MR = 2.83.

(Continued)

Table 6.2. (*Continued*)

Authors	Sample	Patient impairment	Health measures	Psychiatric morbidity	Physical morbidity
Shaw, Patterson, Semple, Grant, et al. (1997)[e]	*n* = 110 Chinese family CGs, recruited from large-scale epidemiologic research program investigating the prevalence of AD and dementia in Shanghai; *M* age = 58 yrs.; 57.3% female; 30.9% spouses & 69.1% other relatives; and *n* = 139 American family CGs, recruited via physician referrals and community support groups; *M* age = 70.7 yrs.; 64.7% female; 100% spouses. For Chinese CGs, *n* = 110 age-, sex- and kinship-matched controls; for American CGs, age-matched married controls.	Alzheimer's disease and dementia	Psychiatric: BSI depression and anxiety; HDRS. Physical: physical symptoms.	Chinese CGs did not differ significantly from their controls on the BSI-depression (.26 vs. .22), BSI-anxiety (.23 vs. .21), or HDRS (2.6 vs. 1.8), whereas American CGs obtained significantly higher scores than their controls on the BSI-depression (.67 vs. .28), BSI-anxiety (.63 vs. .59), and HDRS (5.3 vs. 2.5). InChinese CGs, coping strategies were largely unrelated to depression and anxiety whereas in American caregivers, more behavioral confronting and distancing and cognitive confronting and distancing were related to more depression and anxiety.	Chinese CGs reported significantly more physical symptoms than their controls (5.9 vs. 4.0), whereas American CGs did not differ significantly from their controls (10.5 vs. 8.5). Only among Chines CGs, less cognitive distancing was related to more physical symptoms.
Shaw, Patterson, Semple, Ho, et al.. (1997)[e]	*n* = 150 spouse CGs, recruited from ADRC and through community support groups or physician referrals; *M* age = 70.6 yrs.; 65% female; *M* CG duration = 1.9 yrs.; n = 46 age-matched married controls.	Alzheimer's disease	Psychiatric: None. Physical: 4 indices—extended illness/ disability ($\geq$ 1 mo.); hospitalization (during past 6 mos., as in-patient); nurse's health rating (healthy, borderline, not healthy); no. of mos. before previous three health events were reached.	N/A.	CGs were no more at risk for having an extended illness/ disability, being hospitalized, or being rated as unhealthy by a nurse compared to controls. However, CGs who provided more assistance with ADLs had a greater hazard than controls for reaching at least one of the three health events. More patient problem behaviors were associated with a *lower* risk of CG hospitalization.

Skaff et al. (1996)	n = 456, no recruitment information; 30% continuing CGs, 29% placement CGs (who had place patient in care facility by T2), and 41% bereaved CGs at T2.	Alzheimer's disease	Psychiatric: None. Physical: Single-item self-rated health.	N/A.	For continuing CGs only, self-rated health worsened significantly from T1 to T2.
Smerglia & Deimling (1997)	n = 244, drawn from 2 larger caregiver studies in Cleveland, OH that recruited through formal community agencies, community groups, and self-referrals; 31% spouses, 69% adult children.	Impaired elder requiring assistance with at least one of six personal care tasks	Psychiatric:Zung Depression Scale (ZDS). Physical: Single-item self-rated health.	*M* ZDS significantly higher for spouses (47.2) than for adult children (39.4). Higher ZDS score associated with being the spouse, having less family adaptability and decision-making satisfaction, and poorer self-rated health.	*M* self-rated health signifcantly poorer for spouses (2.7) than for adult children (3.4).
Smith (1996)	n = 225 mothers, recruited through community agencies and other sources; *M* age = 70.3 yrs.; 97.9 white.	Adult children with mental retardation	Psychiatric: None. Physical health: Single-item self-rated health.	N/A.	30.7% rated their health as "fair" or "poor." Better self-rated health was associated with greater ego integrity and less negative affect.
Stephens, Franks, & Atienza (1997)[c]	n = 105 employed CG daughters, recruited via various community sources; *M* age = 48.3 yrs.; 89.5% white; *M* CG duration = 5.6 yrs.	Ill or disabled elderly parent who needed assistance with shopping, personal care, preparing meals, or supervision due to memory problems	Psychiatric: CES-D. Physical: None.	$M = 17$; 44.8% scored ≥ 16 on the CES-D, indicating risk for clinical depression. Greater stress associated with paid employment and caregiving were related to more depressive symptoms.	N/A.
Stephens & Townsend (1997)	n = 296 CG daughters/ daughters-in-law, recruited through newspaper and radio ads, newspaper articles, and notices in local newsletters; *M* age = 43.9 yrs.; 88% white; *M* CG duration = 6.3 yrs.	Ill or disabled elderly parent who needed assistance with one IADL or ADL, or supervision	Psychiatric: CES-D. Physical: None.	$M = 12.5$; 32% scored ≥ 16 on the CES-D, indicating risk for clinical depression. Greater stress associated with caregiving, paid employment, and role of wife were related to more depressive symptoms.	N/A.

(*Continued*)

Table 6.2. (*Continued*)

Authors	Sample	Patient impairment	Health measures	Psychiatric morbidity	Physical morbidity
St.-Onge & Lavoie (1997)	$n = 99$ CG mothers, recruited through general and psychiatric hospitals; M age = 62.1 yrs.	Adult children diagnosed with a psychotic disorder (schizophrenia, schizophreniform or schizoaffective disorder, atypical psychosis)	Psychiatric: Psychiatric Symptom Index (PSI). Physical: four items measuring current, past, and future health, and the extent to which physical discomfort requiring physician consultation was experienced in previous year.	CGs obtained a higher score on the PSI (22.8) than a normative community sample (10.5), and scored above the cut-off representing strong symptoma--tology (i.e., a score of > 20). Psychiatric symptoms declined in CGs whose offspring showed improvement compared to those whose did not. More psychiatric symptoms were associated with poorer social support and poorer self-rated health	$M = 8.6$ (range = 4–16).
Strawbridge et al. (1997)	$n = 44$ grandparents, 44 spouses, 130 adult children, drawn from a longitudinal health and mortality study and who responded to a follow-up in 1994; M age = 60.9, 64.6, and 57.8 yrs., respectively; % female 73.8%, 63.6%, and 66.1%, respectively; nonCGs $n = 1{,}669$.	Based on 1994 responses, grandparents indicated that they were raising a relative < 18 yrs., spouses indicated they were helping to care for a spouse and assisting with instrumental or personal care activities, adult	Psychiatric: ≥1 depressive symptoms. Physical: self-rated health (fair or poor vs. good or excellent); ≥1 chronic conditions; activity limitations.	CGs were more likely to report one or more depressive symptoms than nonCGs.	Grandparents were significantly more likely to rate their health as poor and to have some activity limita-tions compared to nonCGs; spouse and adult child CGs did not differ compared to nonCGs.

		children indicated that they were helping to care for a parent and living or assisting with instrumental and personal care activities			
Vitaliano, Russo, et al. (1996)[f]	n = 81 spouses, recruited using mass mailings to physicians; M age = 69.8 yrs.; 64% female; almost 100% white; M CG duration 40.1 mos. for men and 47.2 mos. for women; age- and sex-matched nonCG group, n = 86.	Alzheimer's disease	Psychiatric: HDRS. Physical: body mass index (BMI), prevalence of obesity, current health problems, medication use, cardiovascular risk factors and health-related behaviors (smoking, estrogen use, exercise, alcohol intake), food diaries.	At T1 and T2, significantly higher HDRS scores were obtained by male CGs than male nonCGs (M = 6.9 vs. 3.6 at T1 and M = 6.4 vs. 4.5 at T2, respectively) and female CGs compared to female nonCGs (M = 8.5 vs. 4.7 at T1 and M = 7.7 vs. 4.6 at T2). Depression ratings were not a significant predictor of weight gain either as a main effect or in interaction with psychological and nutritional variables.	At T1 and T2, male CGs had a higher BMI than male nonCGs (approx. 12 lbs. more); at T1, 20% of the male caregivers vs. 4% of the male nonCGs were classified as obese and at T2, 28% of the male caregivers vs. 4% of the male nonCGs were obese. Female CGs did not differ from female nonCGs on BMI, but by T2 they had gained more weight than their controls (approximately 3 lbs. more). Female CGs with greater anger control and in creased caloric intake and male CGs with lower perceived control and increased fat intake gained the most weight.

(*Continued*)

Table 6.2. (*Continued*)

Authors	Sample	Patient impairment	Health measures	Psychiatric morbidity	Physical morbidity
Vitaliano, Scanlan, et al. (1996)[f]	*n* = 73 spouses, recruited using mass mailings to physicians; *M* age = 69.8 yrs.; 65% female; almost 100% white; *M* CG duration 45.5 mos. for men and 52.7 mos. for women; age- and sex-matched nonCG group, *n* = 69.	Alzheimer's disease	Psychiatric: HDRS, composite measure of psychological distress (HDRS + CG burden + hassles–uplifts). Physical: Fasting insulin/glucose and lipids, cholesterol, hypertension, activity regimen, medications, current health problems, cardio-vascular risk factors, exercise, food intake.	CGs were stratified by gender and hormone-replace therapy (for females). CGs in all groups were rated at T1 and T2 as more depressed on the HDRS(*M* range = 90.8 to 116.0) than their respective controls (*M* range = 33.0 to 49.8). Male CGs had higher lipids than their matched controls and female CGs had less aerobic activity than their controls. On psychologicaldistress, female CGs scored significantly higher than their controls atT1 and T2, and CG males scoredhigher than their controls at T2.	CGs as a group had ignificantly higher insulin at s T1 and T2 than the control group; this effect disap-peared when they statistically controlled for caregivers' psychological distress. No differences on glucose levels between CGs and nonCGs. For both CGs and nonCGs, depressive symptoms were related to insulin and glucose levels.
Vitaliano, Scanlan, Krenz, & Fujimoto (1996)	*n* = 73 spouses, recruited using mass mailings to physicians; *M* age = 69.8 yrs.; 65% female; almost 100% white; *M* CG duration 45.5 mos. for men and 52.7 mos. for women; age- and sex-matched nonCG group, *n* = 69.	Alzheimer's disease	Psychiatric: None. Physical: Fasting insulin/glucose and lipids, cholesterol, hypertension, activity regimen, medications, current health problems, cardio-vascular risk factors, exercise, food intake.	N/A.	Female CGs not on hormone-replacement therapy reported significantly less exercise than their controls. Male CGs were more obese and had higher lipids than their controls. No other differences on health variables between CGs and controls. CGs with high anger-out and hostility had significantly higher glucose levels than CG groups with other combinations of type of anger and hostility.

Williamson & Schulz (1995)	*n* = 82 family CGs, mostly spouses, identified by care recipients recruited from local hospitals. *M* age = 60.2; 95.1% white.	Recurrent cancer patients receiving palliative radiation therapy	Psychiatric: CES-D. Physical: None.	*M* = 11.6. More depressive symptoms associated with low levels of communal behavior prior to the patient's illness. CGs with low past communal behavior scored higher on the CES-D at high and low burden levels whereas CGs with high past communal behaviors scored higher on the CES-D only at high levels of burden. CG burden mediated the links between depression and both past communal behaviors and patient pain.	N/A.

Note: Studies sharing a superscript have overlapping samples.

groups, community samples, and/or population norms (Collins & Jones, 1997; King & Brassington, 1997; Majerovitz, 1995; Malone, Beach, & Zarit, 1995; Miller, Campbell, Farran, Kaufman, & Davis, 1995; Shaw et al., 1997) as well as greater anxiety (King & Brassington, 1997). Studies that have specifically compared dementia caregivers with noncaregiver controls also have obtained support for more self-reported depressive symptoms among caregivers (Haley et al., 1995; Irwin et al., 1997; Kiecolt-Glaser, Glaser, Gravenstein, Malarkey, & Sheridan, 1996). Only a small minority of studies have failed to observe poorer mental health in dementia caregivers compared to noncaregiver controls or community norms (e.g., Fuller-Jonap & Haley, 1995; Meshefedjian, McCusker, Bellavance, & Baumgarten, 1998). Recent studies that have used a psychiatric diagnostic interview also have found a considerable proportion of dementia caregivers to be characterized by mood and/or anxiety disorders (Irwin et al., 1997; Redinbaugh, MacCallum, & Kiecolt-Glaser, 1995; Shaw et al., 1997; Vitaliano, Russo, Scanlan, & Greeno, 1996; Vitaliano, Scanlan, Krenz, Schwartz, & Marcovina, 1996).

Caregivers of Patients with Other Health Conditions

Recently published studies of caregivers of patients with disabilities other than dementia also support the prevalence of poorer mental health in family caregivers compared to noncaregivers (see Table 6.2). This pattern is characteristic of studies that have interviewed caregivers of frail elderly individuals by drawing participants from large, population-based random samples (e.g., Beach, Schulz, Yee, & Jackson, 1998; Bookwala & Schulz, 1998; Cochrane, Goering, & Rogers, 1997; Jutras & Lavoie, 1995; Schulz et al., 1997) as well as studies that have used smaller, nonrandom samples of caregivers to frail elders (Bass, Noelker, & Rechlin, 1996; Fingerman, Gallagher-Thompson, Lovett, & Rose, 1996; Lawrence, Tennstedt, & Assmann; 1998; Martire, Stephens, & Atienza, 1997; Pruchno, Burant, & Peters, 1997; Pruchno, Peters, & Burant; 1995; Stephens, Franks, & Atienza, 1997; Stephens & Townsend, 1997). Researchers also have found significantly greater psychiatric morbidity compared to noncaregivers or community samples among caregivers of handicapped individuals (Pariante et al., 1997), recurrent cancer patients (Williamson & Schulz, 1995), heart transplant patients (Canning, Dew, & Davidson, 1996), and mentally ill adult relatives (Rauktis, Koeske, & Tereshko, 1995; St.-Onge & Lavoie, 1997). Some evidence suggests that caring for an adult child with a mental illness is related to poorer mental health than caring for a child with a developmental disability (Pruchno, Patrick, & Burant, 1996) or mental retardation (Seltzer, Greenberg, & Krauss, 1995).

In summary, the vast majority of recently published studies have found evidence for poorer mental health among caregivers either when compared to community standards or noncaregiving control groups. This trend characterizes both dementia and nondementia caregivers and, as Table 6.2 indicates, crosses kinship lines, with caregiving spouses, adult children, parents, and grandparents reporting elevated depressive and related psychiatric morbidity symptoms.

CAREGIVING AND PHYSICAL MORBIDITY

The link between caregiving and physical morbidity is more equivocal and generally weaker than the association between caregiving and poorer mental health (Schulz et al., 1995). This is perhaps due largely to the use of diverse indicators of physical health in caregiving studies. Among the various physical health indicators that have been used are self-rated global health, the presence of chronic conditions and illness, physical symptom checklists, the use of medications, utilization of health care services, health-related behaviors, and impact on physiological functioning. Based on their review of prior research with dementia caregivers, Schulz and his colleagues (1995) concluded that while caregiving is fairly consistently associated with poorer self-rated health, links with the other, less subjective assessments of physical health have produced mixed results. Since their review, several studies of both dementia and nondementia caregivers have examined physical health outcomes among caregivers, albeit the number of such studies remains smaller than of those examining mental health outcomes among caregivers. And researchers continue to operationalize caregivers' physical health outcomes in various ways. Table 6.2 also presents a summary of recent findings on the relation—or lack thereof—between caregiving and physical morbidity.

Self-Rated Global Health

A common assessment of physical health status in caregiving studies is a single question asking respondents to rate their current health on a scale from poor to excellent. It is estimated that 24–28% of the elderly population rate their health as fair or poor (Schulz et al., 1994). A substantial proportion of caregivers also rate their health to be compromised. For example, Mui (1995) found that 63% of caregiving wives and 50.7% of caregiving husbands reported "fair" or "poor" health, and Smith (1996) found that almost 31% of caregiving mothers to adult children with mental retardation rated their health as "fair" or "poor." Majerovitz (1995) reported that when asked to compare their own health relative to other people their age, 19% of spouse caregivers reported their health to be "fair" or "poor." Consistent with Schulz et al.'s (1995) review, several recent studies also have documented that caregivers generally perceive their health to be worse than noncaregivers or community samples (e.g., Beach et al., 1998; Mui, 1995; Pruchno et al., 1995, Rose-Rego, Strauss, & Smyth, 1998; Schulz et al., 1997; Strawbridge, Wallhagen, Shema, & Kaplan, 1997). Evidence also has been obtained for a decline in caregivers' physical health status over time. In a study that measured self-rated health across time, Skaff, Pearlin, and Mullan (1996) found that self-rated health worsened significantly from baseline to follow-up among continuing caregivers. In general, then, a fairly consistent link has emerged between caregiving and poorer self-rated health, with few studies failing to find this relationship (e.g., Haley et al., 1995; Pruchno et al., 1996).

Chronic Conditions, Illnesses, Physical Symptoms, and Disability

A number of recently published caregiving studies suggest the existence of a link between caregiving and self-reported illnesses, symptomatology, and disability (Bass et al., 1996; Canning et al., 1996; Cochrane et al., 1997; Fuller-Jonap & Haley, 1995; Jutras & Lavoie, 1995). In terms of specific health problems, Fuller-Jonap and Haley (1995) found that caregiving husbands reported more respiratory problems than a comparison group, while Jutras and Lavoie (1995) noted that more caregivers, compared to noncaregivers, reported suffering from diabetes (4.5% vs. 1%) and hay fever (5.8% vs. 3.1%), and more caregivers reported suffering from back problems than individuals who did not reside with an elder (14.0% vs. 9.6%). In terms of disability days, Cochrane et al. (1997) found that although caregivers and noncaregivers did not differ in number of days of total disability, significantly more caregivers reported having "limited-activity days" and "days that required extreme effort" compared to noncaregivers. Other studies, however, have failed to find a link between caregiving and self-reported illnesses or disability (e.g., Irwin et al., 1997; Pruchno et al., 1996; Shaw et al., 1997).

Health-Related Behaviors, Medication Use, and Health Service Utilization

Several studies have assessed caregivers' engagement in health-compromising behaviors (e.g., alcohol use, smoking), use of prescription and over-the-counter medication, and use of health-care services (e.g., number of hospitalizations, physician visits), as proxy indicators of physical morbidity (Burton, Newsom, Schulz, Hirsch, & German, 1997; Glaser & Kiecolt-Glaser, 1997; Haley et al., 1995; Kiecolt-Glaser et al., 1996; King & Brassington, 1997; Schulz et al., 1997). However, findings on the links between caregiving and these health indices are equivocal. Several studies have found that caregivers report less physical activity, sleep, and rest than noncaregivers (Burton et al., 1997; Fuller-Jonap & Haley, 1995; Glaser & Kiecolt-Glaser, 1997; Kiecolt-Glaser et al., 1996; Schulz et al., 1997), but no group differences were apparent in other behaviors such as alcohol consumption, smoking, weight change, finding time to see a doctor, and missing a doctor's appointment. In terms of medication use, Burton et al. (1997) found that caregivers were more likely to report forgetting to take prescription medications than were noncaregivers, but Haley et al. (1995) observed no group differences on medication use. Although caregivers do not appear to use more health care services for themselves than noncaregivers, Bass et al. (1996) reported that a significant number of family caregivers use formal services to assist with caring for the patient, such as assistance with personal-care tasks, household service, health care service, and escort services.

Physiological Indices

Kiecolt-Glaser and her colleagues (1996; Glaser & Kiecolt-Glaser, 1997) have reported several findings relevant to the effects of caregiving on physiological response. They note that, compared to matched controls, caregivers show a poorer immune response after exposure to an influenza virus vaccine and to infection by a latent herpes simplex virus. Similarly, Pariante and her colleagues (1997) observed that caregivers had lower levels of *T* cells, a higher percentage of *T* suppressor/cytotoxic cells, and a lower *T* helper:suppressor ratio compared to matched controls. When they grouped caregivers by age, Pariante et al. found that older caregivers had lower numbers of *T* cells and *T* helper cells, and higher antibody titers for cytomegalovirus compared to matched noncaregivers; they observed no differences between younger caregivers and their controls on these measures. In contrast, Irwin and his colleagues (1997) observed no significant differences between caregivers and noncaregivers on immune, neuroendocrine, and sympathetic measures. In terms of the relative prevalence of cardiovascular risk factors in caregivers versus noncaregivers, Vitaliano, Scanlan, Krenz, et al. (1996) reported that male caregivers had higher lipids than their age- and sex-matched controls, while female caregivers had less aerobic activity than their control counterparts. Caregivers also had significantly higher insulin than the control group, although this effect disappeared after statistically controlling for caregivers' psychological distress. Vitaliano, Russo, et al. (1996) also report that male caregivers had a higher body mass index than their noncaregiving controls at two different time points, with more caregivers than controls being classified as obese. Female caregivers, in comparison, did not differ from their controls on body mass index; however, they had gained more weight than their controls by follow-up.

To summarize, several recent caregiving studies have examined the relationship between caregiving and adverse physical health outcomes using a wide range of health indicators. Overall, a fairly consistent association exists between caregiving and poorer self-rated health. However, consistent with Schulz et al.'s (1995) conclusions, evidence for the relation between caregiving and more objective indicators of physical health (e.g., physical symptomatology, health service utilization, medication use, physiological outcomes) remains mixed. While several studies have obtained an association between caregiving and chronic illnesses, disability, poor immune function, and adverse physiological outcomes, others have reported null findings. We can expect that as the number of studies exploring physical health outcomes among caregivers continues to grow, and more sophisticated measures of physical health are employed, we will be able to determine more reliably the extent to which caregiving is related to objective indicators of physical morbidity.

INTERRELATIONSHIPS BETWEEN CAREGIVERS' PSYCHIATRIC AND PHYSICAL MORBIDITY

Strong relations are known to exist between elderly individuals' mental and physical health status, such that poorer mental health is concomitant with poorer physical health (e.g., Bookwala & Schulz, 1996; Newsom & Schulz, 1996; Schulz et al., 1994). Schulz et al. examined the association between depressive symptoms and self-rated health in a large, population-based sample of older adults participating in the Cardiovascular Health Study. They observed that more depressive symptoms were related to poorer self-rated health. Indeed, among multiple variables from eight domains, including functional status, medication use and medical procedures, self-reported clinical disease, anthropometry (e.g., standing height, waist circumference) and health habits, assessed prevalent disease, assessed subclinical disease, psychosocial measures, and demographic characteristics, depressive symptoms emerged as one of the strongest predictors of poorer self-rated health. In a subsample of elderly married couples drawn from the same sample of older adults, Bookwala and Schulz (1996) found that more depressive symptoms were related to greater functional impairment and use of medication among both women and men.

A significant relation between mental and physical health also has been observed among dementia caregivers, where caregivers' physical health quite consistently emerges as a strong predictor of their mental health (for a review, see Schulz et al., 1995). In caregiving studies, caregivers' physical health frequently has been conceptualized as either a resource variable or a vulnerability factor in shaping caregivers' mental health, and several studies include it as a statistical covariate in data analyses. Schulz et al. (1995) noted that the vast majority of studies that examined the relationship between depression and physical health found a significant association between these variables. Some physical health predictors of greater depression included low self-rated health, declines in self-rated health, more chronic health conditions, and more frequent health service utilization. As Schulz et al. pointed out in their review, most studies have relied on cross-sectional data to examine the relationship between depressed affect and physical health. An exception is a study by Pruchno, Kleban, Michaels, and Dempsey (1990), who used longitudinal data to examine links between depressed affect and physical health. They found evidence for a causal pathway from mental health to physical health only. Specifically, depressive symptoms at Time 1 and Time 2 predicted self-rated health at Time 2 and Time 3, respectively; self-rated health was not a significant predictor of depressive symptoms over time, however. Hence, Pruchno and her colleagues concluded that changes in mental health are likely to bring about changes in assessments of physical health but not vice versa.

As Table 6.2 details, caregiving studies published after Schulz et al.'s (1995) review provide additional support for an association between caregivers' psychiatric morbidity and poorer self-reported health. In general, studies that find no relation between caregivers' mental health and self-reported physical health are few (Draper, Poulos, Poulos, & Ehrlich, 1995). When examined, the vast majority of studies have found that in caregivers of patients with various health conditions, greater psychiatric morbidity is associated with poorer self-rated health (Beach et al., 1998; Bookwala & Schulz, 1998; Hooker et al., 1998; Li, Seltzer, & Greenberg, 1997; Miller et al., 1995; Pruchno et al., 1995; 1996; Seltzer et al., 1995; Smerglia & Deimling, 1997; St.-Onge & Lavoie, 1997) and more self-reported illnesses (Canning et al., 1996; Hooker et al., 1998; Pruchno et al., 1996). When Hooker and her colleagues tested reciprocal relationships between caregivers' mental and physical health using structural equation modeling, they found that poorer mental health significantly predicted poorer physical health, but not the reverse. Only a few of these studies have used longitudinal data, however. For example, Canning et al. (1996) noted that caregivers of heart transplant patients who reported more chronic illnesses soon after the patient's surgery were more likely to have higher psychological distress 7 and 12 months after the surgery. Similarly, Beach et al. (1998) noted that caregivers' poorer self-rated health was associated with more symptoms of anxiety and depression at Time 1 and Time 2. In addition, caregivers' poorer self-reported functioning was associated with more anxiety symptoms at both time points.

Links between depression and more objective indices of caregiver physical health—such as physiological indices and health service utilization—are less common. Kiecolt-Glaser et al. (1996) obtained no significant relationship between caregivers' depressive symptoms and their immune response. Moreover, the inclusion of depressive symptoms (which were higher in caregivers than noncaregivers) as a covariate in analyses of immune response did not nullify the finding of poorer immune response in caregivers compared to noncaregivers, suggesting that the stressful effects of caregiving may impact relatively independently on both caregivers' affect and immune function. Vitaliano, Russo, et al. (1996) found that caregivers' depression ratings were not a significant predictor of weight gain either as a main effect or in interaction with psychological and nutritional variables. Irwin et al. (1997) also found no support for a relationship between depressed affect and caregivers' immune, neuroendocrine, or sympathethic measures. When King and Brassington (1997) examined the relationship between mental health and exercise training among caregivers who were randomly assigned either to an exercise training program or a delayed exercise control condition, they found no difference in depressed affect or anxiety between the two groups. Perhaps the only evidence for a link between mental health and physiological indices was reported by Vitaliano, Scanlan, Krenz, Schwartz, and Marcovina (1996), who found a strong association between depressive symptoms and insulin and glucose levels. However, this link was present in both caregivers and noncaregivers. Indeed, the evidence for higher insulin levels in caregivers compared to noncaregivers disappeared when they controlled for participants' psychological distress.

In summary, significant associations have been reported between older adults' mental and physical health using large, population-based samples. A substantial number of caregiving studies have treated caregivers' self-rated health as an important covariate of mental health and, in general, a fairly robust relationship has emerged between psychiatric morbidity and poorer self-reported health among caregivers. Links between caregivers' psychiatric morbidity and objective physical health indicators—such as immune function, physiological indices, and physical activity—are rare, however. Because most of the studies examining caregivers' health outcomes are cross-sectional, only tentative conclusions can be drawn about causal pathways between caregivers' psychiatric and physical morbidity.

CORRELATES OF PSYCHIATRIC AND PHYSICAL MORBIDITY IN CAREGIVERS

As Table 6.2 indicates, the recent literature is replete with findings that link caregivers' mental health to a variety of factors. By and large, these findings are replications or extensions of the results of previous caregiving studies (see Schulz et al., 1995). Specifically, psychiatric morbidity in caregivers has been linked to sociodemographic variables (Draper et al., 1995; Haley et al., 1995; Li et al., 1997; Miller et al., 1995; Meshefedjian et al., 1998; Rose-Rego et al., 1998), kinship to the patient (Draper et al, 1995; Li et al., 1997, Smerglia & Deimling, 1997; Strawbridge et al., 1997), the patient's disability and behavioral problems (e.g., Bookwala & Schulz, 1998; Draper et al., 1995; Lawrence et al., 1998; Li et al., 1997; Majerovitz, 1995; Meshefedjian et al., 1998; Pruchno et al., 1996; Seltzer et al., 1995), caregiver strain or burden (e.g., Beach et al., 1998; Bookwala & Schulz, 1998; Hooker et al., 1998; Pruchno et al., 1995, 1996; St.-Onge & Lavoie, 1997; Stephens & Townsend, 1997), role overload and role conflict (Lawrence et al., 1998; Martire et al., 1997; Stephens et al., 1997; Stephens & Townsend, 1997), greater perceived stress (Beach et al., 1998; Redinbaugh et al, 1995), personality variables (Bookwala & Schulz, 1998; Fingerman et al., 1996; Hooker et al., 1998), coping patterns (Fingerman et al., 1996; Gignac & Gottlieb, 1996; Seltzer et al., 1995; Shaw et al., 1997), caregiver–patient relationship quality (Canning et al., 1996; Lawrence et al., 1998), family adaptability (Majerovitz, 1995; Smerglia & Deimling, 1997) and family consensus (Pruchno et al., 1997), and social support (Li et al, 1997; MaloneBeach & Zarit, 1995; Redinbaugh et al., 1995; St.-Onge & Lavoie, 1997).

Less common is the examination of correlates of physical health among caregivers. Hence, we devote this section to a more detailed discussion of possible correlates of caregivers' physical morbidity. In general, the sociodemographic variables related to poorer physical health are similar to those associated with poorer mental health in caregivers (see Table 6.2). Specifically, poorer self-rated health in caregivers is associated with being female (Mui, 1995; Rose-Rego et al., 1998), older and less educated (Penning, 1998), and spouse caregivers tend to rate their physical health to be worse than do adult child caregivers (Li et al., 1997; Smerglia & Demling, 1997). In contrast to the nature of the relationship between race and mental health, however, black caregivers report poorer health than do their white counterparts (Miller et al., 1995). Stressful aspects of the caregiving context are related to poorer physical health, as they are to poorer mental health. For example, poorer self-rated health is related to more functional and behavioral impairment in care recipients and more hours of care provision (Penning, 1998). Greater stress experienced by caregivers is related to worse self-rated health (Beach et al., 1998; Hooker et al., 1998; Mui, 1995) and poorer functional status (Mui, 1995), as are greater perceived unmet needs in the patient and longer caregiving duration (Mui, 1995). In addition, Irwin et al. (1997) noted that basal adrenocorticotropic hormone (ACTH) levels were higher among caregivers who reported an extreme mismatch in the amount of care they provided and the amount of respite time they received compared to caregivers who did not experience such an extreme discrepancy. Some evidence suggests, however, that the negative effects of patient disability on caregivers' health deterioration may be alleviated substantially when caregivers use formal care services to assist in care provision (Bass et al., 1996).

As in the case of mental health, personality variables also have emerged as significant correlates of physical health indicators among caregivers. Perceptions of control have been linked to better health by some investigators. For example, Burton and her colleagues (1997) observed that more caregivers with a strong sense of control engaged in preventive health behaviors (e.g., not missing meals, not missing flu shots) compared to caregivers with a weak sense of control. Vitaliano, Russo, et al. (1996) also observed that caregiving husbands with lower perceived control and increased fat intake gained the most weight. In comparison, perceived control was unrelated to wife caregivers' physical health; instead, greater anger control and increased caloric intake was linked to the greatest weight gain among caregiving wives. In another report examining the links between anger and physiological indices among caregivers, Vitaliano and colleagues reported that caregivers with high anger-out and hostility had significantly higher glucose levels than all other combinations of type of anger and level of hostility (Vitaliano, Scanlan, Krenz, & Fujimoto, 1996). Other studies have linked better self-rated health in caregivers with greater ego integrity and less negative affect (Smith, 1996), and poorer self-rated health to higher neuroticism and lower optimism (Hooker et al., 1998).

In summary, findings of several studies point to links between psychiatric morbidity in caregivers and sociodemographic variables, stressful aspects of the caregiving context, personality, coping, and social support. In comparison, a smaller number of studies have examined risk factors of poorer physical health among caregivers. In general, there are fairly parallel links between the correlates of physical health and those associated with mental health.

CONCLUSIONS

Caregiving research documenting health outcomes among caregivers continues to expand at a rapid rate, albeit with more studies assessing the mental health effects of caregiving and significantly fewer examining physical morbidity in caregivers. The quality of the research on caregiving outcomes continues to improve as well. A sizable proportion of the studies reviewed in this chapter compare caregivers' health outcomes with those of noncaregiver groups or community standards, use diagnostic measures of psychiatric morbidity and objective assays of physical morbidity, and draw samples from large, population-based studies. The large majority of caregiving studies, however, continue to rely on cross-sectional data. Hence, the sustained impact of caregiving on psychiatric and physical health remains uncertain. The conduct of longitudinal studies documenting deterioration or improvement in caregivers' health indicators as the caregiving context changes over time remains a pressing need in the area of caregiving research.

Several other issues related to caregiver outcomes demand attention. First, the links between caregiving stress and physical morbidity need to be reliably established through the use of sensitive measures of health indices and sophisticated research methodologies. Second, future research needs to examine systematically the lagged pathways between caregivers' psychiatric and physical morbidity. To date, this issue has received only cursory attention, with the vast majority of studies treating caregivers' physical health status as a statistical covariate in their examination of the links between psychiatric morbidity and psychosocial factors. To the extent that caregivers' psychiatric and physical morbidity are linked over time, such information would offer valuable insights in the development and planning of caregiver interventions that aim at reducing morbidity effects. Finally, it is important to acknowledge that the link between caregiving and negative health outcomes is not inevitable, and that there is considerable variability in the extent of detrimental effects reported by caregivers. Indeed, several studies document that caregivers may experience substantial benefits associated with providing assistance (see Beach et al., 1998; Kramer 1997; Miller & Lawton, 1997; Schulz et al., 1997). Some benefits of care provision include positive role gains, sharing activities with siblings, greater empathy for older adults and disabled individuals requiring assistance, and, in some instances, gains in psychological and physical well-being. In addition, many of the negative effects of caregiving may be alleviated by formal assistance received by the caregiver (e.g., Bass et al., 1996). Negative health outcomes in the caregiving context also may not be limited to the caregiver. Care recipients, too, may experience substantial depressed affect associated with the caregiving context as a result of being the recipients of assistance. For example, Newsom and Schulz (1998) reported that nearly 40% of disabled elderly care recipients reported some emotional distress in response to the help they received from their spouse. In addition, the emotional distress in response to being helped (labeled "helping distress") was associated with greater depressive symptomatology 1 year later, suggesting that care recipients' helping distress may have significant long-term consequences. Each of these issues merits further empirical attention.

In summary, based on the studies that we have reviewed here, we can conclude that caregivers of both dementia and nondementia patient groups evidence significant psychiatric morbidity compared to noncaregivers or community norms. Caregivers also consistently rate their physical health to be poorer than do noncaregivers and, in some cases, exhibit poorer health-related behaviors, physical disability, and immune and physiological functioning. Several studies have documented a link between poorer self-rated health and psychiatric morbidity among caregivers and, based on preliminary evidence, the correlates of psychiatric and physical morbidity among caregivers appear to be parallel. There is a continued need, however, for longitudinal research on the sustained psychiatric and physical morbidity outcomes of caregiving and their interrelationships.

REFERENCES

Administration on Aging. (1997). *Compassion in action: A look at caregiving in America* [On-line]. Available: www.aoa.dhhs.gov/May97/compassion.html.

American Psychological Association. (1997). *What practitioners should know about working with older adults.* Washington, DC: Author.

Bass, D. M., Noelker, L. S., & Rechlin, L. R. (1996). The moderating influence of service use on negative caregiving consequences. *Journal of Gerontology: Social Sciences, 51B,* S121–S131.

Beach, S. R., Schulz, R., Yee, J. L., & Jackson, S. (in press). Negative (and positive) health effects of caring for a disabled spouse: When do they occur? *Psychology and Aging.*

Beck, A. T., Ward, C. H., Mendelson, M., Mock, J., & Erbaugh, J. (1961). An inventory for measuring depression. *Archives of General Psychiatry, 4,* 561–571.

Blazer, D. G. (1994). Geriatric psychiatry. In R. E. Hales, S. C. Yudofsky, & J. A. Talbott (Eds.), *The American Psychiatric Press textbook of psychiatry* (2nd edition, pp. 1405–1421). Washington, DC: American Psychiatric Press.

Bookwala, J., & Schulz, R. (1996). Spousal similarity in subjective well-being: The Cardiovascular Health Study. *Psychology and Aging, 11,* 582–590.

Bookwala, J., & Schulz, R. (1998). The role of neuroticism and mastery in spouse caregivers' assessment of and response to a contextual stressor. *Journal of Gerontology: Psychological Sciences, 53B,* P155–P164.

Burton, L. C., Newsom, J. T., Schulz, R., Hirsch, C. H., & German, P. S. (1997). Preventive health behaviors among spousal caregivers. *Preventive Medicine, 26,* 162–169.

Canning, R. D., Dew, M. A., & Davidson, S. (1996). Psychological distress among caregivers to heart transplant recipients. *Social Science and Medicine, 42,* 599–608.

Cochrane, J. J., Goering, P. N., & Rogers, J. M. (1997). The mental health of informal caregivers in Ontario: An epidemiological survey. *American Journal of Public Health, 87,* 2002–2007.

Collins, C. & Jones, R. (1997). Emotional distress and morbidity in dementia carers: A matched comparison of husbands and wives. *International Journal of Geriatric Psychiatry, 12,* 1168–1173.

Derogatis, L. R. (1983). *SCL-90R administration, scoring, and procedures manual–II* (2nd edition). Towson:, MD: Clinical Psychometrics Research.

Di Nardo, P. A., O'Brien, G. T., Barlow, D. H., Waddell, M. T., & Blanchard, E. B. (1983). Reliability of DSM-III anxiety disorder categories using a new structured interview. *Archives of General Psychiatry, 40,* 1070–1074.

Draper, B. M., Poulos, R. G., Poulos, C. J., & Ehrlich, F. (1995). Risk factors for stress in elderly caregivers. *International Journal of Geriatric Psychiatry, 11,* 227–231.

Fingerman, K. L., Gallagher-Thompson, D., Lovett, S., & Rose, J. (1996). Internal resourcefulness, task demands, coping, and dysphoric affect. *International Journal of Aging and Human Development, 42*(3), 229–248.

Fuller-Jonap, F., & Haley, W. E. (1995). Mental and physical health of male caregivers of a spouse with Alzheimer's disease. *Journal of Aging and Health, 7*(1), 99–118.

General Accounting Office. (1998). *Alzheimer's disease: Estimates of prevalence in the United States.* (GAO/HEHS-98-16). Washington DC: United States General Accounting Office, Health, Education, and Human Services Division.

Gignac, M. A. M., & Gottlieb, B. H. (1996). Caregivers' appraisals of efficacy in coping with dementia. *Psychology and Aging, 11,* 214–225.

Glaser, R., & Kiecolt-Glaser, J. K. (1997). Chronic stress modulates the virus-specific immune response to latent herpes simplex virus type 1. *Annals of Behavioral Medicine, 19,* 78–82.

Goldberg, D. P., & Hillier, V. F. (1979). A scaled version of the General Health Questionnaire. *Psychological Medicine, 9,* 139–145.

Haley, W. E., West, C. A., Wadley, V. G., Ford, G. R., White, F. A., Barrett, J. J., Harrell, L. E., & Roth, D. L.

(1995). Psychological, social, and health impact of caregiving: A comparison of black and white dementia caregivers and noncaregivers. *Psychology and Aging, 10*(4), 540–552.

Hamilton, M. (1967). Development of a rating scale for primary depressive illness. *British Journal of Social and Clinical Psychology, 6*, 278–296.

Hooker, K., Monahan, D. J., Bowman, S. R., Frazier, L. D., & Shifren, K. (1998). Personality counts for a lot: Predictors of mental and physical health of spouse caregivers in two disease groups. *Journal of Gerontology: Psychological Sciences, 53B*, P73–P85.

Irwin, M., Hauger, R., Patterson, T. L., Semple, S., Ziegler, M., & Grant, I. (1997). Alzheimer caregiver stress: Basal natural killer cell activity, pituitary–adrenal–cortical function, and sympathetic tone. *Annals of Behavioral Medicine, 19*, 83–90.

Jutras, S., & Lavoie, J. P. (1995). Living with an impaired elderly person: The informal caregivers' physical and mental health. *Journal of Aging and Health, 7*(1), 46–73.

Kiecolt-Glaser, J. K., Glaser, R., Gravenstein, S., Malarkey, W. B., & Sheridan, J. (1996). Chronic stress alters the immune response to influenza virus vaccine in older adults. *Proceedings of the National Academy of Sciences, USA, 93*(7), 3043–3047.

King, A. C., & Brassington, G. (1997). Enhancing physical and psychological functioning in older family caregivers: The role of regular activity. *Annals of Behavioral Medicine, 19*(2), 91–100.

Kramer, B. J. (1997). Gain in the caregiving experience: Where are we? What next? *The Gerontologist, 37*, 218–232.

Lawrence, R. H., Tennstedt, S. L., & Assmann, S. F. (1998). Quality of the caregiver-care recipient relationship: Does it offset negative consequences of caregiving for family caregivers? *Psychology and Aging, 13*, 150–158.

Li, L. W., Seltzer, M. M., & Greenberg, J. S. (1997). Social support and depressive symptoms: Differential patterns in wife and daughter caregivers. *Journal of Gerontology: Social Sciences, 52B*, S200–S211.

Majerovitz, S. D. (1995). Role of family adaptability in the psychological adjustment of spouse caregivers to patients with dementia. *Psychology and Aging, 10*, 447–457.

Malone Beach, E. E., & Zarit, S. H. (1995). Dimensions of social support and social conflict as predictors of caregiver depression. *International Psychogeriatrics, 7*, 25–38.

Martire, L. M., Stephens, M. A. P., & Atienza, A. A. (1997). The interplay of work and caregiving: Relationships between role satisfaction, role involvement, and caregivers' well-being. *Journal of Gerontology: Social Sciences, 52B*, S279–S289.

Meshefedjian, G., McCusker, J., Bellavance, F., & Baumgarten, M. (1998). Factors associated with symptoms of depression among informal caregivers of demented elders in the community. *Gerontologist, 38*, 247–253.

Miller, B., Campbell, R. T., Farran, C. J., Kaufman, J. R., & Davis, L. (1995). Race, control, mastery, and caregiver distress. *Journal of Gerontology: Social Sciences, 50B*, S374–S382.

Miller, B., & Lawton, M. P. (1997). Introduction: Finding balance in caregiving research. *Gerontologist, 37*, 216–217.

Moritz, D. J., Kasl, S. V., & Berkman, L. F. (1989). The health impact of living with a cognitively impaired spouse: Depressive symptoms and functioning. *Journal of Gerontology: Social Sciences, 44*, S17–S27.

Mui, A. C. (1995). Perceived health and functional status among spouse caregivers of frail older persons. *Journal of Aging and Health, 7*(2), 283–300.

National Alliance for Caregiving and the American Association for Retired Persons (1997). *Family caregiving in the U.S.: Findings from a national survey: Final report.* Bethesda, MD: National Alliance for Caregiving.

Newsom, J. T., & Schulz, R. (1996). Social support as a mediator in the relation between functional status and quality of life in older adults. *Psychology and Aging, 11*, 34–44.

Newsom, J. T., & Schulz, R. (1998). Caregiving from the recipient's perspective: Negative reactions to being helped. *Health Psychology, 17*, 172–181.

Pariante, C. M., Carpiniello, B., Orru, M. G., Sitzia, R., Piras, A., Farci, A. M. G., DelGiacco, G. S., Piludu, G., & Miller, A. H. (1997). Chronic caregiving stress alters peripheral blood immune parameters: The role of age and severity of stress. *Psychotherapy and Psychosomatics, 66*(4), 199–207.

Penning, M. J. (1998). In the middle: Parental caregiving in the context of other roles. *Journal of Gerontology: Social Sciences, 53B*, S188–S197.

Pruchno, R. A., Burant, C. J. & Peters, N. D. (1997). Typologies of caregiving families: Family congruence and individual well-being. *Gerontologist, 37*, 157–168.

Pruchno, R. A., Kleban, M. H., Michaels, J. E., & Dempsey, N. P. (1990). Mental and physical health of caregiving spouses: Development of a causal model. *Journal of Gerontology: Psychological Sciences, 45,* P192–P199.

Pruchno, R., Patrick, J. H., & Burant, C. J. (1996). Mental health of aging women with children who are chronically disabled: Examination of a two-factor model. *Journal of Gerontology: Social Sciences, 51B,* S284–S296.

Pruchno, R., Peters, N. D., & Burant, C. J. (1995). Mental health of coresident family caregivers: Examination of a two-factor model. *Journal of Gerontology: Psychological Sciences, 50B,* P247–P256.

Radloff, L. S. (1977). The CES-D scale: A self-report depression scale for research in the general population. *Applied Psychological Measurement, 1,* 385–401.

Rauktis, M. E., Koeske, G. F., & Tershko, O. (1995). Negative social interactions, distress, and depression among those caring for a serious and persistently mentally ill relative. *American Journal of Community Psychology, 23*(2), 279–299.

Redinbaugh, E. M., MacCullum, R. C., & Kiecolt-Glaser, J. K. (1995). Recurrent syndromal depression in caregivers. *Psychology and Aging, 10*(3), 358–368.

Rose-Rego, S. K., Strauss, M. E., & Smyth, K. A. (1998). Differences in the perceived well-being of wives and husbands caring for persons with Alzheimer's disease. *Gerontologist, 38,* 224–230.

Schulz, R., Mittelmark, M., Kronmal, R., Polak, J. F., Hirsch, C. H., German, P., & Bookwala, J. (1994). Predictors of perceived health status in elderly men and women: The Cardiovascular Health Study. *Journal of Aging and Health, 6,* 419–447.

Schulz, R., Newsom, J. T., Mittelmark, M., Burton, L., Hirsch, C., & Jackson, S. (1997). Health effects of caregiving: The Caregiver Health Effects Study: An ancillary study of the Cardiovascular Health Study. *Annals of Behavioral Medicine, 19,* 110–116.

Schulz, R., O'Brien, A. T., Bookwala, J., & Fleissner, K. (1995). Psychiatric and physical morbidity effects of dementia caregiving: Prevalence, correlates, and causes. *Gerontologist, 35,* 771–791.

Schulz, R., Visintainer, P., & Williamson, G. M. (1990). Psychiatric and physical morbidity effects of caregiving. *Journal of Gerontology: Psychological Sciences, 45,* P181–P191.

Seltzer, M. M., Greenberg, J. S., Krauss, M. W. (1995). A comparison of coping strategies of aging mothers of adults with mental illness or mental retardation. *Psychology and Aging, 10*(1), 64–75.

Shaw, W. S., Patterson, T. L., Semple, Grant, I., Yu, E. S., Zhang, M. Y. He, Y., & Wu W. Y. (1997). A cross-cultural validation of coping strategies and their associations with caregiving distress. *Gerontologist, 37,* 490–504.

Shaw, W. S., Patterson, T. L., Semple, S. J., Ho, S., Irwin, M. R., Hauger, R. L., & Grant, I. (1997). Longitudinal analysis of multiple indicators of health decline among spacial caregivers. *Annals of Behavioral Medicine, 19*(2), 101–109.

Skaff, M. M., Pearlin, L. I., & Mullan, J. T. (1996). Transitions in the caregiving career: Effects on sense of mastery. *Psychology and Aging, 11,* 247–257.

Smerglia, V. L., & Demling, G. T. (1997). Care-related decision-making satisfaction and caregiver well-being in families caring for older members. *Gerontologist, 37*(5), 658–665.

Smith, G. C. (1996). Caregiving outcomes for older mothers of adults with mental retardation: A test of the two-factor model. *Psychology and Aging, 11,* 353–361.

Spitzer, R. L., Williams, J. B. W., Gibbon, M., & First, M. B. (1992). The structured clinical interview for DSM-III-R (SCID): History, rationale, and description. *Archives of General Psychiatry, 49,* 624–629.

St.-Onge, M., & Lavoie, F. (1997). The experience of caregiving among mothers of adults suffering from psychotic disorders: Factors associated with their psychological distress. *American Journal of Community Psychology, 25,* 73–94.

Stephens, S., & Christianson, J. B. (1986). *Informal care of the elderly.* Lexington, MA: Lexington Books.

Stephens, M. A. P., Franks, M. F., & Atienza, A. A. (1997). Where two roles intersect: Spillover between parent care and employment. *Psychology and Aging, 12,* 30–37.

Stephens, M. A. P., & Townsend A. L. (1997). Stress of parent care: Positive and negative effects of women's other roles. *Psychology and Aging, 12,* 376–386.

Stone, R., Cafferata, G. L., & Sangl, J. (1987). Caregivers of the frail elderly: A national profile. *Gerontologist, 27,* 616–626.

Strawbridge, W. J., Wallhagen, M. I., Shema, S. J., & Kaplan, G. A. (1997). New burdens or more of the same? Comparing grandparent, spouse, and adult–child caregivers. *Gerontologist, 37,* 505–510.

Tower, R. B., Kasl, S. V., & Moritz, D. J. (1997). The influence of spouse cognitive impairment on respondents' depressive symptoms: The moderating role of marital closeness. *Journal of Gerontology: Social Sciences, 52B,* S259–S269.

United States Bureau of the Census. (1995). *Statistical brief: 65+ in the United States*. Washington, DC: Department of Commerce.

Vitaliano, P. P., Russo, J., Scanlan, J. M., & Greeno, C. G. (1996). Weight changes in caregivers of Alzheimer's care recipients: Psychobehavioral predictors. *Psychology and Aging, 11*, 155–163.

Vitaliano, P. P., Scanlan, J. M., Krenz, C., & Fujimoto, W. (1996). Insulin and glucose: Relationships with hassles, anger, and hostility in nondiabetic older adults. *Psychosomatic Medicine, 58*, 489–499.

Vitaliano, P. P., Scanlan, J. M., Krenz, C., Schwartz, R. S., & Marcovina, S. M. (1996). Psychological distress, caregiving, and metabolic variables. *Journal of Gerontology: Psychological Sciences, 51B*, P290–P299.

Williamson, G. M., & Schulz, R. (1995). Caring for a family member with cancer: Past communal behavior and affective reactions. *Journal of Applied Social Psychology, 25*, 93–116.

Wright, L. K., Clipp, E. C., & George, L. K. (1993). Health consequences of caregiver stress. *Medicine, Exercise, Nutrition, and Health, 2*, 181–195.

Zung, W. K. (1965). A self-rating depression scale. *Archives of General Psychiatry, 12*, 63–70.

Zung, W. K. (1975). *The measurement of depression*. Columbus, OH: Merrill.

II

Conditioning Variables and Outcomes

7

Depression, Immune Function, and Health in Older Adults

KATHERINE L. APPLEGATE, JANICE K. KIECOLT-GLASER, and RONALD GLASER

In this chapter, we summarize research on the relationships among age, depression (or depressed mood), and immune function in older adults. These studies suggest that depression, stress, and an immune system declining naturally with age yield greater down-regulation of immune function among older than younger adults. Therefore, we consider possible physical health consequences, with an emphasis on recent work addressing vaccine responses and wound healing. We also include a discussion of the implications for recovery from surgery, rates of infectious illness, and general physical well-being.

DEPRESSION AND IMMUNE FUNCTION

Major depressive disorder (MDD) has been associated with decrements in immune function in a variety of adult populations; however, this effect has not been as well explored among the elderly. Existing evidence suggests that declines in immune function may actually be larger among distressed older adults than younger individuals because of the added complications of an age-impaired immune system (Herbert & Cohen, 1993a; Kiecolt-Glaser, Glaser, Gravenstein, Malarkey, & Sheridan, 1996; Schleifer, Keller, Bond, Cohen, & Stein, 1989).

Immunological research reviewed in this chapter refers to two common assessment methods: quantitative and qualitative assays (Kiecolt-Glaser & Glaser, 1988). Quantitative assays identify the percentage and number of specific types of lympho-

---

KATHERINE L. APPLEGATE and JANICE K. KIECOLT-GLASER • Department of Psychiatry, Ohio State University College of Medicine, Columbus, Ohio 43210. **RONALD GLASER** • Department of Medical Microbiology and Immunology, Ohio State University College of Medicine, Columbus, Ohio 43210.

Physical Illness and Depression in Older Adults: A Handbook of Theory, Research, and Practice, edited by Gail M. Williamson, David R. Shaffer, and Patricia A. Parmelee. Kluwer Academic/Plenum Publishers, New York, 2000.

cytes. For example, one technique employs monoclonal antibodies bound to fluorescent dyes to detect specific surface proteins on lymphocytes. The percent of cells that fluoresce can be measured by a flow cytometer yielding a count of the specific cells being investigated. By obtaining a differential blood count, one can also determine the absolute number of each subpopulation of cells. However, such quantitative assays may be difficult to interpret because the number of circulating immune cells may have little correlation with the protective capabilities of the cells (Herbert & Cohen, 1993b).

Qualitative measures are more strongly associated with psychological stressors than are quantitative measures (Kiecolt-Glaser & Glaser, 1995); one type of qualitative (or functional) assay is blastogenesis, which assesses lymphocyte proliferation in response to stimulation by a mitogen (Kiecolt-Glaser & Glaser, 1995; O'Leary, 1990). Lymphocytes are generally found in a resting condition, but they become activated when in contact with an infectious agent and reproduce to combat the infection. Blastogenesis provides a way to assess this proliferation in a laboratory setting. Mitogens are added to media in which peripheral blood lymphocytes are cultured to induce proliferation *in vitro*; two common *T*-cell mitogens are phytohemagglutinin (PHA) and concanavalin A (Con A), which each target different subsets of lymphocytes. Radioactive isotopes are included in the media, and when the cells divide, the isotope is incorporated into cellular DNA. Proliferation can be quantified by measuring the amount of radioactive emission expressed as counts per minute. Interestingly, both enumerative and functional immune assays show depression-associated declines in elderly subjects (Herbert & Cohen, 1993a); however, age-related immune declines are more notable among functional assays.

Clinical depression is correlated with a decrease in the number of several leukocyte subsets. For example, elderly depressed women had fewer total *T* lymphocytes and *T* helper cells than elderly nondepressed women (Targum, Marshall, Fischman, & Martin, 1989). In another study, community subjects without a history of depression had more *T* suppressor cells associated with aging, while unipolar depressed patients did not show a comparable response (Schleifer et al., 1989). These data provide evidence of differences in immune cell numbers between depressed and nondepressed older adults. However, to be included in this latter study, elderly subjects could not be taking any medication, a criterion that may have selected for those in exceptionally good health, thus producing a nonrepresentative sample.

Not only are there dissimilarities in cell number between depressed and nondepressed older adults, but there are also studies that suggest differences in the qualitative aspects of immune function; a meta-analysis suggested that there were reliable variations among depressed elderly subjects compared to nondepressed peers (Herbert & Cohen, 1993a). Proliferative responses to three mitogens, phytohemagglutinin (PHA), concanavalin A (Con A), and pokeweed, as well as natural killer (NK) cell activity, showed moderate effect sizes for depression. Mitogen-induced lymphocyte proliferation, or blastogenesis, is an *in vitro* assay of cell function that examines the ability of cells to replicate when stimulated. NK cell activity, determined by a measurement of target cell killing or lysis, is an important antiviral and antitumor defense (Kiecolt-Glaser & Glaser, 1995).

Severity of depression may be related to qualitative differences in immune function between depressed and nondepressed subjects (Schleifer et al., 1989). For example, elderly participants with higher scores on the Hamilton Rating Scale for Depression (HRSD; Hamilton, 1960) and the Beck Depression Inventory (BDI; Beck, Ward, Mendelson, Mock, & Erbaugh, 1961) had lower T-lymphocyte responses to mitogen stimulation than those who were less depressed (Darko et al., 1988). Participants having MDD with melancholic features, or psychotic depressives, showed lower lymphocyte responses to PHA and pokeweed than those with minor depression (Maes, Bosmans, Suy, Minner, & Raus, 1989). Thus, changes in immune function may become more problematic as the severity of depression increases.

Two proposed pathways link depression and down-regulation of the immune system: neuroendocrine and behavioral (Herbert & Cohen, 1993a). Support for neuroendocrine mediation comes from work showing activation of the hypothalamic–pituitary–adrenal (HPA) axis and the sympathetic nervous system (SNS) in depressed patients versus nondepressed controls (Ritchie & Nemeroff, 1991; Stokes, 1987). Both cortisol and catecholamines can have adverse effects on immune cells that have receptors for these hormones (Rabin, Cohen, Ganguli, Lysle, & Cunnick, 1989). Modulation of the immune system via the neuroendocrine pathway has received some empirical support in studies that have linked depression to increases in cortisol, as well as further work showing that elevations in cortisol can down-regulate the immune system. Unfortunately, studies incorporating all three elements simultaneously have not provided convincing evidence for this mechanism. However, there are two methodological reasons that may explain this difficulty: an overreliance on depressed populations has restricted the range of cortisol data available, and single cortisol assessments in many studies have limited the assessment of cortisol variability (Herbert & Cohen, 1993a).

Changes in health behaviors provide a second pathway for depression-associated immune down-regulation (Kiecolt-Glaser & Glaser, 1988). Compared to their nondepressed counterparts, those suffering from depression exercise less, sleep less, maintain poorer diets, and engage in greater use of cigarettes, alcohol, and other substances (Grunberg & Baum, 1985). Indeed, there is ample evidence that these behaviors can alter immune function (Friedman, Klein, & Specter, 1991; Irwin, Smith, & Gillin, 1992; Simon, 1991). Thus, the health habits of depressed older adults may also contribute to the observed changes in immune function.

Although many studies have found immunological differences between depressed and nondepressed samples, not all investigations find support for such a relationship (e.g., Brambilla, Maggioni, Cenacchi, Sacerdote, & Panerai, 1995). Perhaps depression-related decrements in immune function are not inevitable; some individuals may be more susceptible than others. Even among the aged, diminished function is not uniform. It is thus important to examine what factors best predict changes in immune function, and which individuals are at greatest risk.

STRESS AND IMMUNE FUNCTION

Although syndromal depression is associated with diminished immune function, meeting criteria for clinical depression is not a necessary condition for immune change; individuals undergoing stressful life experiences may also be at risk. A meta-analysis of stress and immune function revealed significant effect sizes for several immune parameters including stress-related increases in the number of white blood cells, and decreases in the number of circulating *B* cells, *T* cells, *T* helper cells, and *T* suppressor/cytotoxic cells (Herbert & Cohen, 1993b). On functional assays, stress was associated with a decreased proliferative response to Con A and PHA, as well as lower NK cell activity.

The elderly are likely to be at higher risk than younger individuals following stress-related immunological changes because aging itself weakens the immune system. Qualitative immunological assays show the strongest age-related immune alterations. For example, lymphocyte proliferative responses to PHA, Con A, and pokeweed mitogen decline with age (Maes et al., 1989). In addition, although NK-cell cytotoxicity does not appear to reliably diminish with age, lymphokine-activated NK-cell killing is lower in older populations (Kutza, Kaye, & Murasko, 1995).

As a consequence of these age-related changes, stressful experiences for older adults may have particularly important immunological consequences. If both stress and age can independently weaken the immune system, the interaction of these factors may produce even greater decline. A number of studies have examined these relationships in more detail, particularly studies addressing the physical and psychological effect of one long-term stressor, caregiving for a family member with a progressive dementia.

Caregiving

Alzheimer's disease (AD) caregivers are faced with an uncontrollable and unpredictable disease course in which the patient requires ever-increasing amounts of care. While the patient's survival time after disease onset varies, the modal length is between 5 and 10 years (Hay & Ernst, 1987). Therefore, caregiving has been conceptualized as a chronic stressor. In the studies that follow, caregivers are providing 5 or more hours of care for a family member per week, while controls have no current caregiving responsibilities.

A number of immunological alterations have been related to the stress of caregiving. For example, dementia family caregivers had lower percentages of *T* lymphocytes and lower *T* helper/suppressor cell ratios than noncaregivers (Kiecolt-Glaser et al., 1987; Pariante et al., 1997). Other studies suggest that caregivers have fewer NK cells (Castle, Wilkins, Heck, Tanzy, & Fahey, 1995) and *T* helper cells (Kiecolt-Glaser et al., 1987) compared to noncaregivers. Levels of antibody to Epstein–Barr virus (EBV) virus capsid antigen (VCA) immunoglobulin G (IgG) were higher among 34 family caregivers than 34 matched noncaregivers (Kiecolt-Glaser et al., 1987). Normally, after an active herpesvirus infection, the virus is repressed in a latent state within specific host cells. Immunosuppressed individuals often show elevated levels of antibody to herpesviruses. *Higher* antibody levels to a latent herpesvirus, including EBV, suggest *poorer* cellular immune control over viral latency; for example, patients on immunosuppressive therapies often have elevated herpesvirus antibody levels (Glaser & Jones, 1994). In a longitudinal study with 69 spousal caregivers and 69 matched noncaregivers, caregivers showed greater increases in their antibody levels to EBV VCA between intake and the 1-year follow-up compared to controls (Kiecolt-Glaser, Dura, Speicher, Trask, & Glaser, 1991). At the second assessment, the former also had lower lymphocyte proliferative responses to two mitogens, Con A and PHA, than the latter.

Modulation of another latent herpesvirus, herpes simplex virus Type 1 (HSV-1), may also differ between chronically stressed caregivers and their noncaregiving peers (Glaser & Kiecolt-Glaser, 1997). In comparisons between 71 family caregivers and 58 noncaregivers, caregivers had higher antibody levels to latent HSV-1 and poorer HSV-1-specific *T* cell responses; however, there were no significant differences in neutralizing antibody titers between groups. These data support the hypothesis that psychological stress may alter the immune system's control over herpesvirus latency.

Delayed hypersensitivity skin tests also differentiated spousal caregivers and noncaregiving participants (McCann, 1991). Used to assess cell-mediated immune responses in the body, delayed hypersensitivity skin test reactions are provoked by placing sensitizing chemicals under the participants' skin (Brock & Madigan, 1991). In healthy individuals with an intact immune system, a response to the antigens should be visible within a few hours to a few days. Longer reaction times, or the absence of a reaction, suggest poorer immune function. Fifty percent of caregivers were categorized as totally or relatively anergic compared to only 12% of noncaregivers; these immunological deficits were shown in comparison not only to controls but also age-based norms (McCann, 1991).

Caregiver studies have explored psychological concomitants of downward change in immune function by identifying individuals who show immunological change over a period of time. In one such study, approximately 32% of caregivers were classified in an "at risk" category because of downward change on two out of three functional immune measures over 1 year compared to only 14% of noncaregivers (Kiecolt-Glaser et al., 1991). At the initial assessment, these "at risk" caregivers had reported more distress in response to dementia-related patient behaviors and less social support than caregivers who had not shown such uniform declines. Although these two caregiver groups did not differ in amount of time per day they provided care, the total length of time they had provided care, or the extent of the patient's cognitive impairment, the former were more likely to have institutionalized their spouse between the initial and follow-up assessments.

What do these immunological changes mean for caregivers? Evidence suggests that caregivers show a dampened but stable pattern of immune responses to this chronic stressor; they do not exhibit continued decline (Townsend, Noelker, Deimling, & Bass, 1989). Therefore, immunological status does not simply reflect how long the person has been caregiving but, rather, seems to be affected by the new challenges and struggles caregivers continue to face as part of providing care for their loved one.

Bereavement

The loss of a spouse often leads to a stressful and difficult adjustment period. Work examining this experience has suggested that immune function may decline soon after bereavement. For example, compared to prebereavement levels, husbands of cancer patients exhibited significantly lower proliferative response to PHA, Con A, and pokeweed mitogen stimulation 2-months postbereavement; however, 4 to 14 months postbereavement, immune function had improved somewhat (Schleifer, Keller, Camerino, Thornton, & Stein, 1983).

Bereavement studies that included depressed and nondepressed widows have examined the effects of both syndromal depression and stress on immune function. For example, depressed widows showed larger decreases in NK cell activity and poorer response to mitogen stimulation than nondepressed widows (Zisook et al., 1994). Data such as these suggest that the presence of several factors may produce increasingly more severe decrements in immune function.

How does bereavement differ for caregivers? Unlike the rebound observed in "normal" bereavement (Harlow, Goldberg, & Comstock, 1991), early evidence suggests that caregivers' depressed mood and immunological function may not substantially improve following the spouse's death. Using structured psychiatric interviews with noncaregivers, continuing caregivers, and bereaved (former) caregivers, the latter had rates of syndromal depression that were comparable to those of continuing caregivers, even though an average of 19.8 months passed between the initial assessment and bereavement (Bodnar & Kiecolt-Glaser, 1994). Both groups had elevated rates of depression compared to noncaregivers.

The long-term immunological consequences of bereavement and caregiving were addressed in a study with 14 continuing AD caregivers, 17 bereaved caregivers, and 31 noncaregivers (Esterling, Kiecolt-Glaser, Bodnar, & Glaser, 1994). Continuing caregivers had been providing care for all 5 years between the two assessments; former caregivers had been providing care at intake but had lost their spouse in the interim. On average, approximately 2 years had passed since the death of their spouse. Researchers assessed NK-cell activity, NK-cell activity after enhancement with stimulation by recombinant interferon-gamma (rIFN-gamma), and stimulation by recombinant interleukin-2 (rIL-2); these two cytokines increase NK-cell cytotoxicity *in vitro* (Herberman & Ortaldo, 1981). Current caregivers, former caregivers, and noncaregivers showed no differences on NK-cell cytotoxicity, consistent with data from Irwin et al. (1991). However, both continuing and former caregivers demonstrated poorer responses to both rIFN-gamma and rIL-2 in comparison to controls, although the two caregiving groups were not significantly different from one another. Similar patterns of responding for continuing and bereaved caregivers suggest that physiological consequences of this chronic stressor may persist well after the death of the patient.

Continuing and bereaved caregivers were then divided into two groups (Esterling et al., 1994). High responders were those above the median in their response to one or both cytokines, while low responders were below the median for both cytokines. In comparison to high responders, low responders reported less positive social support from their social networks, and less closeness from the individuals in their network. Importantly, the groups did not differ on depression or health behaviors. These results highlight the importance of strong social networks for individuals experiencing chronic stress; those without such support may be susceptible to additional immunological decline.

Summary

The mechanisms through which stress alters immune function are similar to those discussed in the depression literature—neuroendocrine and behavioral pathways (Herbert & Cohen, 1993a, 1993b). Emotional stress can activate the SNS, modifying levels of hormones and neurotransmitters in the body. For example, caregivers who reported higher life stress had higher plasma levels of norepinephrine (Mills et al., 1997), and older caregivers had higher levels of neuropeptide Y (NPY) than older controls (Irwin et al., 1991). NPY is a sympathetic neurotransmitter thought to regulate immune function during stress. In addition, behavioral changes such as diet, exercise, or alcohol consumption may also alter immune function.

Caregiving for a family member with a progressive dementia and bereavement are both associated with immunological consequences for older adults. Changes in both the number and function of peripheral blood leukocytes suggest significant declines in an aging system. Thus, individuals under stress, even if they do not meet syndromal depression criteria, may show clear deficits in immune function.

AGE AND IMMUNE FUNCTION

As discussed earlier, distress-related declines in immune function are larger among the elderly compared to younger individuals (Herbert & Cohen, 1993a). A meta-analysis of the depression and immune function literature found stronger effect sizes for older versus younger subjects on both enumerative and qualitative assays (Herbert & Cohen, 1993a). The combination of distress and an aging immune system may produce more significant deficits in the immune system's ability to function effectively.

Mechanisms for age-related changes in immune function are still under investigation. In earlier literature, pathways thought to account for age-related declines in *T*-cell function involved a decrease in thymic function or production of thymic hormone (Thompson et al., 1984). These declines would impair the maturation of helper/inducer *T* lymphocytes, thereby inhibiting their ability to proliferate rapidly in response to antigen. In contrast, more recent work has highlighted the effects of aging on *T*-cell function. As individuals grow older, there are decreases in the number of cytotoxic-effector cells, *T*-cell proliferation after mitogen stimulation, *T*-cell responses to new antigens, and *T*-cell stimulation of *B*-cell proliferation and maturation (Miller, 1996). These age-related declines may result from alterations in the signal transduction pathways, specifically, defects in calcium mobilization and protein phosphorylation.

Declines in immune function have been linked to increased morbidity and mortality. In fact, these downward changes may actually serve as general markers of physiological aging (Murasko, Weiner, & Kaye, 1988). For example, older adults whose lymphocytes did not proliferate in response to three mitogens were twice as likely to die over a 2-year period as their counterparts who demonstrated a more robust response. Similarly, a study of older adults showed that poorer cell-mediated immunity was linked with greater morbidity and mortality (Wayne, Rhyne, Garry, & Goodwin, 1990). In another longitudinal study examining the 3-year period prior to death, decreases in the absolute number of peripheral blood lymphocytes were associated with mortality (Bender, Nagel, Adler, & Andres, 1986). In addition to marking more general physical decline, these alterations in immune function place individuals at greater risk for immunologically mediated diseases.

HEALTH IMPLICATIONS

Early investigations showing distress-related declines in immune function were unclear about the significance of such changes on overall physical health. Because down-regulation of immune function does not necessarily indicate poorer health status among young and healthy adults, small immunological decrements do not provide clear information on health risks. To assess these effects, responses of chronically stressed populations, such as caregivers, have been compared to nonstressed controls on indices including vaccine response, length of wound healing time, and recovery from surgery.

To determine whether chronic stress was associated with impaired immune responses to influenza virus inoculation, 32 caregivers and 32 well-matched noncaregivers received a vaccine (Kiecolt-Glaser et al., 1996). The timing and strength of antibody and *T*-cell or cytokine response following vaccination provide data on the body's response to challenge by pathogens. Although participants in this study had similar vaccine histories, rates of chronic illnesses, and medication usage, caregivers had poorer cellular and humoral immune responses to influenza vaccine than controls. Four weeks after vaccination, caregivers were less likely to show a significant increase in antibody to the vaccine; they had lower levels of interleukin-1-ß production (IL-1ß; an important promoter of immunological activities including antibody responses), and their peripheral blood lymphocytes produced lower levels of IL-2 in response to vaccine stimulation than controls. The immune system must utilize the protective element of vaccines for them to be effective; these data suggest that chronically stressed subjects show deficits in their immune response after vaccination compared to nonstressed peers.

These vaccination results are particularly relevant for older adults. Influenza and pneumonia together are the fourth leading cause of death among adults over the age of 75 (Yoshikawa, 1983), and mortality from influenza infection is four times higher among those over 60 years of age than those under 40 (Burns, Lum, Seigneuret, Giddings, & Goodwin, 1990). Declines in immune function are thought to be related to this increased morbidity and mortality (Phair, Kaufmann, Bjornson, Adams, & Linnemann, 1978). Vaccine response data such as those described here (Kiecolt-Glaser et al., 1996) can provide a window on the body's response to other pathogens such as viruses or bacteria; individuals who show a delayed or blunted vaccine response could be at greater risk for more severe illness.

Because the immune system plays a central role in wound repair, wound healing may also be affected by distress-related immunological changes (Kiecolt-Glaser, Marucha, Malarkey, Mercado, & Glaser, 1995). Thirteen female caregivers and 13 female noncaregivers underwent a 3.5 mm punch biopsy wound on the forearm; healing was assessed every few days by photography and response to hydrogen peroxide until the wound healed (healing was defined as no foaming when covered with peroxide). Caregivers took an average of 48.7 days to heal, compared to only 39.3 days in controls, a 24% difference between groups. The peripheral blood lymphocytes of caregivers also produced significantly less IL-1ß messenger RNA (mRNA) in response to lipopolysaccharide stimulation in relation to controls. During wound repair, IL-1ß regulates enzymes involved with the reconstruction of damaged connective tissue matrices, and it also stimulates production of other cytokines needed for healing. The group difference in IL-1ß production suggests that this may be one mechanism related to differences in wound healing.

Older adults undergo surgical procedures more often than do younger individuals, and age itself has been linked to a greater number of postsurgical complications (Linn, Linn, & Jensen, 1983). In one study, younger and older participants were divided into groups based on their preoperative anxiety. After surgery, the older anxious group had significantly more complications than the other three groups. Postoperative morbidity and mortality are substantially higher among older adults compared to younger individuals (Thomas & Ritchie, 1995); further suppression of immune function by depression or chronic stress may place the elderly at even greater risk (Kiecolt-Glaser, Page, Marucha, MacCallum, & Glaser, 1998).

CONCLUSION

In summary, there are clearly significant immunological declines associated with depression, depressed mood, and stress. Unfortunately, these distress-related effects are magnified in elderly populations because the aging process itself contributes to downward change in immune function. The interaction of distress and aging leaves older adults open to physical health complications. Alterations of immune function have been linked to poorer vaccine response and slower wound healing time, as well as increased mortality and morbidity. These trends reveal greater risks for distressed older adults, who are particularly vulnerable to the consequences of immune decline.

This review highlights several significant issues for practitioners working with older adults. First, the treatment of depression in older adults is critical not only for psychological reasons but also to minimize the potential physical health complications of depression. Second, identifying older adults undergoing a chronic stressor, such as those caregiving for an ill family member, may facilitate a search for additional resources from their family or community. Such individuals need emotional outlets and social support to buffer the effects of these stressful experiences. Finally, evidence on vaccine response among elderly caregivers emphasizes the need for older adults to receive regular influenza vaccinations to protect against infection.

ACKNOWLEDGMENTS: Work on this paper was supported in part by National Institute of Health Grant Nos. K02 MH01467, R37 MH42096, and P01 AG11585.

REFERENCES

Beck, A. T., Ward, C. H., Mendelson, M., Mock, J. E., & Erbaugh, J. K. (1961). An inventory for measuring depression. *Archives of General Psychiatry, 4,* 561–571.

Bender, B. S., Nagel, J. E., Adler, W. H., & Andres, R. (1986). Absolute peripheral blood lymphocyte count and subsequent mortality of elderly men. *Journal of the American Geriatric Society, 34,* 649–654.

Bodnar, J., & Kiecolt-Glaser, J. K. (1994). Caregiver depression after bereavement: Chronic stress isn't over when it's over. *Psychology and Aging, 9,* 372–380.

Brambilla, F., Maggioni, M., Cenacchi, T., Sacerdote, P., & Panerai, A. R. (1995). *T*-lymphocyte proliferative response to mitogen stimulation in elderly depressed patients. *Journal of Affective Disorders, 36,* 51–56.

Brock, T. D., & Madigan, M. T. (1991). *Biology of microorganisms.* Engelwood Cliffs, NJ: Prentice Hall.

Burns, E. A., Lum, L. G., Seigneuret, M. C., Giddings, B. R., & Goodwin, J. S. (1990). Decreased specific antibody synthesis in old adults: Decreased potency of antigen-specific B cells with aging. *Mechanisms of Ageing and Development, 53,* 229–241.

Castle, S., Wilkins, S., Heck, E., Tanzy, K., & Fahey, J. (1995). Depression in caregivers of demented patients is associated with altered immunity: Impaired proliferative capacity, increased CD8+, and a decline in lymphocytes with surface signal transduction molecules (CD38+) and a cytotoxicity marker (CD56+ CD8+). *Clinical Experimental Immunology, 101*, 487–493.

Darko, D. F., Lucas, A. H., Gillin, J. C., Rish, S. C., Golshan, S., Hamburger, R. N., Silverman, M. B., & Janowsky, D. S. (1988). Age, cellular immunity and the HP axis in major depression. *Progress in Neuro-Psychopharmacology and Biological Psychiatry, 12*, 713–720.

Esterling, B., Kiecolt-Glaser, J. K., Bodnar, J., & Glaser, R. (1994). Chronic stress, social support and persistent alterations in the natural killer cell response to cytokines in older adults. *Health Psychology, 13*, 291–299.

Friedman, H., Klein, T., & Specter, S. (1991). Immunosuppression by marijuana and components. In R. Ader, D. L. Felten, & N. Cohen (Eds.), *Psychoneuroimmunology* (pp. 931–953). San Diego: Academic Press.

Glaser, R., & Jones, J. (Eds.). (1994). *Human herpesvirus infections*. New York: Dekker.

Glaser, R., & Kiecolt-Glaser, J. K. (1997). Chronic stress modulates the virus-specific immune response to latent herpes simplex virus Type 1. *Annals of Behavioral Medicine, 19*, 78–82.

Grunberg, N. E., & Baum, A. (1985). Biological commonalities of stress and substance abuse. In S. Shiffman & T. A. Wills (Eds.), *Coping and substance use* (pp. 25–62). San Diego: Academic Press.

Hamilton, M. (1960). A rating scale for depression. *Journal of Neurology, Neurosurgery, and Psychiatry, 23*, 56–62.

Harlow, S. D., Goldberg, E. L., & Comstock, G. W. (1991). A longitudinal study of risk factors for depressive symptomology in elderly widowed and married women. *American Journal of Epidemiology, 134*, 526–538.

Hay, J., & Ernst, R. L. (1987). The economic costs of Alzheimer's disease. *American Journal of Public Health, 77*, 1169–1175.

Herberman, R. B., & Ortaldo, J. R. (1981). Natural killer cells: Their role in defenses against disease. *Science, 214*, 24–30.

Herbert, T. B., & Cohen, S. (1993a). Depression and immunity: A meta-analytic review. *Psychological Bulletin, 113*, 472–486.

Herbert, T. B., & Cohen, S. (1993b). Stress and immunity in humans: A meta-analytic review. *Psychosomatic Medicine, 55*, 364–379.

Irwin, M., Brown, M., Patterson, T., Hauger, R., Mascovich, A., & Grant, I. (1991). Neuropeptide Y and natural killer cell activity: Findings in depression and Alzheimer's caregiver stress. *FESEB Journal, 5*, 3100–3107.

Irwin, M., Smith, T. L., & Gillin, J. C. (1992). Electroencephalographic sleep and natural killer activity in depressed patients and control subjects. *Psychosomatic Medicine, 54*, 10–21.

Kiecolt-Glaser, J. K., Dura, J. R., Speicher, C. E., Trask, O. J., & Glaser, R. (1991). Spousal caregivers of dementia victims: Longitudinal changes in immunity and health. *Psychosomatic Medicine, 53*, 345–362.

Kiecolt-Glaser, J. K., & Glaser, R. (1988). Methodological issues in behavioral immunology research with humans. *Brain, Behavior, and Immunity, 2*, 67–78.

Kiecolt-Glaser, J. K., & Glaser, R. (1995). Measurement of immune response. In S. Cohen, R. C. Kessler, & L. U. Gordon (Eds.), *Measuring stress: A guide for health and social scientists* (pp. 215–229). New York: Oxford University Press.

Kiecolt-Glaser, J. K., Glaser, R., Gravenstein, S., Malarkey, W. B., & Sheridan, J. (1996). Chronic stress alters the immune response to influenza virus vaccine in older adults. *Proceedings of the National Academy of Sciences, 93*, 3043–3047.

Kiecolt-Glaser, J. K., Glaser, R., Shuttleworth, E. C., Dyer, C. S., Ogrocki, P., & Speicher, C. E. (1987). Chronic stress and immunity in family caregivers for Alzheimer's disease victims. *Psychosomatic Medicine, 49*, 523–535.

Kiecolt-Glaser, J. K., Marucha, P. T., Malarkey, W. B., Mercado, A. M., & Glaser, R. (1995). Slowing of wound healing by psychological stress. *Lancet, 346*, 1194–1196.

Kiecolt-Glaser, J. K., Page, G. G., Marucha, P. T., MacCallum, R. C., & Glaser, R. (in press). Psychological influences on surgical recovery: Perspectives from psychoneuroimmunology. *American Psychologist, 0*, 000–000.

Kutza, J., Kaye, D., & Murasko, D. M. (1995). Basal natural killer cell activity of young versus elderly humans. *Journal of Gerontology: Biological Sciences, 50A*, B110–B116.

Linn, B. S., Linn, M. W., & Jensen, J. (1983). Surgical stress in the healthy elderly. *Journal of the American Geriatric Society, 31,* 544–568.
Maes, M., Bosmans, E., Suy, E., Minner, B., & Raus, J. (1989). Impaired lymphocyte stimulation by mitogens in severely depressed patients: A complex interface with HPA-axis hyperfunction, noradrenergic activity and the ageing process. *British Journal of Psychiatry, 155,* 793–798.
McCann, J. J. (1991). *Effects of stress on spouse caregivers' psychological health and cellular immunity.* Unpublished doctoral dissertation, Rush University College of Nursing, Chicago, IL.
McNerlan, S. E., Rea, I. M., Alexander, H. D., & Morris, T. C. M. (1998). Changes in natural killer cells, the CD57CD8 subset, and related cytokines in healthy aging. *Journal of Clinical Immunology, 18,* 31–38.
Miller, R. A. (1996). The aging immune system: Primer and prospectus. *Science, 273,* 70–74.
Mills, P. J., Ziegler, M. G., Patterson, T., Dimsdale, J. E., Hauger, R., Irwin, M., & Grant, I. (1997). Plasma catecholamine and lymphocyte beta 2-adrenergic receptor alterations in elderly Alzheimer caregivers under stress. *Psychosomatic Medicine, 59,* 251–256.
Murasko, D. M., Weiner, P., & Kaye, D. (1988). Association of lack of mitogen-induced lymphocyte proliferation with increased mortality in the elderly. *Aging: Immunology and Infectious Disease, 1,* 1–6.
O'Leary, A. (1990). Stress, emotion, and human immune function. *Psychological Bulletin, 108,* 363–382.
Pariante, C. M., Carpiniello, B., Orrù, M. G., Sitzia, R., Piras, A., Farci, A. M. G., Del Giacco, G. S., Piludu, G., & Miller, A. H. (1997). Chronic caregiving stress alters peripheral blood immune parameters: The role of age and severity of stress. *Psychotherapy and Psychosomatics, 66,* 199–207.
Phair, J., Kauffmann, C. A., Bjornson, A., Adams, L., & Linnemann, C. (1978). Failure to respond to influenza vaccine in the aged: Correlation with B-cell number and function. *Journal of Laboratory and Clinical Medicine, 92,* 822–828.
Rabin, B. S., Cohen, S., Ganguli, R., Lysle, D. T., & Cunnick, J. E. (1989). Bidirectional interaction between the central nervous system and the immune system. *Critical Reviews in Immunology, 9,* 279–312.
Ritchie, J. C., & Nemeroff, C. B. (1991). Stress, the hypothalamic–pituitary–adrenal axis, and depression. In J. A. McCubbin, P. G. Kaufmann, & C. B. Nemeroff (Eds.), *Stress, neuropeptides, and systemic disease* (pp. 181–197). San Diego: Academic Press.
Schleifer, S. J., Keller, S. E., Bond, R. N., Cohen, J., & Stein, M. (1989). Major depressive disorder and immunity: Role of age, sex, severity, and hospitalization. *Archives of General Psychiatry, 46,* 81–87.
Schleifer, S. J., Keller, S. E., Camerino, M., Thornton, J. C., & Stein, M. (1983). Suppression of lymphocyte stimulation following bereavement. *Journal of the American Medical Association, 250,* 374–377.
Simon, H. B. (1991). Exercise and human immune function. In R. Ader, D. L. Felten, & N. Cohen (Eds.), *Psychoneuroimmunology* (pp. 869–895). San Diego: Academic Press.
Stokes, P. E. (1987). The neuroendocrine measurement of depression. In A. J. Marsella, R. M. A. Hirschfeld, & M. M.. Katz (Eds.), *The measurement of depression* (pp. 153–195). New York: Guilford Press.
Targum, S. D., Marshall, L. E., Fischman, P., & Martin, D. (1989). Lymphocyte subpopulations in depressed elderly woman. *Biological Psychiatry, 26,* 581–589.
Thomas, D. R., & Ritchie, C. S. (1995). Preoperative assessment of older adults. *Journal of the American Geriatrics Society, 43,* 811–821.
Thompson, J. S., Wekstein, D. R., Rhoades, J. L., Kirkpatrick, C., Brown, S. A., Roszman, T., Straus, R., & Tietz, N. (1984). The immune status of healthy centenarians. *Journal of the American Geriatrics Society, 32,* 274–281.
Townsend, A., Noelker, L., Deimling, G., & Bass, D. (1989). Longitudinal impact of interhousehold caregiving on adult children's mental health. *Psychology of Aging, 4,* 393–401.
Wayne, S. J., Rhyne, R. L., Garry, P. J., & Goodwin, J. S. (1990). Cell mediated immunity as a predictor of morbidity and mortality in subjects over sixty. *Journal of Gerontology, Medical Sciences, 45,* M45–M48.
Yoshikawa, T. T. (1983). Geriatric infectious diseases: An emerging problem. *Journal of the American Geriatrics Society, 31,* 34–39.
Zisook, S., Shuchter, S. R., Irwin, M., Darko, D. F., Sledge, P., & Resovsky, K. (1994). Bereavement, depression, and immune function. *Psychiatry Research, 52,* 1–10.

8

Quality of Life, Depression, and End-of-Life Attitudes and Behaviors

M. POWELL LAWTON

The time of chronic illness preceding the end of life has become a major focus of research in recent years. Concern over both the quality of life (QOL) of the individual and the allocation of health care resources is a major reason for the large amount of research in this area. The discussion in this chapter has the broad goal of examining depression and chronic illness in relation to the end of life. Depression and physical illness are analyzed as antecedents of treatment preferences near the end of life, explicitly treatment whose purpose is to extend life. Inasmuch as little is known about how such individual characteristics determine actual treatment received, the outcomes to be examined are usually expectations or preferences for treatment or, alternatively, wishes for length of life given hypothetical conditions under which life would exist. The overall guiding hypothesis whose tenability will be examined in a critical literature and conceptual review is that the wish to live a shorter time is partially explainable on the basis of depression (and other negative aspects of life), but that such a relationship is also conditioned in the opposite direction by the positive qualities of the person's life. Valuation of life (VOL) is offered as a construct proposed to mediate the relationship between negative and positive aspects of mental health and end-of-life wishes and behaviors.

The first section of this chapter describes briefly the concept of health-related quality of life (HRQOL) and the place given depression in the operationalization of HRQOL. The next section sketches my conception of QOL and mental health, and provides definitions of several important constructs. The four major sections are then devoted to evidence concerning the prevalence of depression as life nears its end,

M. POWELL LAWTON • Polisher Research Institute, Philadelphia Geriatric Center, Philadelphia, Pennsylvania 19141.

Physical Illness and Depression in Older Adults: A Handbook of Theory, Research, and Practice, edited by Gail M. Williamson, David R. Shaffer, and Patricia A. Parmelee. Kluwer Academic/Plenum Publishers, New York, 2000.

wishes for length of life and end-of-life treatment by elders, the relationship between depression and cognitions regarding end-of-life issues, and my conception of VOL as a mediator between QOL and end-of-life attitudes and behavior.

HEALTH-RELATED QUALITY OF LIFE

The most inclusive definition of HRQOL is the evaluation of any or all aspects of a person's life that originate in or are influenced by the person's health. "Evaluation" denotes the desirable versus undesirable dimension that is inherent in the concept. The evaluative aspect is debated among investigators in terms of whether it is properly limited to the individual's subjective evaluation or whether consensual, social-normatively-based evaluation ("objective QOL") is also included. Among all possible facets of a person's life that might be evaluated, "health-related" delimits the facets that are included. The facets studied in HRQOL research may be either components of a state of health (e.g., mortality risk, symptoms, disabilities) or phenomena secondarily influenced by health (e.g., anxiety over health state, occupational disability, social withdrawal).

HRQOL research relies on the observation that the worth of a person's life may be judged not only by the number of years of its duration but also by the quality of those years as affected by health. The concept, "quality-adjusted life years" (QALYS; Patrick, Bush, & Chen, 1973) expresses the consistent conclusion that people's preference for good health extends to a hypothetical willingness to trade some chronological longevity of poor quality time for a shorter period of good health. Typically, in HRQOL research, participants are given a scenario that portrays a cluster of indicators of good health and poor health, and asked to make judgments that lead (through a variety of methodological operations) to a bottom line translatable into how great a "discount" on normal life expectancy a person would accept for a shorter time in good health. Terms used in the literature include *health preference*, which refers to the express wishes people voice regarding how long they would like to live or which treatments they would like, under which conditions. The discounted number of years or proportion of chronological years the person chooses in return for good is "health utility." Some of the methods used to estimate discount rates, such as the standard gamble or the time trade-off (Torrance, Thomas, & Sackett, 1972) are tedious and, in the case of poorly-educated or deprived people, difficult to accomplish. Therefore, more easily usable observer or self-response instruments measuring health and HRQOL without any trade-off content have been developed and calibrated to a standardization group's direct health utility judgments. For example, the Quality of Well-Being Scale (QOWBS; Fanshel & Bush, 1970) details 23 symptoms and 3 role impairments, and yields a single score based on weights that were established through health-preference judgments of an adult normative group.

Similarly, the McMaster Health Utilities Index (Boyle, Furlong, Feeny, Torrance, & Hatcher, 1995) consists of questions answered by the person on sensory, communicative, functional, social and psychological states. Like the QOWBS, the McMaster was developed to be correlated with time-trade-off utility judgments. Measurement of health utility for an individual presumably provides care-relevant information useful to caregivers and clinicians and, conceivably, counseling with the person regarding end-of-life care issues. Aggregated across people, health utility judgments provide societywide data that are useful for health planning purposes, for example, in allocating services or paying for health care (Kaplan & Bush, 1982).

Most research on HRQOL has not been yoked so systematically to health utility, however. Rather, health-related quality has been estimated as "health status" in freestanding fashion, that is, without regard to its correlation with utility or with actual or estimated longevity. In such health-related measures, QOL in its more general form (see next section) is estimated. Its health-related quality is defined by the attribution of QOL judgments to health as an explanation for any perceived disability. For example, one of the most frequently used HRQOL measures, the Sickness Impact Profile (Bergner, Bobbitt, Carter, & Gilson, 1981), prefaces its questions regarding one's present state with the lead-in, " . . . because of your health." HRQOL has been operationalized even without the health attribution. The measure of health status now in most frequent use is the Short Form 36 (SF-36), originally developed by the RAND Corporation from its Medical Outcomes Study (Stewart & Ware, 1992; Ware & Sherbourne, 1992). Its 36 items include 7 health estimates and 12 items relevant to mental health. The other 17 items all include the attribution of some functional disability to one's physical health.

The health attribution is often implicit, as seen in many disease-specific (Patrick & Deyo, 1989) measures of QOL (see Spilker, 1996, for an exhaustive review of both generic and disease-specific QOL measures). This type of HRQOL measure strives to represent the bothersome symptoms and disabilities characteristic of a single illness as, for example, the Arthritis Impact Measure (AIM; Meenan, Gertman, & Mason, 1980).

In overview, measures of HRQOL vary widely both in their degree of health attribution and the choice of domains of life that are to be evaluated. Depression is one of many domains used as a criterion in HRQOL research. Unquestionably, the most ubiquitous domains included in HRQOL research are those of physical pain/distress and functional disability (activities of daily living (ADL), mobility, physical performance). Depression was estimated in Sackett and Torrance's (1978) time-trade-off judgment process to occupy a midway position among other primarily physical symptoms as a discounting factor in QOL; it was given a weight of .45 on a scale where .00 represented death and 1.00, perfect health. Depression is not a component of every HRQOL measure. The QOWBS, the McMaster Scale, the SF-36, and many others contain "mental health" sections where depressive symptoms appear in individual items but not as a separate scale. An informal review of QOL reprints in my files indicated that the great majority included some assessment of mental health, but only about one-third of the total reported depression measured separately and analyzed in relation to other measures. Very few reported DSM diagnoses of depression. A number used the Profile of Mood States (POMS; McNair, Lorr, & Droppelman, 1971) or the Center for Epidemiological Studies–Depression Scale (CES-D, Radloff, 1977). This voluminous literature would offer an opportunity for a meta-analysis to be performed to estimate how different illnesses compare in their absolute levels of depression or the extent to which depression is associated with illness-specific symptoms, behavioral, or treatment characteristics, and other aspects of QOL.

Three Problems with the HRQOL concept

There are three problems with the HRQOL concept. The basic rationale for HRQOL research is to understand how HRQOL influences people's readiness to seek life-extending treatment for illness, particularly treatments that undermine QOL, and their wishes about living in the face of distressing illnesses. A first problem is that many writers specifically exclude from HRQOL areas of life such as affectional ties, leisure-time pursuits, or self-growth efforts, unless they are used to represent outcomes of poor health or treatment side effects. Such an artificial boundary between health and nonhealth aspects of QOL seems not only philosophically problematic but also very difficult to operationalize satisfactorily. Who is to say whether the continuance of a life role devoted to helping others is or is not vulnerable to the effects of illness? It also seems psychometrically naive to think that simply bracketing a set of questions with ". . . because of your health" will in fact succeed in leading people to screen out veridically all explanations for that state other than those in the health domain.

A second problem is that the core of HRQOL research is based on a decremental model of QOL: Illness and some treatments erode QOL, but excellent and ordinary health are merged in a category that is essentially a "no pathology" category. The scenarios from which health utility is directly measured are invariably decrements; they never portray positive states. In the broader realm where health status (without the utility aspect) is assessed across several domains, positive measures do appear. For example, the SF-36 includes four positive items (e.g., "pep," "happy") but, like the explicit depression items, they are embedded in a single mental health score. Other studies have occasionally used such measures as the Bradburn (1969) positive affect scale (Revenson & Felton, 1989) and a number have used the CES-D, which contains a 4-item group of positive items that could be, but rarely are, used as a separate factor. In summary, it seems essential to recognize that end-of-life decisions may be made by balancing losses in some areas (e.g., pain, disability) with gains in others (e.g., more time with family, completion of a subjectively defined life plan). Recognition of this hedonic calculus constitutes the basis for including positive states as well as depression and other negative states among the concepts discussed in this chapter. It is entirely possible that anticipating positive affect may compensate for or override the poor aspects of life engendered by enduring a painful illness or treatment.

The third problem in most of the HRQOL research to date lies in the assumption that the value attached to different health states is the same for all people. The QOWBS, for example, was calibrated on the responses of a large number of adults in good health, under the assumption that good and poor health are defined in social-normative, rather than individual or subjective terms (Robert Kaplan, personal communication, August 12, 1996). In my view, the judgment processes applied to both health status and other types of QOL yield differing preference levels as a person's age, health, and prospective remaining life span changes. To ignore the possibility that subgroups' preferences may differ, or that one's own changing health state can lead to a rearrangement of preferences, is to deny the adaptive potential of human beings. Empirical evidence regarding group-specific utilities is scant, but what there is will be reviewed later.

In addition, the concepts used in QOL research are often poorly differentiated from one another, resulting in frequent disagreement among investigators regarding what qualities of life should be measured. In the next section, I discuss my conception of QOL and offer definitions of its components (for a more extended discussion, see Lawton, 1991).

A MODEL OF QUALITY OF LIFE

Quality of life is the multidimensional evaluation, by intrapersonal and social-normative standards, of the total life of the person and her or his environmental context. There are four sectors of QOL, each with its own structural differentiation into domains. Two sectors are objective in the sense that the attributes to be evaluated are directly observable and capable of being assessed in terms of absolute or social-normative standards. The first, *behavioral competence*, subsumes successively more complex levels of health, functional health, cognition, time use, and social behavior. The second, *environment*, is suggested as being composed of physical, personal, suprapersonal (the characteristics of the aggregate of people in the person's proximity), and social-institutional domains. Two sectors are subjective, depending on people's own evaluations of their quality or lack of quality. *Perceived quality of life* (PQOL) is the person's satisfaction with, or other evaluation of, specific domains of life; there is no finite list of such domains, but frequently included (e.g., in Andrews & Withey, 1976; Campbell, Converse, & Rodgers, 1976) are such domains as family, marriage, housing, economic state, and spare-time activities. *Psychological well-being* (PWB) represents the person's most global evaluation of self in environment, including both affective and cognitive manifestations of positive and negative mental health. This conception thus views each sector and each domain within each sector as potentially composing a facet of QOL. Objective facets (e.g., high cognitive performance or a resource-rich environment) are aspects of high QOL because they are socially valued and rewarded. Although such evaluations may not reflect a single individual's evaluation, in the aggregate, people in general prize these characteristics and are more likely to experience higher-quality subjective QOL if they have them. PQOL is what is frequently referred to simply as quality of life—the subjective evaluation of limited aspects of a person's life. Just as frequently, however, psychological well-being is also used to represent QOL when it is conceived (erroneously, in my view) as a single overall entity, as in life satisfaction, or in the many measures yielding a single index named QOL but in fact composed of a mixture of many domains that are not separately scored (QOWBS, McMaster scale; e.g., functional health, pain, depression, social isolation).

The present view is that any domain within the four sectors of QOL may be used legitimately as an indicator of QOL in the appropriate circumstance, but is important to recognize the domain specificity of each measure and to choose the right indicator for the specific circumstance of QOL assessment. Although I prefer a model whereby the objective sectors are antecedents of both PQOL and PWB, and PQOL an antecedent of PWB (thereby defining PWB as an ultimate outcome criterion, as in Figure 8.1), such a causal model is not essential to the assertion that we may look anywhere among the domains of QOL for appropriate measurement indicators depending on the need of the research.

Health status, health preference, and health utility may be located within this framework. Health status, when defined behaviorally as in a clinical diagnosis, a functional ability, use of health services, or length of life, is a domain of behavioral competence. Residence in a nursing home or the community, as well as consensually established measures of the quality of these environments, denotes a domain of objective environment. Pain, distress, bothersome symptoms, treatments with bothersome side

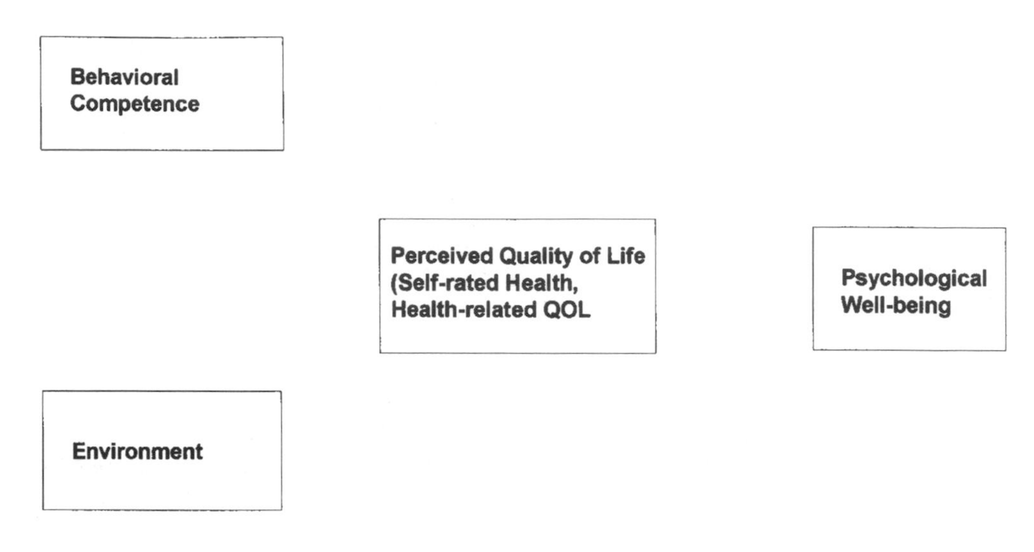

Figure 8.1. Elements of a model of quality of life (health-related and nonhealth-related).

effects, disease-specific QOL, and general self-rated health are domains within the PQOL sector, each reflecting the individual's idiosyncratic assessment of that single domain. A special instance of PQOL is represented by health preference and its derivative, health utility, in that the basis for stating how long one wishes to live under specified conditions is a subjective perception of the domains of life specified by the conditions enumerated in the question. My conception of QOL as a whole suggests that a health preference or utility judgment that is not limited to health-related domains of QOL properly belongs in the sector of PWB; that is, if people can perform the cognitive-affective calculus required to assess all domains of PQOL and PWB in terms of their net cost–benefit utility, a condition-free estimate of health utility would emerge, whose broad representation of the many facets of QOL would warrant being used as an indicator of PWB. In practice, measures of general mental health, depression, and, occasionally, anxiety represent PWB in most HRQOL research.

With the conceptual relationships between QOL and HRQOL having been detailed, depression, the specific focus of this book, is clearly located in the PWB sector. To the extent the limited literature in some areas allows, the QOL outcomes pertinent to the end of life that will be reviewed here are depression and its counterpart positive effect.

PREVALENCE OF DEPRESSION IN RELATION TO THE END OF LIFE

An obvious assumption might be that as life nears its end, depressive affect should increase, whether because depression is an inherent component of some fatal diseases and their symptoms, or because people's perception that their time of death is near evokes feelings of hopelessness or situational depression. Should this association hold true, we might expect depression and death anxiety to increase with age and depression to be more present in some diseases than in others. The literature in these areas is voluminous and cannot be reviewed here. However, none of these expectations has been established clearly as yet.

One source of information on the relationship between depression and end of life is the study of people's attitudes toward life-sustaining measures given different hypothetical scenarios. Cicirelli (1997) constructed 17 such scenarios and had healthy older community residents rate each in terms of seven questions about what to do in each case. These responses were converted into factor scores denoting for each person the extent to which the ending of life, leaving the decision to others, or maintaining life and life supports were advocated. On a univariate level, those with high self-esteem and low depression were more likely to wish to maintain life; neither state was related to the active wish to end life. These small relationships remained in some but not all multivariate tests.

In terms of clinical depression, unipolar depression appears to diminish with age (Regier et al., 1988) and the prevalence of dysphoric affect has been shown to diminish from young-adult to young-old range but to show some sign of increasing after age 75 (Gatz & Hurwicz, 1990; Newmann, 1989). The weight of the evidence (Kalish, 1985) suggests that death anxiety does not increase with age. Whether fatal diseases are more likely to be accompanied by depression is uncertain. Some of the evidence in this regard may be found in other chapters of this book. Examples of negative evidence in this regard are found in several studies of cancer patients (predominantly middle-aged and older) with several types of cancer who reported lower total scores and lower depression and anxiety scores on the POMS (McNair et al., 1971) than available comparison groups of college-age students and psychotherapy patients both after treatment (Cassileth, Lusk, Brown, & Cross, 1985) and after diagnosis but before treatment (Cella et al., 1989). Cassileth et al. (1984), in another study, compared RAND Mental Health Index Depression and Anxiety scale (Brook at al., 1979) scores from patients with arthritis, diabetes, cancer, kidney disease, and dermatological disorders. Despite the broad differences in presumed symptom severity and mortality risk across these groups, depression and anxiety scores did not differ by disease group or between any disease group and a large sample of the general population. By contrast, outpatients who were being treated for depression were, in every case, more depressed and anxious than each of the physical illness groups.

At best, evidence of this type is indirect and subject to error because participants were not characterized in terms of their death risk or nearness to death. Chochinov et al. (1995) used DSM-III-R criteria to classify 200 palliative-care hospitalized cancer patients. The combined major depression and dysphoria prevalence rate was 12.5%, clearly higher than epidemiologically determined base rates (Regier et al., 1988). In the large-scale National Hospice Study, Mor (1986) found the POMS Depression score to be elevated among hospice patients in comparison with the physical-illness groups mentioned in the preceding paragraph, but in contrast to these late-stage patients, depression scores of newly diagnosed hospice cancer patients were quite low, as noted by Cella et al. (1989).

The only two studies located that attempted to estimate the prevalence of depression among those known to have been on a death trajectory were both based on the retrospective reports of survivors (usually from family members) of deceased elders. Lawton, Moss, and Glicksman (1990) interviewed 200 survivors recruited from follow-back contacts through death certificates in the greater Philadelphia area. Although names were sampled randomly, the high incidence of refusals or inability to locate the informant (51%) rendered the 49% who responded a clearly nonpresentative sample. A comparison group of 150 informants for still-living elders was also interviewed. Face-to-face structured interviews were conducted, asking the informant to report both objective (event-related) and subjective aspects of the last year of the decedent's life. The core of this interview asked the informant to rate the extent to which the person evidenced or experienced 19 phenomena broadly fitting into the model of QOL previously described: pain, ADL (7 activities), energy, alertness, social interaction (4 indicators), doing nothing, interest in outside world, satisfaction with time use, depression, and hope. Each rating was performed with time referents of 12, 3, and 1 month prior to death. The details of the findings are reported in Lawton et al. (1990). Depression was significantly greater at all three times for the decedent group as compared to a group of relatively healthy community residents. Even so, 62% of decedents were reported "never" or "seldom" to have been depressed in the last month of life. Figure 8.2 shows survivors' estimates of the trajectories of depression over the last year of their relative's life up to the last month. For 41%, depression remained constant or decreased at a low level, while for 29% it increased in the last year. Given that survey recruitment is known to be selectively unsuccessful with the most disturbed, it is probable that the percentage who experienced some depression was underestimated in our study.

Lynn et al. (1997) utilized subsamples of two large, national multisite projects, the Study to Understand Prognosis and Preferences for Outcomes and Risks of Treatments (SUPPORT; Lynn & Kraus, 1990) and the Hospitalized Elderly Longitudinal Project (HELP; Tsevat et al., 1998). Severely ill and very-old hospitalized patients were recruited for the two studies. Surrogates of 3,357 patients were interviewed either 6 months (SUPPORT) or 12 months (HELP) after the baseline interview; the successfully interviewed group represented 83% of those eligible. Short subsets from the POMS Depression and Anxiety scales were included in the interview, which inquired about the decedent's status 3 days prior to death. Almost half (45%) were unconscious at the time. The percentages reported were based on persons who were conscious 3 days prior to death. Neither percentage distributions of depression and anxiety scores nor means are reported. However, the range of "average" item ratings was 0.7 to 1.7, which would be at the "a little" point on item scales for each depression or anxiety item, ranging from 0 (Not at all) to 4 (Extremely). The nearest estimate of depression prevalence in the published report is that "about one fourth of patients . . . had scores greater than 2 (indicating at least moderate anxiety or depressed affect in the last 3 days of life" (p. 101). The closest comparison in the Lawton et al. data indicated that in the last month of life, 34% were depressed often or all the time. Regrettably, the only study of any representative sample of elders near the end of life did not include such symptoms in its inquiry of survivors (Brock & Foley, 1998).

Thus, the seemingly simple question—Do older people get depressed as their lives move toward the end?—in fact may be answered only in very preliminary fashion. The data available are based on nonrepresentative samples, filtered through the reports of informants and measured by scales that are marginally comparable with any standard comparison groups. Such comparisons do suggest that depression is more prevalent at the very end of life. Nonetheless, the evidence at hand suggests a relatively low rate of depressive affect, considering the magnitude of the threat and the distressing states

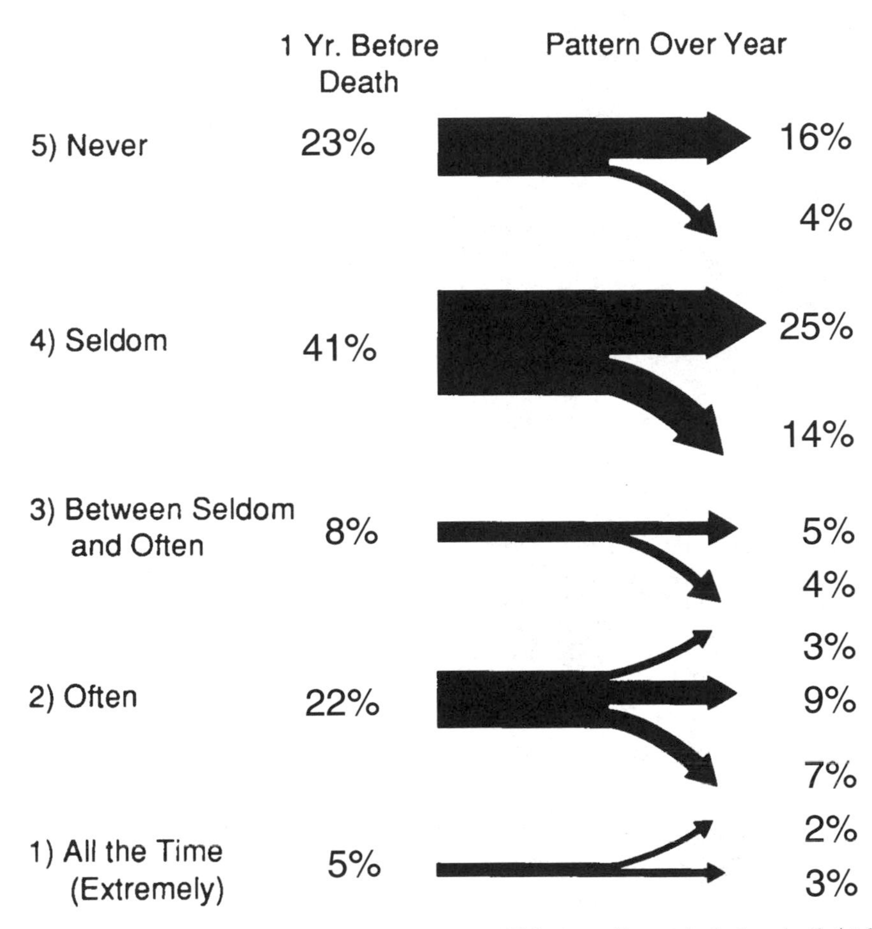

Figure 8.2. Trajectories of depression over the last year of life, from 12 months before death (rightmost percentages). Reprinted from Lawton, Moss, & Glicksman (1990), Figure 5, page 19. Copyright Blackwell Publishers and reprinted with permission.

associated with fatal disease—one-fourth to one-third of elders in the last days of their lives. There are also hints that the direct results of the illness may be more important than the cognitive and affective recognition of the imminence of death. This suggestion is based on the lack of depressive feelings among the newly diagnosed patients studied by Cella et al. (1989). In the latter study, one subtype, pancreatic cancer patients, did exhibit relatively elevated depression scores, which may well reflect a direct somatic connection associated with the site of the cancer. As the illness moves on toward its final phase, the documented pain and distress clearly are correlated with depressive symptoms for some, while unconsciousness overtakes others. Yet one cannot help being impressed that so many manage to adapt to the threat of death to overcome what would seem to be very appropriate depression over the impending loss of life. Research is needed to shed light on the mechanism by which this occurs. My hypothesis is that a subjective weighting process occurs in which the gains and losses of one's present life and the resources and liabilities of one's current state are converted into a judgment of valuation of life. This process is discussed later.

HEALTH UTILITIES AND END-OF-LIFE TREATMENT PREFERENCES OF ELDERS

Before addressing the critical questions of the place of depression and other determinants of end-of-life issues, it is necessary to review what is known regarding older people's responses to inquiries regarding their health utility and treatment preferences. Relatively little information exists regarding these issues, especially older people's willingness to trade years of chronological life in poor health for fewer years of life in good health. The scarcity of such knowledge is not surprising in light of the complexity of most methods for determining health utilities. Methods requiring paired-comparison judgments, ratio scaling, or complex health status by treatment matrices are very difficult for most older people, and no studies of this type have been found in the literature. All published studies have utilized simplifications of the classic methods, which makes comparisons across studies difficult. Nonetheless, it is important to review such estimates in the interest of identifying particular knowledge gaps. Studies that have attempted to estimate health utilities and preferences (usually in the form of years of desired life, preferences for particular treatments, or changes in either of these wishes over time) are reviewed here. Their link to depression is examined in the next section.

Health Utility

Two of the few studies of health utility of older people measured in time-trade-off form were the SUPPORT and the HELP studies (Tsevat et al., 1995; 1998). Rather than presenting a series of scenarios depicting a variety of conditions, a single anchoring point was used: the patient's current state which, because all HELP subjects were hospitalized, represents a compromised health state. The respondent was asked, "Would you prefer living 1 year in your current state of health or living only 11 months in excellent health? (if yes, the question was repeated for successively fewer months until the point of indifference was reached. The fraction of a year represented by the indifference point is the health utility).

In the SUPPORT study (Tsevat et al., 1995) one criterion for recruiting the primarily older sample was that they were not expected to live more than 6 months. In this highly vulnerable group, the mean health utility was .73 (i.e., subjects were willing to give up about 3.2 months if they had a choice based on their current state). In the HELP study, consisting only of elders, of the total 1,226 enrollees, 43% were ineligible for various reasons, but of the 622 remaining, 67% understood and completed the time-trade-off question. This degree of attrition displays the unlikelihood that total population utility estimates might be obtained in this manner but does show that this form of inquiry is usable among some proportion of very-old people. The mean health utility was 9.7 months, that is, a willingness to forego 2.3 months for good health in the remainder. However, 68% were willing to trade no time or only 1 month; the strong central tendency in this old-old population was thus to discount years of desired life relatively little in the face of their current health limitations.

In a study of 600 healthy and chronically ill elders, Lawton et al. (1999) used the Quality of Life Questionnaire (QOLQ, Schneiderman, Pearlman, Kaplan, Duderson & Rosenberg, 1992) in slightly revised form in order to portray several health-compromised conditions more systematically: no impairment, impaired function in three domains (ADL, cognition, and pain), as well as the contrast between being at home and in an institution. The QOLQ asks people to estimate the number of years (or months) they would like to live in each of 10 conditions. Although the response form is easy, the cognitive challenge of the questions is substantial. Across the 10 conditions, from 32% to 61% of subjects were unable to respond in years or days. Cognitive impairment occasioned the greatest discounting of chronological life (willingness to relinquish time alive). If unconscious, 61% of subjects would not wish to live at all; in the event of extreme pain, controllable only by narcotics, and being confused in a nursing home, the percentages who would not wish to live were 41% and 43%, respectively. Pain was more tolerable. Faced with mild pain, no health limitations, and one's present health 22%, 30%, and 25%, respectively, wished to live as long as possible. Functional impairment was intermediate between low impairment and cognitive impairment in reducing the number of years respondents wished to live.

One form of health utility could be represented by attitudes toward euthanasia. In an all-ages well sample, Ward (1980) found that 62% approved of euthanasia in incurable patients, but among respondents age 70 and over, only 49% approved. A different form of health utility may be expressed by the overt wish to die, whose metric is not in time but could be thought of as a wish to live zero to a small number of days. Chochinov et al. (1995) inquired about this possibility in a sample of 200 palliative-care cancer patients of mean age 71. Their inquiry was in the form of the question, "Do you ever wish that your illness would progress more rapidly so that your suffering could be over sooner?," followed by open-ended elaboration to determine the seriousness and pervasiveness of the wish to die. Reliable coding converted responses into a 6-point scale ranging from *No desire* to *Extreme desire for death*. Scale points of 4, 5, and 6 (*Moderate and genuine*, *Strong*, and *Extreme*) were selected as indicators of the desire to die. Although 45% acknowledged at least fleeting thoughts of this type (scale points 2–6), only 8.5% met the desire-to-die criterion of 4 or higher.

Research has clearly just begun to address the issue of direct estimation of health utility by older people The most obvious need is for more usable methods to elicit such judgments. It is entirely possible that there may be no very simple way to make such inquiries. In that case, there are two possible approaches to add to our knowledge. The first is to increase the proportion of persons capable of responding to such an inquiry through qualitative interviewing, where people's own terms and frames of reference may be better able to elicit relevant information. This approach would provide complementary information but would not afford qualitative subgroup comparisons. The second approach is for the researcher to be resigned to the reality that only intellectually relatively well-endowed people can formulate responses to such hypothetical questions and will therefore be the main providers of such information. In this case, subgroup variations (e.g., racial, cultural, health-related) would have to be investigated within privileged groups, with corresponding bias and the danger that the homogeneity of perspective consequent to social privilege would mask real differences among the less-privileged subgroups.

Treatment Preference

More data are available regarding older people's preferences for various treatments because such questions can be posed about a single condition or a single treatment. Even those that specify more than one condition can usually be answered in dichotomous or short rating-scale form. No standard form exists, however, making it very difficult to search for convergence among different studies. Table 8.1 shows some of the studies of treatment preference with largest numbers of subjects or particularly interesting features. Although the major features of each study are characterized in the columns, each study has unique features that can be ascertained only by referring to the original article. It is immediately apparent that a number of factors may be capable of influencing the estimates:

1. *Respondent characteristics*. Major differences would be expected depending on whether healthy elderly volunteers, primary practice patients, chronically ill community residents, nursing home residents, or acute-hospital patients are questioned.
2. *The complexity of the question, the response format, and the care taken to assure that respondents comprehend the question*. The latter is a particular problem because many people do not comprehend exactly what is involved in procedures such as cardiopulmonary resuscitation (CPR), tube feeding, or artificial ventilatory treatment. Most investigators have striven to portray these procedures. The study of O'Brien et al. (1995) is notable for not accepting subjects unless they "passed" a comprehension test of the procedures after receiving an explanation.
3. *Physical condition under which life-extending treatment is applied.* In fact, this aspect has two sources of variability that are not always specified or differentiated. It is possible to specify a new condition, such as a heart attack, superimposed on persons as they are now, or when totally healthy, or confused (cognitively impaired). Most frequently, a single future state is specified. In addition some researchers specify a prognosis (no chance of improvement, temporary, etc.).
4. *Treatment.* CPR, tube feeding, and ventilation are the most frequently targeted treatments. They may also be specified as permanent or temporary. Hospitalization and surgery usually are characterized in terms of the condition under treatment.
5. *Response format*. Dichotomous (yes–no) and multiscale-point ratings are equally prevalent. "Don't know" and other nonresponses are very prevalent, and sometimes it is unclear whether they were counted or included in percentages. Sometimes response categories offered to the respondent represent alternatives unclassifiable in preferential terms (e.g., "The doctor should decide").

It is clear from Table 8.1 that cognitive impairment and unconsciousness elicit very low treatment preferences. Hospitalization is most acceptable, but the question wording does not always specify for what. Although the treatment wishes for nursing home residents are sometimes characterized as "high" or higher than expected, no standard or central tendency is discernible from Table 8.1 that would enable such a comparative judgment to be made. Neither is there a basis in published data for judging whether the estimates shown in Table 8.1 are systematically different from those

Table 8.1. Treatment Preferences Estimated under Various Conditions by Nine Studies

Study	Subjects	*N*	Condition	Percentage "yes" or mean score				
				Hospital	CPR	Surgery	Tube feeding	Ventilation
Cohen-Mansfield et al. (1991)	Nursing home patients	103	Intact		3.6[a]		3.1	3.6
			Confused		3.2		3.0	3.1
			Unconscious		3.1		2.9	3.0
Danis et al. (1991)	Competent nursing home residents	126	"Critical illness"	58%[b]	47	32		35
			"Terminal illness"	29	20	14		20
			"Permanently unconscious"	20	19	8	16	13
Emanuel et al. (1991)	Primary practice all ages	405	Coma, chance of recovery		29[c]	27	20	20
			Vegetative state		10	9	8	7
			Dementia		16	13	11	7
			Dementia and terminal illness		8	8	9	8
Uhlmann & Pearlman (1991)	Outpatients 65+	258	Now		75%[d]			23[e]
			Severe chronic lung disease, no improvement	61				21
			Stroke, ADL deficit	11				14
			(all 3 with "heart stoppage")					
Cohen-Mansfield et al. (1992)	Hospital patients 65+	97	Now		4.9[a]		3.6	3.8
			Confused		3.0		3.0	3.0
			Unconscious		2.6		1.7	2.8
Garrett et al. (1993)	Primary practice 65+	2,536	"Terminal illness"	47%[f]	27	24	18	19
Gerety et al. (1993)	VA nursing home residents	52	Current health	98%[g]	67		81	65
			ADL impairment only		60		38	
			Cognitive impairment only		33		25	
			(all of above "with life-threatening Illness")					

(Continued)

Table 8.1. (*Continued*)

Study	Subjects	*N*	Condition	Hospital	Percentage "yes" or mean score			
					CPR	Surgery	Tube feeding	Ventilation
O'Brien et al. (1995)	Cognitively intact nursing home residents	424 – 380	"Serious illness"	89%[f]			33	
			"Permanent brain damage"	73				
			"Sudden cardiac arrest"		60			
Gramelspacher et al. (1997)	Medical practice patients 75+ or chronic illness and 50+	831	"Terminal Illness"	58%[f]	44	15	37	27

Note. Cell entries are % "yes" as indicated by first entry for each study. Other decimal entries are multiscale-point means.
[a] Mean of 7-point scale (1 = Absolutely no, 7 = Absolutely yes).
[b] Other alternatives: No treatment, doctor decide, family decide, don't know.
[c] Other alternatives: "No," wants a trial, undecided.
[d] Definitely or probably "yes." Other alternatives: uncertain, probably "no," definitely "no."
[e] CPR and ventilator
[f] Other alternatives: No, don't know.
[g] Other alternatives: Refuse, uncertain.

made by younger, healthy adults. Such comparative estimates would be very useful and worth deriving if identical methods could be applied to multiple respondent groups. What is clear, however, is the wide variability across individuals. The measurement of treatment preferences and health utilities is important in the study of depression near the end of life because these are the major attitudinal indicators of the wish to live that are available as outcomes against which to test the effect of depression. Determinants of treatment preferences associated with depression are examined later in this chapter.

Change in Health Utility and Treatment Preference over Time

A critical question, the third of the problems with HRQOL raised at the beginning of this chapter, is whether the respondent's present or changing health state alters wishes regarding length of life or end-of-life care. Several studies yielded information on change over time. Change over a short term was assessed by Llewellen-Thomas, Sutherland, and Thiel (1993), who studied laryngeal cancer patients (ages not specified) before and after invasive radiation therapy that resulted in clearly measurable onset of distressing symptoms. Yet the health utilities measured for these symptoms at baseline (hypothetically) did not change significantly when measured while being actually experienced at the end of treatment. In the SUPPORT study (Tsevat et al., 1995) 732 seriously ill, hospitalized patients were reinterviewed 2 months after baseline and 593 were measured 6 months after baseline. Mean health utilities increased significantly at both time intervals, and the increase was correlated with improvement in health status. Tsevat et al. (1998) interviewed surviving HELP subjects (176 of the original 414) 1 year after their original hospital-based interview. Their time-trade-off scores increased significantly; the amount of time they were willing to forego decreased by about 2 weeks, a small amount of time but statistically significant and possibly indicative of a process of change in the frame of reference. Decreased "physiological reserve" was associated with decreased health utility but ADL change was not. Was change evidenced in both the SUPPORT and HELP studies simply a function of the fact that, having survived for a year, these privileged people were heartened and allowed themselves the luxury of feeling that they could tolerate a longer time of life, even if compromised by distress? Or was it a decision-theory-based effect whereby the perceived remaining time was shorter and their wishes reflected an enhanced attachment to remaining time? Clearly, more research is needed regarding such questions.

Another study was a 2-year follow-up of 2,073 (representing 82% of the Garrett, Harris, Norbrun, Patrick, & Davis, 1993, baseline sample) primary-practice community residents ages 65+ who indicated treatment preferences (*Yes*, *No*, or *Don't know*) for six types of treatment under the single stated condition of "having a terminal illness." In general, the judgments were most stable (66–75%) over 2 years among all those who did not wish a treatment at baseline but less stable (18% to 43%) among those who did wish treatment at baseline. The *Don't know* percentages were relatively high at baseline (18–46%) and displayed lowest stability (23–29%). There was a slight decrease in overall prevalence of wish for treatment. Factors associated with a change in the direction of an increased wish for treatment were hospitalization, having an accident, lessened mobility, and increased physical problems on the QWBS (Kaplan & Bush, 1982), but not self-rated health. Decreased social support was also associated with increased wish for treatment.

Despite a very low number of subjects, one other study deserves mention because of the urgency of the research questions asked. This study of change began with only 6 cancer patients receiving palliative care in a general hospital, who were followed up 2 weeks later (Chochinov et al., 1995). All had been characterized as wishing to die on the Desire for Death scale. When followed up, 4 of the 6 no longer met the Desire for Death criterion rating, while 2 continued to do so.

Information relevant to this question may also be obtained from comparisons of the health utilities of people in poor health with those of healthy people. Kaplan (August 12, 1996, personal communication) has suggested that people's judgments about health states mirror social norms and do not vary with the judge's personal health. Corroborating information for this position came from a study (Balaban, Sagi, Goldfarb, & Nettler, 1986) of arthritis patients (mean age 50) in which the category scaling procedure used to derive the original QOWB Scale (Patrick et al., 1973) was repeated. In this task of judging the health preferences for differing scenarios, the original weights obtained from healthy people were replicated. On the other hand, Sackett and Torrance (1978) found that end-stage renal-disease patients displayed higher time-trade-off utilities for their own functional and dialysis conditions then did members of the general public judging those conditions.

None of these studies have dealt explicitly with chronically ill elders extending (and therefore changing) over a period of time. The small increase in utilities reported from the SUPPORT and HELP study at the very least marks this question as one requiring continued investigation of these and other factors' influences on health utility. The effects of selective survival on these findings remain to be adequately explored (i.e., having survived longer than many in the sample, did subjects simply adjust their expectations upward in a realistic manner?).

DEPRESSION AND LIFE-PROLONGATION

It has often been suggested that because clinical depression is associated with suicidal and other life-denying wishes, a person's request for euthanasia or physician-assisted suicide may be motivated by psychopathology. Viewed in combination with the instability of depression and its amenability to pharmacological and psychological treatment, a conservative view on the appropriateness of legalizing such measures seems warranted (Lee, 1990). Relatively little empirical evidence on this assertion exists, however. Even extending the question to ask whether health utilities or treatment preferences are associated with depression, evidence is surprisingly scant.

Several early studies found no association between depressive symptoms and wishes for single or aggregated life-extending treatment measures (Cohen-Mansfield et al., 1991; Michelson, Mulvhill, Hsu, & Olson, 1991; Uhlmann & Pearlman, 1991). A much larger study of community-resident primary care elders (Danis, Garrett, Harris, & Patrick, 1994; Garrett et al., 1993) utilized an abridged Center for Epidemiological Studies–Depression Scale (CES-D, Radloff, 1977) and six treatments in a scenario of "terminal illness." At the time of the first study, people who were more depressed wished more treatment (Garrett et al., 1993). Stability over 2 years was reported by Danis et al. (1994), indicating that people whose depression had increased over the 2-year period were more likely to have increased their wish for treatment over the same period. Thus among these reports, depression was associated with a greater wish for treatment, the opposite of the direction predicted by the psychopathological explanation for a reduced wish for treatment..

Another nursing home study, which paid special attention to ensuring that residents comprehend the phenomena about which they were questioned, found that depressed residents were slightly less likely to wish for CPR (O'Brien et al., 1995) but equally likely to wish for tube feeding (O'Brien et al., 1997) compared to nondepressed people.

Lee and Ganzini (1992) took special pains to diagnose depression by both psychometric and clinical criteria, excluding Veterans Administration inpatients who met one, but not the other, criterion. They inquired about seven treatments using five scenarios: the patient as he was at the time ("now"), conditions with a relatively favorable prognosis (pneumonia, iatrogenic renal failure), and conditions with very poor prognosis (devastating stroke, metastatic cancer). Comparing 50 depressed and 50 nondepressed patients, they found that depressed patients would be less likely to wish for life-sustaining treatment as they are now and in the favorable conditions, but they did not differ from nondepressed patients in their wish for treatment for the poor-prognosis conditions. It seems likely that the strength of the poor prognosis was sufficient in both groups to override the effect of the difference in depression between the two groups.

In the cross-sectional baseline analysis of both the SUPPORT (Tsevat et al., 1995) and the HELP data, Tsevat et al. (1998) found that people who were more depressed on the POMS were more willing to give up months of life. The correlations between depression score and health utility were –.36 and –.27, respectively. Similar relationships were observed for POMS anxiety and a single quality-of-life question. In the Chochinov et al. (1995) study of palliative care patients, those indicating depression on the Beck Depression Index (Beck, Ward, Mendelson. Mock, & Erbaugh., 1951) expressed a stronger wish to die ($r = .29$). Furthermore, 59% of those clearly wishing to die were depressed, while only 8% of those who did not wish to die were depressed.

Cross-sectional relationships reveal little about causal mechanisms. The SUPPORT study followed people 2 months and 6 months, while the HELP study (Tsevat et al., 1998) followed subjects for 1 year. The cross-sectional relationships between depression and willingness to relinquish more months of life were repeated longitudinally, in that increased depression was associated with a decrease in health utility; increase in depression and decrease in health utility were correlated .24 in both the SUPPORT and HELP studies. By contrast, a 6-month longitudinal evaluation of the Lee and Ganzini (depressed) patients (Lee & Ganzini, 1992) failed to show any overall change in life-enhancing treatment wishes among patients who recovered from their depression ($N = 19$). Their number of desired treatments remained constant at 12, both before and after recovery, although the 15 patients who remained depressed exhibited a slight (though nonsignificant) decrease from 14 at the beginning of the study to 12 at follow-up. Nondepressed patients showed greater overall stability of choice than did depressed patients.

The strongest test of causality was provided in another study by Ganzini, Lee, Heintz, Bloom, and Fenn (1994), who studied the therapeutic outcomes of 43 inpatients with major depression. There were 24 patients discharged with a remission of their depression; their number of desired treatments remained constant before (8.9) and after (8.8) treatments, despite a major improvement in Geriatric Depression Scale score (Yesavage et al., 1983). These authors formed subgroups of subjects whose treatment preferences did not change ($N = 32$) and those that increased by three or more ($N = 11$). The increased-wish group was originally considerably more depressed, hopeless, and pessimistic about treatment possibilities than the stable group had been before treatment, however. The authors thus concluded that it may be people with the more severe forms of depression who are at risk of inappropriately rejecting treatment, while milder depression (as in Lee & Ganzini, 1992) seems to be associated with treatment preferences not notably different from those made by nondepressed people. Ganzini et al. (1994) provide a thoughtful discussion of ways of thinking about treatment decisions among depressed people and how these fit in with more general attitudes regarding giving and withholding treatment.

Looking over this scattered set of empirical findings, the lack of consensus as to whether depression leads to rejection of treatment is evident. Hints are given, however, regarding possible topics for future research. First, although depression is by no means inevitable during terminal illness, those who are depressed clearly value their lives less: Chochinov et al. (1995) observed a strong co-occurrence of depression and wish for death. Tsevat et al. (1995; 1998) demonstrated in two major studies that depressed persons were willing to give up more months of life in a health utility judgment. A second tentative conclusion alerts us to the real difference between two of the major concepts reviewed here: first, the value placed on one's life and, second, readiness to endure distressing treatment. The results reviewed here suggest that depression may affect judgments such as health utility more readily than it does more concrete readiness to accept specific treatments, even though utility and treatment acceptance are highly correlated. Low expressed wish to live shows many indications of being a sign of depression. Third, the relationship between depression and desired treatment may be moderated by the severity of the condition. In more severe conditions where treatment success is unlikely, pessimism is realistic, and this factor overshadows depression as a determinant of treatment wish (Lee & Ganzini, 1992). Where less severe and therefore more hopeful conditions are depicted, judgments are at greater risk of being irrationally conditioned by depressive cognition. Finally, the hypothesis of Ganzini et al. (1994), that treatment of more severe depression may lead to greater revisions of treatment pessimism than does treatment of milder depression, deserves further attention. If upheld by other research, such a finding suggests that a diminished wish to live may resist being instigated by mild depressive feelings.

In summary, the exact nature of health preferences, the realistic chances of success, and the severity of depression appear to be simultaneous and possibly interacting causal factors whose effects need to be probed in the effort to understand how depression effects end-of-life treatment decisions. There is enough support for the idea that depression may explain some significant proportion of variance among elders' wishes to live. Therefore, caution is clearly in order regarding the meaning and permanence of people's expressed wishes to curtail their lives.

VALUATION OF LIFE

The psychological processes that mediate conversion of the many internal and external inputs relevant to attitudes and behaviors regarding the end of life have been investigated very little. My QOL model discussed earlier provides structure to account for the inputs. The global assertion is that any domain within any of the four QOL sectors may contribute to the wish to shorten or prolong life, or to engage in behaviors serving those purposes. The work to be described in this section originated in the hypothesis that QOL-related inputs need to be processed subjectively into a schema more directly relevant to end-of-life attitudes and behavior than are the many generalized QOL factors. Our work posited that a VOL schema was the internal psychological force that was the proximal determinant of end-of-life attitudes and behaviors. VOL in its positive sense denotes purpose, meaning, hope, a sense of futurity, persistence, and self-efficacy. It is operationalized with a minimum of content directly related to mental health or pathology and is free of any direct reference to health, longevity, or death. Our hypothesis regarding the dynamics of end-of-life behavior suggests that a mix of QOL elements (including both positive and negative aspects of QOL) contribute to valuation of life, which in turn is the dominant influence on outcomes such as health utility, treatment preference, and treatment choice.

An instrument to assess VOL was constructed utilizing items expressing the major aspects of this construct. Most existing measures of positive mental health could not be used, because the content of some of their items included frank psychopathology. A group of borrowed and newly constructed items, including 5 items from the Hope Scale (Snyder, et al., 1991), was assembled. In several stages of analysis, a 13-item Positive Valuation of Life factor and a 6-item Negative Valuation of Life factor emerged (Lawton et al., in press). Illustrative items include "I feel able to accomplish my life goals," "Each day I have much to look forward to," and (Negative Valuation of Life), "The real enjoyments of my life are in the past." Although the Positive Valuation of Life scale is preferred, because the Negative scale introduces greater response error for elders with low educational achievement, both factors and their sum exhibited favorable psychometric characteristics. In particular, VOL demonstrated moderate (.30 to .60) correlation with positive affect and negative affect, and a number of standard measures of positive mental health. A structural model of VOL, however, indicated that the relationships of health and depression to VOL were completely accounted for by a factor measuring perceived quality of time use (Lawton, 1997) and positive affect (Pearlman et al., 1992). To measure health utility, the Quality of Life Questionnaire (Schneiderman et al., 1992) was slightly simplified for use with our age 70+ sample of healthy and chronically ill elders. Ten scenarios are presented, including ideal health, health as in the present, and various depictions of physical and cognitive disability, and home versus nursing home. The respondent is asked to state how long she or he would wish to live given the condition (Years of Desired Life). The task was difficult for elders; about one-half could respond only with "As long as possible," "Only God knows that," or provide no estimate at all. Nonetheless, among those who did provide valid responses in chronological time, despite some significant zero-order correlations between QOL measures and Years of Desired Life, when the full model was tested using Years as the dependent variable and Positive Valuation of Life as the most proximal independent variable, only VOL remained a significant predictor of Years of Desired Life for most conditions (Lawton et al., 1999). In three conditions, there were no significant predictors: confused in a nursing home, severe pain controllable only by heavy narcotics, and unconscious with no hope of recovery. The strength of the relationship between Positive VOL and years of Desired Life diminished as negative scenarios became more intense (significant standardized betas ranged from .40, with "no limitation," to .16 for confused at home).

First of all, these results are consistent with most previous research in showing that the physical and mental disabilities associated with poor health undermine people's attachment to life. Second, our research on VOL added an intrapsychic dimension to the measurement of attachment to life that is relatively free of contamination with either health, psychopathology, or health utility. VOL in multivariate analysis, however, was strongly related to positive states but not health or depression. Yet we found that VOL strongly mediated both positive and negative features of QOL in affecting the most direct prospective measure of personal health utility, Years of Desired Life.

The final point to be made from this research is the light it sheds on the three problematic aspects of HROL. It demonstrates that health is by no means the only determinant of the seemingly health-related outcome, personal health utility. People do, in fact, process multiple inputs, especially those from the fulfilling, stimulating, happy aspects of their lives as they think about their psychological "net worth."

The place of depression in end-of-life phenomena is most central to this book's theme. Although psychopathology is clearly the dominant end point in most mental health research, the two-factor view of mental health has come into its own and reinstated positive affect and positive mental health as equal components of overall mental health (Ryff, 1989; Watson & Tellegen, 1985). Our data showed that depression lowered, and positive affect elevated to a small degree, Years of Desired Life given the respondent's current condition; the effects of both were diminished greatly as VOL was introduced as the mediator. This suggests an important conclusion: People may be depressed without going the further step and wishing for a shorter life. The two may covary, as shown in the preceding section. But our data suggest that some people process depression in terms of reduced attachment to life, while others call up nonhealth-related (mental or physical) factors to counteract the eroding effect of poor health and depression or their wish to live. In our data, quality of time use and positive affect level contributed to VOL.

More direct evidence supporting the ability of positive features of life to contribute to life-sustaining motivation came in one study of cancer patients (Yellen & Cella, 1995) and another of normal college students and elders (Ditto, Druley, Moore, Danks, & Smucker, 1996). Among the cancer patients, three factors—a measure of social QOL, sharing a household, and having children at home—increased people's willingness to endure toxic and invasive end-stage treatments. Ditto et al. inquired about the role of valued activities in people's hypothetical preference for living or dying under 28 conditions of compromised health. The degree to which each health condition was rated as interfering with a most valued activity was the strongest and often the only predictor of a 9-point preference-to-live-or-die scale. Although the Ditto et al. study does not separate health-related from nonhealth-related influences (because all interferences were attributed to health by the way the questions were asked), their results underline the centrality of the favored activity, a major component of our QOL model. It is worth noting that Williamson and Shaffer (Chapter 9, this volume) build a strong empirical case for the salience of activities related to the person's sense of control and the enjoyable quality of the activities.

Our own data speak minimally to the third HRQOL problem area, the effect of proximity to death or to major health problems as a factor in health utility judgments.

Current health was related to few of the Years of Desired Life ratings; in fact, the literature seems to indicate a generally small association between present health and health utility (see Tsevat et al., 1998). The critical issue is longitudinal change among individuals as they move from good to poor health and the end of life. The few studies of this type that have come to light, however, do suggest that there may be a process of accommodation whereby people stretch the upper range tolerability of health-related intrusions on their QOL as the lower end becomes constricted, as suggested by prospect theory (Kahneman & Tversky, 1984). These authors' general theory of decision, judgment, and preferences, if applied to people's estimates of health preferences, indicates that the relationship of preference judgments ("prospects") to loss–gain probabilities (i.e., state of health) is not linear (it is an ogive function), but is partly dependent on a third factor, the person's position on the health dimension. For a person in good health, the difference between the worst state (death) and a state only incrementally better (very poor health but still alive) may seem small. For the person in poor health, that same health-state difference may be associated with a much larger preference improvement (Winter, Lawton, & Ruckdeschel, 1997). The increase in health utility over time in the HELP study (Tsevat et al., 1998) and the SUPPORT study (Tsevat et al., 1995) supports this idea.

CONCLUSION

This overview of literature dealing with the end of life makes clear that depression is an important state at this time of life. At the same time, it is clear that depression is neither inevitable in the terminal phase of life nor a necessary motivation for shortening or ending one's life. Despite the well-documented association between depression and physical illness, and the erosion of quality of life engendered by poor health, it seems virtually certain that non-health-related aspects of life enter into people's judgment processes in thinking about life prolongation and treatment. This process, which we have termed VOL, is influenced directly and indirectly by both depression and positive affect, as well as health and other aspects of QOL, but was dominated in our early empirical research by the positive inputs. VOL, in turn, is the primary determinant of elders' judgments of how long they might wish to live under some health-compromised conditions.

Throughout the chapter, I have emphasized the major gaps in knowledge or application of methods that can yield definitive answers to some of the questions. As in many areas of the behavioral sciences, longitudinal studies of people as they move across the entire trajectory of good health to poor health to terminal illness are needed in order to answer many questions. Such questions include the following:

1. Do people become more depressed over time as illness intrudes? Is there some nonlinear time or event function of depression marked by early recognition of illness, treatment, chronicity, disability, and nearness of death?
2. How do positive countervailing resources, events, relationships, affects, and cognitions exert their effects on VOL, health utilities, and preferences for and choices of, treatment? Concepts such as the meaningfulness of activities, the qualities and functions of social relationships, and personal goals should be

introduced into research on end-of-life phenomena to take us beyond the simple statement that good things can moderate the effects of bad things in one's life.

3. Methods of estimating health utility and treatment preference in older, frail, and poorly educated people are needed. Too much of what has been ascertained to date is based on samples that are elite not by design but by virtue of excluding those who are too ill, cognitively limited, or educationally deprived to respond to typical modes of inquiry used in this research.
4. Greater order needs to be brought into the specification of treatment preferences and the factors that condition them. Some such factors are the health state of respondents, their residential locus, their degree of comprehension of the questions, the specific conditions portrayed, and the hypothetical treatment procedures used in the inquiry. In addition to these substantive influences, methodological variations, such as the form of the choices, the alternatives provided, and the handling of nonresponse, have been so varied in research to date as to defy any attempt to characterize central tendencies.
5. The critical lack is comparison of earlier-life wishes with actual treatment (already approached to some degree in research not reviewed here; e.g., Teno et al., 1997) and comparison of end-of-life outcomes in the affective arena with people's earlier hypothetical judgments of what they would and would not find tolerable.

With such a long list of unanswered questions and a short list of solid conclusions, further research in this area is clearly indicated. The psychological and mental health aspects of the end of life are as important as physical health for the well-being of both the individual and society in the aggregate.

ACKNOWLEDGMENT: Some research reported in this chapter was supported by Grant No. MH37707 from the National Institute of Mental Health, and Grant No. AG11995 from the National Institute on Aging.

REFERENCES

Andrews, F. M., & Withey, S. B. (1976). *Social indicators of well-being.* New York: Plenum Press.

Balaban, D. J., Sagi, P. C., Goldfarb, N. E., & Nettler, S. (1986). Weights for scoring the quality of well-being instrument among arthritic patients. *Medical Care, 24*, 973–980.

Beck, A. T., Ward, C. H., Mendelson, M., Mock, J., & Erbaugh, J. (1951). An inventory for measuring depression. *Archives of General Psychiatry, 4*, 561–571.

Bergner, M. B., Bobbitt, R. A., Carter, W. B., & Gilson, B. S. (1981). The Sickness Impact Profile: Development and final revision of a health status measure. *Medical Care, 19*, 787–798.

Boyle, M. H., Furlong, W., Feeny, D., Torrance, G. W., & Hatcher, J. (1995). Reliability of the Health Utilities Index–Mark III used in the 1991/cycle 6 Canadian General Social Survey Health Questionnaire. *Quality of Life Research, 4*, 249–257.

Bradburn, N. (1969). *The structure of psychological well-being.* Chicago: Aldine.

Brock, D. B., & Foley, D. J. (1998). Demography and epidemiology of dying in the U.S. with emphasis on deaths of older persons. *Hospice Journal, 13*, 49–60.

Brook, R. H., Goldberg, G. A., Harris, L. J., Applegate, K., Rosenthal, M., & Lohr, K. N. (1979). *Conceptualization in measurement of physiologic health in the Health Insurance Study.* Santa Monica CA: Rand Corporation.

Campbell, A., Converse, P., & Rodgers, W. (1976). *Quality of life in America.* New York: Russell Sage Foundation.

Cassileth, B., Lusk, E .J., Brown, L. L., & Cross, P. A. (1985). Psychosocial status of cancer patients and next of kin: Normative data from the Profile of Mood States. *Journal of Psychosocial Oncology, 3,* 99–105.

Cassileth, B. R., Lusk, E. J., Strauss, T. B., Miller, D. S., Brown, L. L., Miller, C., Cross, P. A., & Tenaglia, A. N. (1984). Psychosocial status in chronic illness? *New England Journal of Medicine, 311,* 506–511.

Cella, D .F., Tross, S., Orav, E. J., Holland, J. C., Silverfarb, P. M., & Rafla, S. (1989). Mood states of patients after the diagnosis of cancer. *Journal of Psychosocial Oncology, 7,* 45–54.

Chochinov, H. M., Wilson, K. G., Enns, M., Mowchum, N., Lander, S., Levitt, N., & Clinch, J. J. (1995). Desire for death in the terminally ill. *American Journal of Psychiatry, 152,* 1185–1191.

Cicirelli, V. G. (1997). Relationship of psychosocial and background variables to older adults' end-of-life decisions. *Psychology and Aging, 12,* 72–83.

Cohen-Mansfield, J., Droge, J. A., & Billig, N. (1992). Factors influencing hospital patients' preferences in the utilization of life-sustaining treatments. *Gerontologist, 32,* 89–95.

Cohen-Mansfield, J., Rabinovitch, B. A., Lipson, S., Fein, A., Gerber, B., Weisman, S., & Pawlson, G. (1991). The decision to execute a durable power of attorney for health care and preferences regarding the utilization of life-sustaining treatments in nursing home residents. *Archives of Internal Medicine, 151,* 289–294.

Danis, M., Garrett, J., Harris, R., & Patrick, D. L. (1994). Stability of choices about life-sustaining treatments. *Annals of Internal Medicine, 120,* 567–573.

Danis, M., Southerland, L. I., Garrett, J. M., Smith, J. L., Hielema, F., Pickard, C. G., Egner, D. M., & Patrick, D. L. (1991). A prospective study of advance directives for life-sustaining care. *New England Journal of Medicine, 324,* 882–888.

Ditto, P. H., Druley, J. A., Moore, K. A., Danks, J. H., & Smucker, W. D. (1996). Fates worse than death: The role of valued life activities in health-state evaluations. *Health Psychology, 15,* 332–343.

Emanuel, L. L., Barry, M. J., Stoeckle, J. D., Ettelson, L. M., & Emanuel, E. (1991). Advance directives for medical care—a case for greater use. *New England Journal of Medicine, 324,* 889–895.

Fanshel, S., & Bush, J. W. (1970). A health-status index and its applications to health-services outcomes. *Operations Research, 18,* 1021–1066.

Frankl, D., Oye, R. K., & Bellamy, P. E. (1989). Attitudes of hospitalized patients toward life support: A survey of 200 medical inpatients. *American Journal of Medicine, 86,* 645–648.

Ganzini, L., Lee, M. A., Heintz, R. T., Bloom, J. D., & Fenn, D. S. (1994). The effect of depression treatment on elderly patients' preferences for life-sustaining medical therapy. *American Journal of Psychiatry, 151,* 1631–1636.

Ganzini, L., Smith, D. M., Fenn, D. S., & Lee, M. A. (1997). Depression and mortality in medically ill older adults. *Journal of the American Geriatrics Society, 45,* 307--312.

Garrett, J. M., Harris, R. P., Norbrun, J. K., Patrick, D. L., & Danis, M. (1993). Life sustaining treatments during terminal illness: Who wants what? *Journal of General Internal Medicine, 8,* 361–368.

Gatz, M., & Hurwicz, M. L. (1990). Are old people more depressed? Cross-sectional data on Center for Epidemiological Studies Depression scale factors. *Psychology and Aging, 5,* 284–290.

Gerety, M. B., Chiodo, L. K., Kantea, D. N., Tuley, M. R., & Cornell, J. E. (1993). Medical preferences of nursing home residents. *Journal of the American Geriatrics Society, 41,* 953–960.

Gramelspacher, G. P., Zhou, X. H., Hanna, M. P., & Tierney, N. M. (1997). Preferences of physicians and their patients for end-of-life care. *Journal of General Internal Medicine, 12,* 346–351.

Kahneman, D., & Tversky, A. (1984). Choices, values, and frames. *American Psychologist, 39,* 341–350.

Kalish, R. A. (1985). The social context of death and dying. In R. H. Binstock & E. Shanas (Eds.), *Handbook of aging and the social sciences* (2nd ed., pp. 149–170). New York: Van Nostrand Reinhold.

Kaplan, R. M., & Bush, J. W. (1982). Health-related quality of life measurement for evaluation research and policy analysis. *Health Psychology, 1,* 61–80.

Lawton, M. P. (1991). A multidimensional view of quality of life in frail elders. In J. E. Birren, J. E. Lubben, J. C. Rowe, & D. E. Deutchman (Eds.), *The concept and measurement of quality of life* (pp. 3–27). New York: Academic Press.

Lawton, M. P. (1997). *Quality of life in health and illness.* Philadelphia: Philadelphia Geriatric Center.

Lawton, M. P., Moss, M., & Glicksman, A. (1990). The quality of the last year of life of older persons. *Milbank Quarterly, 68,* 1–28.

Lawton, M.P., Moss, M., Hoffman, C., Grant, R., Ten Háve, T., & Kleban, M. H. (1999). Health, valuation of life, and the wish to live. *Gerontologist, 39,* 406–416.

Lawton, M. P., Moss, M., Hoffman, C., Kleban, M. H., Ruckdeschel, K., & Winter, L. (in press). Valuation of life: A concept and a scale. *Aging and Health.*

Lee, M. A. (1990). Depression and refusal of life support in older people: An ethical dilemma. *Journal of the American Geriatrics Society, 38,* 710–714.

Lee, M. A., & Ganzini, L. (1992). Depression in the elderly: Effect on patient attitudes toward life-sustaining therapy. *Journal of the American Geriatrics Society, 40,* 983–988.

Llewellyn-Thomas, H. A., Sutherland, J. J., & Thiel, E. C. (1993). Do patients' evaluations of a future health state change when they actually enter that state? *Medical Care, 31,* 1002–1012.

Lynn, J., Teno, J. M., Phillips, R. S., Wu, A.W., Desbiens, N., Harrold, J., Claessens, M. T., Wenger, N., Kreling, B., & Connors, A. F. (1997). Perceptions by family members of the dying experience of older and seriously ill patients. *Annals of Internal Medicine, 126,* 97–106.

McNair, D. M., Lorr, M., & Droppelman, L. F. (1971). *Profile of Mood States manual.* San Diego: Educational and Industrial Testing Service.

Meenan, R. F., Gertman, P. M., & Mason, J. R. (1980). Measuring health status in arthritis: The Arthritis Impact Measurement Scales. *Arthritis and Rheumatism, 23,* 146–152.

Michelson, C., Mulvhill, M., Hsu, M. A., & Olsen, E. (1991). Eliciting medical care preferences from nursing home residents. *Gerontologist, 31,* 358–363.

Mor, V. (1986). Assessing patient outcomes in hospice: What to measure? *Hospice Journal, 2,* 17–35.

Newmann, J. P. (1989). Aging and depression. *Psychology and Aging, 4,* 150–165.

O'Brien, L. A., Grisso, J. A., Maislin, G., LaPann, K., Krotki, K. P., Gress, P. J., Siegert, E. A., & Evans, L. K. (1995). Nursing home residents' preferences for life-sustaining treatments. *Journal of the American Medical Association, 274,* 1775–1779.

O'Brien, L. A., Siegert, E. A., Grisso, J. A., Maislin, G., LaPaun, K., Evans, L. K., & Krotki, K. P. (1997). Tube feeding preferences among nursing home residents. *Journal of General Internal Medicine, 12,* 364–371.

Patrick, D. L., Bush, J. W., & Chen, M. M. (1973). Toward an operational definition of health. *Journal of Health and Social Behavior, 14,* 6–23.

Patrick, D. L., & Deyo, R. A. (1989). Generic and disease-specific measures in assessing health status and quality of life. *Medical Care, 27*(Suppl. 3), S217–S232.

Radloff, L. S. (1977). The CES-D Scale: A self-report depression scale for research in the general population. *Applied Psychological Measurement, 1,* 385–401.

Regier, D. A., Boyd, J. H., Burke, J. D., Rae, D. S., Myers, J. K., Kramer, M., George, L. N., Karns, M., & Locke, B. Z. (1988). One-month prevalence of mental disorders in the United States. *Archives of General Psychiatry, 45,* 977–986.

Revenson, T. A., & Felton, B. J. (1989). Disability and coping as predictors of psychological adjustment to rheumatoid arthritis. *Journal of Consulting and Clinical Psychology, 57,* 344–348.

Ryff, C. D. (1989). Happiness is everything, or is it? Explorations on the meaning of psychological well-being. *Journal of Personality and Social Psychology, 57,* 1069–1081.

Sackett, D. L., & Torrance, G. W. (1978). The utility of different health states as perceived by the general public. *Journal of Chronic Diseases, 31,* 697–704.

Schneiderman, L.J ., Pearlman, R. A., Kaplan, R. M., Anderson, J. P., & Rosenberg, E. M. (1992). Relationship of general advance directive instructions to specific life-sustaining treatment preferences in patients with serious illness. *Archives of Inernal Medicine, 152,* 2114–2122.

Snyder, C. R., Harris, C., Anderson, J. R., Holleran, S. A., Irving, L. M., Sigmon, S.T ., Yosinobu, L., Gibb, J., Langelle, C., & Havney, P. (1991). The will and ways: Development and valuation of an individual differences measure of hope. *Journal of Personality and Social Psychology, 60,* 570–585.

Spilker, B. (Ed.) (1996). *Quality of life and pharmacoeconomics in clinical trails.* (2nd ed.). Philadelphia: Lippincott-Raven.

Stewart, A. L., & Ware, J. E. (Eds.). (1992). *Measuring functioning and well-being*: The Medical Outcomes Study approach. Durham, NC: Duke University Press.

Teno, J. M., Licks, S., Lynn, J., Wenger, N., Connors, A. F., Phillips, R. S., O'Connor, M. A., Murphy, D. P., Fulkerson, W. J., Desbiens, N., & Kraus, W. A. (1997). Do advance directives provide instructions that direct care? *Journal of the American Geriatrics Society , 45,* 508–512.

Torrance, G. W., Thomas, W. H., & Sackett, D. L. (1972). A utility maximization model for evaluation of health care programs. *Health Services Sesearch, 7,* 118–133.

Tsevat, J., Cook, E. F., Green, M. L, Matchar, D. B., Dawson, H. V., Broste, S. K., Wu, A.W., Phillips, R. S.,

Aye, R. K., & Goldman, L. (1995). Health values of the seriously ill. *Annals of Internal Medicine, 122,* 514–520.

Tsevat, J., Dawson, N. V., Wu, A.W., Lynn, J., Soukop, J. R., Cook, E. F., Vidaillet, H., & Phillips, R. S. (1998). Health values of hospitalized patients 80 years or older. *Journal of the American Medical Association, 279,* 370–375.

Uhlmann, R. F., & Pearlman, R. A. (1991). Perceived quality of life and preferences for life-sustaining treatment in older adults. *Archives of Internal Medicine, 151,* 495–497.

Ward, R. A. (1980). Age and acceptance of euthanasia. *Journal of Gerontology, 35,* 421–431.

Ware, J. E., & Sherbourne, C. D. (1992). The MOS 36-item Short-Form Health Survey (SF-36). *Medical Care, 30,* 473–483.

Watson, D., & Tellegen, A. (1985). Toward a consensual structure of mood. *Psychological Bulletin, 98,* 219–235.

Winter, L., Lawton, M. P., & Ruckdeschel, K. (1997, November). *End-of-life decision-making: A prospect theory approach.* Paper presented at the annual meeting of the Gerontological Society of America, Cincinnati, OH.

Yellen, S. B., & Cella, D. F. (1995). Someone to live for: Social well-being, parenthood status, and decision-making in oncology. *Journal of Clinical Oncology, 13,* 1255–1264.

Yesavage, J.A ., Brink, T. L., Rose, T. L., Lum, O., Huang, V., Adey, M. B., & Leirer, V. O. (1983). Development and validation of a geriatric depressing rating scale. A preliminary report. *Journal of Psychiatric Research, 17,* 37–49.

9

The Activity Restriction Model of Depressed Affect

Antecedents and Consequences of Restricted Normal Activities

GAIL M. WILLIAMSON and DAVID R. SHAFFER

Gerontological researchers have long been interested in the extent to which aspects of illness and disability affect the well-being of older adults. One such aspect is whether people are able to maintain satisfactory levels of normal activities in the face of debilitating illness conditions. Activity theory (Lemon, Bengtson, & Peterson, 1972; Longino & Kart, 1982; Reitzes, Mutran, & Verrill, 1995) fostered much of the research in this area. However, our approach is somewhat different from that of other researchers, most of whom have relied on traditional measures of health status and disability and also have focused on identifying main effect predictors of distress. For several years, we have been investigating the role of perceptions of restriction of normal, routine activities in adjustment to various illness- and disability-related factors. We are social psychologists by training and, as such, our orientation is toward identifying not only direct (main effect) predictors of distress but also indirect (i.e., qualifying and buffering) effects of predictor variables. As a result, the Activity Restriction Model of Depressed Affect has evolved over time in response to consistent findings from a systematic program of research on restriction of normal activities in a variety of medically compromised populations.

WHAT IS THE MODEL?

The Activity Restriction Model of Depressed Affect proposes that the extent to which one's normal activities are restricted by a major life stressor plays a central role in

GAIL M. WILLIAMSON and DAVID R. SHAFFER • Department of Psychology, University of Georgia, Athens, Georgia 30602.

Physical Illness and Depression in Older Adults: A Handbook of Theory, Research, and Practice, edited by Gail M. Williamson, David R. Shaffer, and Patricia A. Parmelee. Kluwer Academic/Plenum Publishers, New York, 2000.

psychological adjustment, with major disruptions in normal activities resulting in poorer mental health outcomes (e.g., Williamson, 1996, 1998; Williamson, Shaffer, & Schulz, 1998). Consistent with a portion of an integrative model of depression proposed by Lewinsohn, Hoberman, Teri, and Hautzinger (1985; Zeiss, Lewinsohn, Rohde, & Seeley, 1996), we conceptualize illness and disability as life stressors, depression as a possible reaction to these stressors, and activity restriction as a *mediator* of associations between health-related stressors and emotional well-being. As discussed in a previous chapter (Williamson, Chapter 4, this volume) and later in this chapter, data from a number of studies support these predictions. Like earlier research (e.g., Parmelee, Katz, & Lawton, 1991), our research indicates that health-related stressors (e.g., illness severity, pain) are positively associated with depressed affect. Moreover, activity restriction consistently has been shown to mediate associations between a variety of illness-related variables and depressed affect (Walters & Williamson, 1999; Williamson, in press; Williamson & Schulz, 1992a, 1995a; Williamson, Schulz, Bridges, & Behan, 1994; Williamson et al., 1998). Stated explicitly, health-related stressors appear to affect depressive symptoms largely (and sometimes, only) to the extent that they restrict ability to conduct routine activities.

We have focused on the role of restricted routine activities for methodological, conceptual, and practical reasons. Methodologically, decreased activity level often is studied, not as a distinct construct in its own right, but rather, as one indicator of illness severity. For example, instruments often include decreased activity as one assessment of health status (e.g., Herzog, Franks, Markus, & Holmberg, 1998). The same has been true for measures of depression; an indicator of depression is a decline in previous activity levels. In other words, the extent to which a person foregoes usual activities has been operationalized as an indicator of health status, an indicator of depression, or both.

Yet, conceptually, the constructs of illness severity, activity restriction, and depression, although interrelated, are clearly distinguishable from each other (e.g., Bernard et al., 1997; Kaplan, Barell, & Lusky, 1988; Liang, 1986; Zeiss, Lewinsohn, & Rohde, 1996). To illustrate this point, consider that the levels of association typically found among these variables are, at best, in the modest to moderate range; that is, all individuals who are seriously ill do not become depressed nor do they become functionally disabled, even in the presence of severe pain. Similarly, all individuals who are functionally disabled by illness do not become depressed. In fact, widespread variability in adaptation to serious illness and disability is so common a finding that, some time ago, researchers began looking for other variables that predict which people are likely to adapt well and which are not. Activity restriction is one such variable, and we believe that studying its predictors and outcomes can add substantially to understanding the link between illness and depression.

From a practical standpoint, activity restriction and its antecedents represent additional points of intervention. Specifically, in addition to treating illness symptoms and/or treating depression, interventions can be designed to increase participation in routine activities—even in the presence of illness symptoms, depression, or both. At the end of this chapter, we return to some practical implications of the activity restriction studies. First, however, in the next section, we summarize these studies and their results, paying particular attention to the trajectory of events that led to the development of the Activity Restriction Model.

THE ACTIVITY RESTRICTION STUDIES

Background

Williamson (Chapter 4, this volume) describes the onset of this program of research which began with an interest, prompted by Parmelee et al. (1991), in gathering more information about the association between indicators of illness severity (e.g., pain), functional disability, and symptoms of depression. Contrary to Parmelee et al.'s results for institutionalized older adults, we predicted, and found, that activity restriction mediated the relation between pain and depressed affect in community-residing older adults (see Williamson, Chapter 4, this volume; Williamson & Schulz, 1992a). In addition to finding no mediating effect for functional disability, Parmelee et al. reported that severity of physical illness (as assessed by physician summaries of patient health status) did not mediate the association between pain and depression in their sample. This lack of results is consistent with evidence that physical illness may not, in and of itself, have detrimental effects on either normal activities or emotional states. In fact, illness may not compromise quality of life if it is not accompanied by disability (Bernard et al., 1997; George & Landerman, 1984; Zeiss et al., 1996). Like Parmelee et al., we found no mediating effect of illness in the relation between pain and depression among older adults residing in the community (Williamson & Schulz, 1992a). However, we suspected that illness severity might occupy a role analogous to that of chronic pain in predisposing people to experience both functional disability (activity restriction) and depressed affect. If our reasoning was correct, then activity restriction should mediate the impact of illness severity (as well as pain) on symptoms of depression such that both illness and pain predict depression primarily to the extent that they interfere with normal activities. This proposition was tested in the first of our activity restriction studies.

The First Study

The sample consisted of 228 geriatric outpatients with a wide variety of illness conditions (Williamson & Schulz, 1992a). A major advantage was that, like Parmelee et al. (1991), rather than relying on self-reports, physical illness severity was assessed objectively; that is, physicians' summaries were derived from comprehensive data on patient histories, physical examinations, and laboratory tests. Patients provided self-reports of pain, symptoms of depression (Center for Epidemiological Studies—Depression Scale [CES-D]; Radloff, 1977), and restriction of normal activities (Activity Restriction Scale [ARS]; see Williamson, Chapter 4, this volume; Williamson & Schulz, 1992a)

As shown in Figure 9.1, results of path analyses confirmed our hypotheses. There was no direct effect of illness on depression (and only a small direct effect of pain). Instead, both pain and illness directly affected activity restriction, and activity restriction directly affected CES-D scores. In additional analyses, activity restriction totally mediated the association between illness severity and depressed affect and partially mediated the impact of pain on depressed affect (for the criteria for partial and total mediation, see Darlington, 1990). As we expected, in community-residing older adults, effects of illness severity on depression depended on the extent to which patients were no longer able to conduct routine activities.

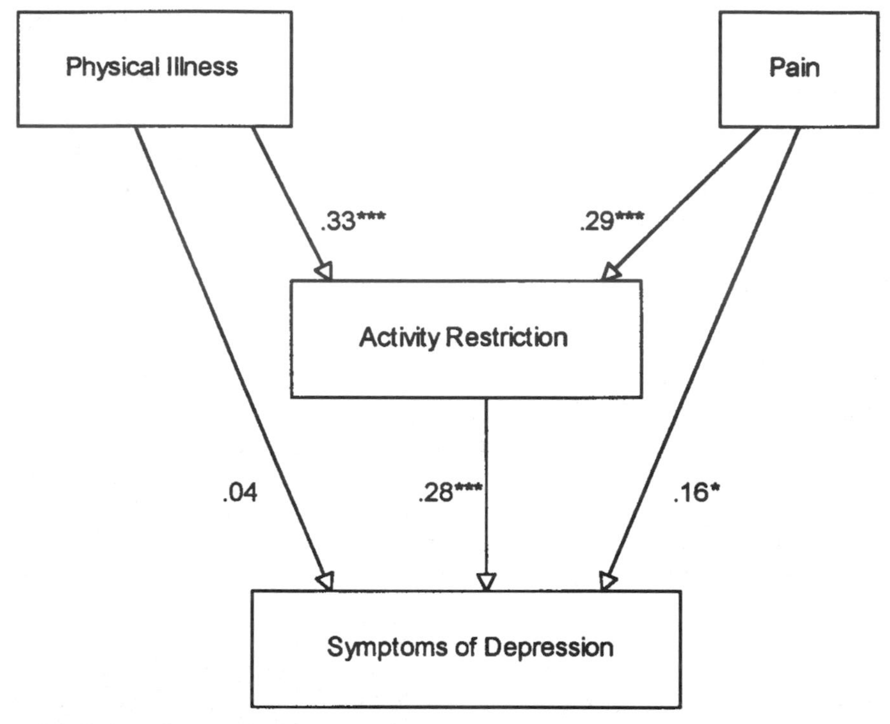

Figure 9.1. Path analysis results for geriatric outpatients: pain, physical illness, activity restriction, and depressed affect. (Derived from data reported in Williamson & Schulz, 1992a).

This study provided evidence that the relation between illness severity (in addition to pain) and depression in community-residing older adults is attributable to functional impairment. Illness severity and pain are sources of activity restriction which, in turn, contributes to depressed affect. We suspected that other indicators of illness severity would be important in this context, a suspicion fostered by findings showing that the presence of multiple chronic illness conditions can influence the extent to which activities are perceived as restricted by a particular illness such as recurrent cancer (see Williamson, Chapter 4, this volume; Williamson & Schulz, 1995a). In subsequent research, we have conceptualized pain as one of many indicators of illness severity and focused on identifying other illness-related variables that contribute to increased activity restriction. Most often, participants in each of these studies were suffering from a particular category of illness or disease. Thus, we have assessed both general indicators of illness severity and those specific to the condition that characterizes the sample. These studies are summarized in the following sections.

Subsequent Studies

Limb Amputation. Our second study looked at adjustment to loss of a limb (Williamson et al., 1994). Not surprisingly, ample evidence indicates that amputation poses serious threats to psychological well-being (e.g., Frank et al., 1984). Moreover, it is primarily the domain of older adults, with people over age 65 accounting for approximately 75% of all amputations, 85–90% of which involve removal of a lower extremity (e.g., Bradway, Malone, Racy, Leal, & Poole, 1984; Goldberg, 1984). Loss of a lower extremity places an individual at high risk for functional disability (LaPlante, 1989; also see Lawrence & Jette, 1996; Stump, Clark, Johnson, & Wolinsky, 1997). Major lower-extremity amputations usually (i.e., in 75-80% of cases) result from peripheral vascular disease (Clark, Blue, & Bearer, 1983; Moore & Malone, 1989; Pinzur, Graham, & Osterman, 1988; Stewart, Jain, & Ogston, 1992), often as a complication of diabetes (Keyser, 1993; Pinzur et al., 1988; Reiber, Pecoraro, & Koepsell, 1992). Thus, lower-extremity amputation is a source of activity restriction among older adults. Because preservation of the knee joint is critical in learning to use a prosthesis (Goldberg, 1984; Thornhill, Jones, Brodzka, & VanBockstaele, 1986), and because many older adults who undergo amputation are never able to walk with a prosthesis (Goldberg, 1984; Pell, Donnan, Fowkes, & Ruckley, 1993; Pinzur et al., 1988), activity restriction is especially likely when a leg is removed above the knee.

Based on results of the first study, we expected that activity restriction would play a central role in adjustment to amputation among older adults. In face-to-face interviews, we used the same measures of activity restriction (the ARS) and depressed affect (the CES-D) as in the first study. We also measured phantom limb pain and pain of any nature in the last week. Illness severity was operationalized as factors specific to amputation (type of amputation, cause of amputation, time since amputation, and total number of amputations) and prosthesis use (number of hours/day worn and satisfaction with prosthesis).

Pain dropped out of the equation in this sample, most likely due to a ceiling effect; that is, over 94% of the 160 participants were in some pain, usually in the moderate range, and almost 80% reported experiencing phantom limb pain. In path analysis, one amputation-related variable (removal of a leg above the knee) and one prosthesis-related variable (wearing a prosthesis fewer hours per day) emerged as direct predictors of increased activity restriction. However, these two variables were not directly related to symptoms of depression. Instead, above-knee amputation was directly related to activity restriction, which in turn predicted depressed affect, and additional analyses indicated that activity restriction totally mediated the effects of prosthesis use on depression. In other words, patients were depressed by being less frequently able (or totally unable) to use a prosthesis to the extent that this limitation interfered with their ability to conduct routine activities.

Thus, our second study provided evidence that activity restriction mediates the association between illness-related variables other than pain and physician ratings of health status, and symptoms of depression. Results of additional studies extend this finding to include two other populations: (1) breast cancer patients and (2) caregivers of spouses with recurrent cancer.

Breast Cancer. The 95 women in this study had been diagnosed with either Stage 1, Stage 2, or Stage 3 breast cancer (for details, see Williamson, in press). Severity of illness was operationalized multidimensionally with three indicators specific to breast cancer (time since diagnosis, tumor stage, and recurrence/metastasis) and two indicators not specific to breast cancer (number of other chronic conditions and general bodily pain). As would be expected, pain predicted depression, and activity restriction totally mediated this association (see Williamson, Chapter 4, this volume; Williamson, in press). Tumor stage, recurrence, and other chronic health conditions were not directly related to either activity restriction or depression. Time since diagnosis was not related to depressed affect, but less time elapsed since diagnosis predicted more activity restriction, which in turn was a strong predictor of more depressed affect.

Taken together, results of these studies indicate that the intervening effect of activity restriction applies to an, albeit selective, variety of operational definitions of severity of illness. However, the mediating effect of activity restriction may depend on the population under consideration and the ways in which illness severity is operationalized. Data reviewed in a previous chapter (Williamson, Chapter 4, this volume) suggest that activity restriction may most consistently mediate the impact of illness severity when it is operationalized as degree of ongoing pain. On the other hand, our amputation study (Williamson et al., 1994) indicates that this may not be the case when there is little variability in the sample (i.e., when the vast majority of participants are subject to experiencing a great deal of chronic pain). Still, we have evidence that when illness severity is operationalized in various other ways (e.g., physician summaries and self-reports specific to a particular condition), intervening effects of activity restriction are found.

For theoretical reasons, we have also asked whether the activity restriction effect applies only to patients. Recall that the Activity Restriction Model proposes that the extent to which normal activities are restricted by a major life stressor plays a central role in psychological adjustment. Major life stressors encompass many events in addition to becoming seriously ill or disabled oneself. Because of our long-standing interest in family caregivers, and because a large literature clearly indicates that the illness and disability of one family member can be highly stressful for other family members (see Bookwala, Yee, & Schulz, Chapter 6, this volume, for an excellent summary), our next step was to extend our previous findings to include caregivers. We hypothesized that greater severity of patient illness would cause caregivers' activities to be more restricted (e.g., by increasing the time and effort involved in providing care) and that more caregiver activity restriction would then lead to higher levels of caregiver depressed affect. It was also expected that caregiver activity restriction would mediate the association between patient illness severity and caregiver symptoms of depression. In this study, we included caregiver resentment as another indicator of psychological adjustment and predicted a similar pattern of results.

Spousal Caregivers of Recurrent Cancer Patients. The sample in this study (Williamson et al., 1998) consisted of 75 cancer patients (51 men and 24 women, *M* age = 63.4 years) and their spousal caregivers (*M* age = 62.2 years). Severity of illness was measured by asking patients whether they had experienced any of 15 symptoms (e.g., fatigue, pain, loss of appetite) in the past month as a result of cancer or treatment. Caregivers indicated the extent to which their own activities were restricted by caregiving responsibilities. Caregiver depression was assessed with the CES-D, and caregiver resentment was operationalized as perceptions of excessive care recipient dependence (see Williamson et al., 1998, for a detailed description). Depressed affect and resentment were not correlated, indicating that they may be independent reactions to the caregiving experience.

As we expected, results for caregivers were much the same as the pattern observed in previous studies of patients; that is, caregiver activity restriction totally mediated the association between patient-reported illness severity and caregiver symptoms of depression. The same effect was found for caregiver resentment. Thus, the Activity Restriction Model extends to reactions to life stressors beyond one's own health problems to include psychological distress (both depression and resentment) associated with dealing with the health problems of a spouse.

Other Contributors to Activity Restriction

We found the results of our initial studies to be both encouraging and interesting, but it also should be noted that in our first study (Williamson & Schulz, 1992a), illness severity and pain explained only about 25% of the variance in activity restriction, indicating that factors other than those related to physical health status are important. Therefore, our goals in subsequent research have not merely been to replicate our original finding with other indicators of illness severity. Rather, we also have focused on identifying other variables that contribute to more restriction of usual activities. As reported in Williamson (Chapter 4, this volume), age and the interaction between age and multiple illness conditions appear to be major players in this context. In addition, the large stress and coping literature indicates that psychosocial factors are important in adaptation to stressful life events (e.g., Lewinsohn, Rohde, Seeley, & Fischer, 1991). We have drawn on this literature in our efforts to identify other contributors to activity restriction, and the findings to date are summarized in the following sections.

Financial Resources. Inadequate income may deter ability to conduct normal activities in the presence of illness and disability (Merluzzi & Martinez Sanchez, 1997). However, we have rarely found financial resources to be related to activity restriction, and the results we have obtained appear to depend on sampling and measurement techniques; that is, when the measure of financial resources is based solely on dollars of annual household income, we have not found that less income contributes to more activity restriction. For example, this was the case in our sample of breast cancer patients (Williamson, in press), but it should also be noted that median income in this study was in excess of $50,000/year with little variability (*SD* < $10,000). Thus, the lack of results could be explained by the measure of financial status we used, the restricted (high) income range in this sample, or both.

Indeed, some evidence supports this notion. Specifically, in our research on limb amputation (Williamson et al., 1994), participants were considerably less affluent (*Mdn* = $15,000–20,000/year), with more variability (*SD* > $10,000) than the sample of breast cancer patients. In addition, rather than relying only on actual income in dollars/year, in the amputation sample, we asked how adequate income was to meet their needs. The average participant reported household income as less than adequate, and financial resources perceived as less than adequate (but not absolute income) emerged as a predictor of more restricted activities. We suspected this was the case, at least in part, because adequate financial resources can make life considerably less stressful for persons with amputations by providing special equipment, home modifications, and assistance with activities that may be difficult to perform alone. Breast cancer patients, on the other hand, are unlikely to have these kinds of ongoing needs. In fact, the vast majority of women in that study reported receiving the types of treatment and services they needed and preferred, most of which were covered by insurance.

Personality. Individuals differ in many respects other than demographic characteristics such as age and household income. We have also begun investigating stable dispositions for their association with activity restriction. Two studies have focused on a particular personality trait, public self-consciousness, because of its potential importance in adaptation to disfiguring illness conditions. Public self-consciousness is a dispositional "tendency to think about those self-aspects that are matters of public display, qualities of the self from which impressions are formed in other people's eyes" (Scheier & Carver, 1985, p. 687). Compared to people low in public self-consciousness, those high in public self-consciousness are more concerned about their physical appearance, more motivated to present themselves to others favorably, more desirous of avoiding disapproval and rejection, and more likely to be disturbed if their appearance does not conform to a model endorsed by society (e.g., Buss, 1980; Fenigstein, 1979; Fenigstein, Scheier, & Buss, 1975; Ryckman et al., 1991). We expected public self-consciousness to be an important predictor of activity restriction in both limb amputation and breast cancer patients, and we have obtained results supporting this expectation.

More than patients low in public self-consciousness, amputation among those high in public self-consciousness can result in fear of rejection because they feel others may perceive them negatively. Consistent with this idea, Rybarczyk et al. (1992) reported that after controlling for age, gender, social support, and a variety of health- and amputation-related variables, social discomfort remained a strong predictor of depression. In the wake of losing a limb, being high in public self-consciousness should lead to more social discomfort due to sensitivity about how others will react. This reasoning led to the prediction that, in order to avoid social discomfort, high public self-consciousness would predict higher levels of activity restriction—especially activities conducted in public situations (e.g., shopping, visiting friends).

These hypotheses were tested in a sample of 89 men and 41 women (M age = 69.2 years) who had undergone limb amputation (Williamson, 1995). They completed the ARS and the 7-item Public Self-Consciousness Scale (PSC; Scheier & Carver, 1985). Participants also indicated the extent to which their amputation caused them to feel (1) less comfortable being out in public and (2) more vulnerable and less able to defend themselves. They were categorized according to whether they had experienced above-knee amputation, and path analyses revealed that both above-knee amputation and high public self-consciousness were independently related to more restriction of normal activities. Both variables also predicted feeling more vulnerable and less able to defend oneself which, in turn, predicted increased activity restriction. In addition, as shown in Figure 9.2, public self-consciousness interacted with above-knee amputation such that among those low in self-consciousness, activities were more restricted in the presence of above-knee amputation. In contrast, those high in public self-consciousness reported high levels of activity restriction regardless of type of amputation.

The importance of public self-consciousness in the face of disfiguring health conditions was confirmed by data from breast cancer patients (Williamson, in press). The literature indicates that self-consciousness and related concerns about body image are associated with a variety of reactions to a diagnosis of breast cancer, most notably decisions about and adjustment to having a mastectomy or a lumpectomy (e.g., Ashcroft, Leinster, & Slade, 1985; Blichert-Toft, 1992; Margolis, Goodman, & Rubin, 1990; Moyer, 1997; Schain, 1991; Sneeuw et al., 1992; Wolberg, Tanner, Romsaas, Trump, & Malec, 1987). In our own research, of a host of demographic, personality, and medical factors,

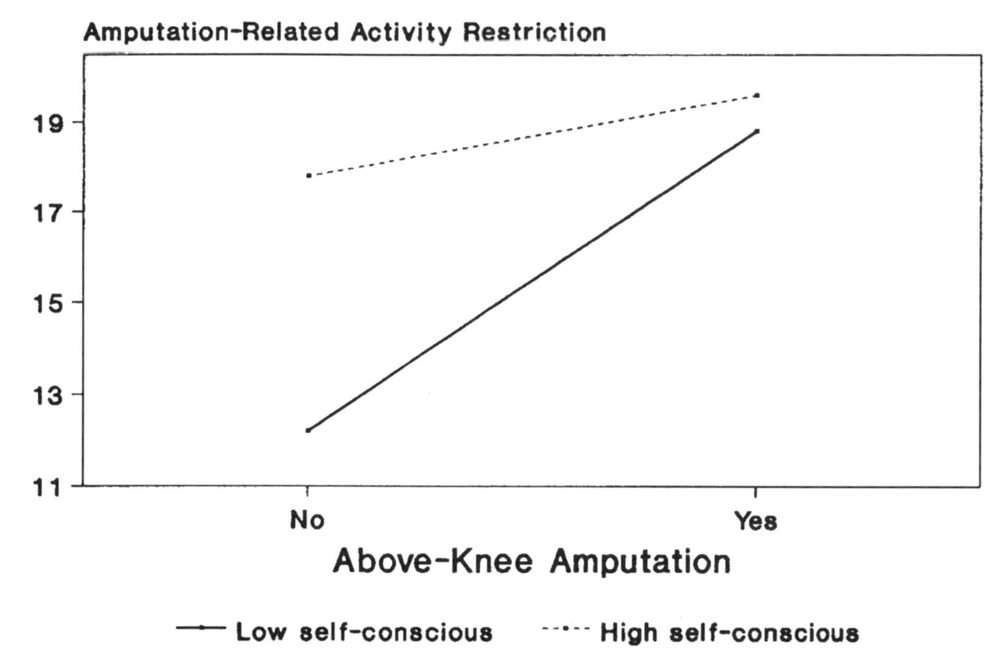

Figure 9.2. Limb amputation: Activity restriction as a function of above-knee amputation and public self-consciousness. (Source: Williamson, 1995).

only public self-consciousness emerged as a significant predictor of reconstructive surgery following mastectomy (Williamson, Jones, & Ingram, 1999). Among breast cancer patients, the concerns about physical appearance and desire to avoid disapproval that accompany high public self-consciousness may lead to restricted social and sports activities because they fear that clothing appropriate to the activity will reveal their surgical disfigurement to others. And, indeed, as reported in Williamson (in press), activities were more restricted among women high in public self-consciousness, an association not qualified by indicators of illness severity (whether general or specific to breast cancer).

Because of its hypothesized relation to disfiguring conditions, these two studies focused on one particular personality variable, public self-consciousness. This is not to say that other personality traits do not matter. Indeed, our own research provides some clues about the potential importance of other personality characteristics in adjustment to major life stressors. For example, caregivers of Alzheimer's disease (AD) patients who are high in communal orientation (a characteristic belief that people should help each other; e.g., Clark, Ouellette, Powell, & Milberg, 1987; Williamson & Clark, 1989) are less distressed by the costs associated with providing care than are those low in communal orientation (Williamson & Schulz, 1990). We suspect that a viable explanation for these results is that, because they believe that helping each other is appropriate behavior, people high in communal orientation have better social support systems than do those low in communal orientation. The relation between social support and psychological adjustment to stressful life events is well documented (e.g., Billings & Moos, 1984; Cohen & Wills, 1985; Irvine, Brown, Crooks, Roberts, & Browne, 1991; Petrisek, Laliberte, Allen, & Mor, 1997; Stanton & Snider, 1993), and in the next section, we turn to evidence indicating that less social support contributes to activity restriction.

Social Support. The large social support literature can be summarized by saying that social support is a multifaceted construct, and different operationalizations produce different results (e.g., Cohen & Wills, 1985; Li, Seltzer, & Greenberg, 1997; Oxman & Hull, 1997; Sarason, Sarason, & Pierce, 1990). In our research, we have chosen to bypass structural measures of social networks because of their generally poor utility for predicting reactions to stressful life events (e.g., Kaplan & Toshima, 1990) and, instead, to focus on the functional and psychological aspects of social support. It seems that it is *perceptions* of available help that buffer the impact of stress that has already occurred (e.g., Cohen & Wills, 1985; Kaplan, Sallis, & Patterson, 1993; Oxman & Hull, 1997; Wethington & Kessler, 1986), as is the case in our populations of interest. We have operationalized perceptions of social support in two ways: (1) perceived help available and (2) satisfaction with social contacts. In both cases, our hypothesis was that individuals who perceive that they are lacking in support resources experience more activity restriction as a result of their medical conditions and treatment.

In the sample of breast cancer patients (Williamson, in press), perceptions of available social support were assessed with a 6-item version of the Interpersonal Support Evaluation List (ISEL; Cohen, Mermelstein, Kamarck, & Hoberman, 1985; Williamson & Schulz, 1992b). Participants rated such statements as "When I need suggestions on how to deal with a personal problem, I know someone I can turn to" and "There is at least one person I know whose advice I really trust." In multivariate analyses, perceived social support explained significant additional variance in activity restriction after controlling for the effects of demographics, public self-consciousness, and illness severity. Activities were more restricted among breast cancer patients who perceived they had less social support.

In our first limb amputation study, we included a measure of satisfaction with social contacts (see Williamson et al., 1994, for details). Less satisfaction with social contacts emerged as a direct predictor of activity restriction. This was the case after controlling for demographics (age, income adequacy), amputation- and prosthesis-related factors, and pain. Those who were less satisfied with their social contacts reported more restricted activities.

Results of these two studies are consistent with evidence that social support is a critical component of coping successfully with stressful life events (e.g., Russell & Cutrona, 1991; for reviews, also see Cohen & Wills, 1985; House, Landis, & Umberson, 1988). These findings also confirmed our expectation that social support resources facilitate ability to conduct routine activities in the presence of serious illness and disability (for similar findings, see Mutran, Reitzes, Mossey, & Fernandez, 1995; Oxman & Hull, 1997).

Given our results, and those of others, indicating how important social support is in adjusting to major stressful life events, we have begun giving serious attention to the antecedents of adequate levels of social support. One approach is to consider that social support is most likely to be given in the context of close interpersonal relationships, and in the next section, we turn to evidence supporting the critical nature of the quality of such relationships. Our work in this area relies heavily on the Theory of Communal Relationships, formulated by Clark and her colleagues (e.g., Clark & Mills, 1979, 1993; Mills & Clark, 1982).

Interpersonal Relationship Quality. Thus far, our research in this domain has focused on how good relationships between caregivers and care recipients were before onset of the illness that led to the care recipient's need for assistance. It has long been assumed that affection and closeness influence willingness to become a caregiver (e.g., Cantor, 1983; Given, Collins, & Given, 1988; Jarrett, 1985). However, regardless of the quality of relationships between family members, cultural norms dictate that family members are *expected* to help each other in times of need (e.g., Aldous, 1977; Williamson & Clark, 1989; Williamson & Schulz, 1990). These societal expectations lead many people–including those who have not had a good relationship with the care recipient in the past–to assume the burden of caregiving. Still, distress should be higher among these caregivers than among those who have enjoyed a good relationship with their care recipient. And, indeed, this is what we found in our study of 174 family caregivers of AD patients (Williamson & Schulz, 1990). Those who reported having a close relationship with the patient before illness onset were less burdened by caregiving responsibilities than were those whose relationship had not been close. Among the items constituting the measure of "burden" in this study were several that assessed aspects of caregiver activity restriction (e.g., effects of caregiving on one's social life).

The Theory of Communal Relationships (e.g., Clark & Mills, 1979, 1993; Mills & Clark, 1982) has much to say about which individuals should adapt well to providing care (e.g., feel less burdened) and which should not. Consequently, before describing the results of our subsequent research, we highlight major components of the theory and research supporting predictions derived from it.

Communal relationships are characterized by behaviors on the part of both partners that are responsive to (or indicative of a desire to respond to) their partner's needs. Such relationships are most likely to be found among close friends, romantic partners, and family members (e.g., Clark & Mills, 1979, 1993; Mills & Clark, 1982). However, close relationships between friends, spouses, and family members can vary in exactly how communal they are (e.g., Clark & Mills, 1993). In highly communal relationships, partners routinely are concerned about and attend to each other's needs as these needs arise. Less communal relationships are characterized by low levels of feelings of responsibility for the other's welfare and less responsiveness to one another's needs. The Theory of Communal Relationships has fostered interesting predictions about how people allocate benefits and how they react to helping situations, and empirical studies support these predictions (e.g., Clark & Mills, 1979; Clark, Mills, & Corcoran, 1989; Clark, Mills, & Powell, 1986). Several findings are especially relevant to reactions to caring for an ill or disabled family member. Specifically, communal partners do not feel exploited when their partner cannot reciprocate their aid (Clark & Waddell, 1985). Moreover, in the context of communal relationships, elevated affect follows having helped a partner (Williamson & Clark, 1989, 1992), with analogous declines after failing to help (Williamson, Clark, Pegalis, & Behan, 1996).

These data imply that the psychological impact of providing care may vary dramatically depending on how mutually communal the preillness relationship was between caregiver and care recipient. In historically communal relationships, responding to care recipient needs is a continuation of "in-role" behavior rather than a distasteful responsibility to be endured. These caregivers most likely display genuine concern for the recipient's welfare but still miss the intimacy and mutual concern that may no longer be apparent. Consistent with this notion, Williamson and Shaffer (1998) reported that depressed affect among caregivers in highly communal relationships was directly related to deterioration in the couples' interpersonal behavior and interactions.

In contrast, caregivers whose relationship with the care recipient has been historically characterized by few mutually communal behaviors may provide care more out of obligation than concern for promoting the recipient's welfare (Williamson & Schulz, 1995b). Providing care, then, may be perceived as burdensome and induce resentment because it qualifies as a highly communal but "out-of-role" activity for people unaccustomed to placing their partners' needs ahead of their own (or having their partners behave in kind toward them). Indeed, Williamson and Shaffer (1998) found that depressed affect among caregivers in less communal relationships was more related to perceived burdens than to interpersonal loss.

How does all this relate to activity restriction? Recall that in our study of spousal caregivers of patients with recurrent cancer (Williamson et al., 1998), caregiver activity restriction mediated associations between severity of patient illness and caregiver depression and resentment. This effect was obtained with data from the total sample. However, we also predicted that a different pattern of results would emerge when the sample was divided according to quality of preillness relationship. Restricted activities of caregivers in highly communal relationships should be predicted by intimacy and affectional loss, while activity restriction among those in less communal relationships should be predicted by the amount of care they must provide and should focus on restriction of more idiosyncratic activities.

Relationship history was assessed with the 10-item Mutual Communal Behaviors Scale (MCBS; Williamson & Schulz, 1995b), a measure of frequency of expressions of communal feelings between caregiver and care recipient prior to illness onset. Caregivers were instructed to think about "the types of interactions you had with the patient *before* his/her illness." Five items evaluated caregiver communal behaviors toward the patient, and five items evaluated patient communal behaviors toward the caregiver (for details of this measure, see Williamson & Schulz, 1995b). To evaluate hypotheses specific to high and low communal relationships, the sample was divided at the MCBS median. Caregivers also responded to items measuring perceived loss of affection and physical intimacy resulting from care recipient illness.

Results were quite consistent with predictions derived from communal relationships theory. Among caregivers in high MCBS relationships, intimacy/affectional loss (but not severity of patient symptoms) was a strong predictor of caregiver activity restriction. In the low MCBS group, restriction of routine activities was predicted by severity of patient symptoms and not by loss of intimacy and affection.

Summary

Having shown that activity restriction mediates the association between various indicators of illness severity and depression, we turned our attention to identifying other variables that contribute to more restricted routine activities. As reviewed in the preceding sections, we have found that, in addition to indicators of illness severity, psychosocial factors are important. Among these are age, income adequacy, social support, personality (e.g., public self-consciousness), and aspects of interpersonal relationships (e.g., quality of past relationship, loss of a close relationship). Our interest in this endeavor was fostered by the observation in our first study (Williamson &

Schulz, 1992a) that pain and illness severity together explained only 25% of the variance in activity restriction. We now know that this finding is reasonably consistent. Depending largely on how comprehensively it is measured, illness severity explains as little as 7% of the variance in activity restriction (Williamson, 1995) to as much as 32% (Williamson, in press), but the average across studies is about 20%.

Adding psychosocial variables to the equation has in every case allowed us to account for significant portions of the variance in activity restriction beyond the effects of illness severity—ranging from 8% (Williamson, 1999) to 22% (Williamson, 1995), averaging 14% across studies. Combining illness severity and psychosocial variables, we have been able to explain from 24% of the variance in activity restriction among caregivers of cancer patients (Williamson et al., 1998) to 35% among breast cancer patients (Williamson, in press).

Thus, we feel comfortable saying that psychosocial factors contribute to activity restriction over and above illness severity. Overall, this program of research seems to indicate nothing so strongly as that adjustment to a major life stressor is a complex and multifaceted endeavor, a process influenced by many factors. However, our research makes the case that activity restriction is one such factor, perhaps a critical one. Although we have consistently shown that restriction of normal activities is a highly proximal predictor of symptoms of depression and, in additional research, identified psychosocial factors that contribute to activity restriction beyond the effects of illness severity, it remained possible that activity restriction is merely a surrogate for various other factors known to contribute to depression in the wake of major life stress. Demonstrating that this was not true was viewed as a critical step in our program of research, and in the next section, we turn to evidence supporting the importance of the construct of activity restriction.

HOW IMPORTANT IS ACTIVITY RESTRICTION IN PREDICTING EMOTIONAL DISTRESS?

There were several ways to confront this issue. One was to show that, beyond the effects of known correlates of depression, activity restriction explains significant variance in depressed affect. Another was to show that activity restriction intervenes in associations between depression and variables not related to illness severity. And, finally, we thought it was important to provide evidence that activity restriction can predict affective reactions other than depression.

Does Activity Restriction Explain Significant Variance After Controlling for Other Variables?

Above and beyond effects of the variety of variables included in each of our studies, activity restriction had explained a significant portion of the variance in depressed affect, ranging from 6% ($p < .001$) among geriatric outpatients and amputees (Williamson & Schulz, 1992a; Williamson et al., 1994) to 13% ($p < .001$) among cancer patient caregivers (Williamson et al., 1998). Depending largely on how many other variables were included, we had been able to account for between 18% (Williamson & Schulz, 1992a) and 40% (Williamson et al., 1994) of the total variance in depression (all $p <$

.001) when activity restriction was added to the equation. The study of breast cancer patients (Williamson, in press) was designed to provide a strong test of how important activity restriction is in predicting depression; that is, the goal was to demonstrate that, after controlling for the effects of known predictors of depressed affect, having one's usual activities restricted accounts for significant additional variance in symptoms of depression. And, indeed, convincing evidence came from analyses showing that after controlling for factors (age, public self-consciousness, indicators of illness severity, and perceived social support) that explained exactly half of the variance in depressed affect among breast cancer patients, being less able to conduct routine activities explained a highly (and clinically) significant additional 19% ($p < .0001$).

Does Activity Restriction Qualify the Impact of Non-illness-Related Variables on Depression?

Other evidence for the importance of activity restriction in predicting depression would be obtained if activity restriction could be shown to qualify (i.e., mediate and/or moderate) the effects of non-illness-related variables on symptoms of depression. And, in fact, results of the breast cancer study yielded new information about why public self-consciousness predicts depressed affect (Williamson, in press). As expected, activities were more restricted and depressed affect was higher among women who were more self-conscious. Also, as predicted, activity restriction qualified the effects of self-consciousness on symptoms of depression; that is, restriction of normal activities both moderated and totally mediated the impact of public self-consciousness. As shown in Figure 9.3, symptoms of depression were much higher among highly self-conscious

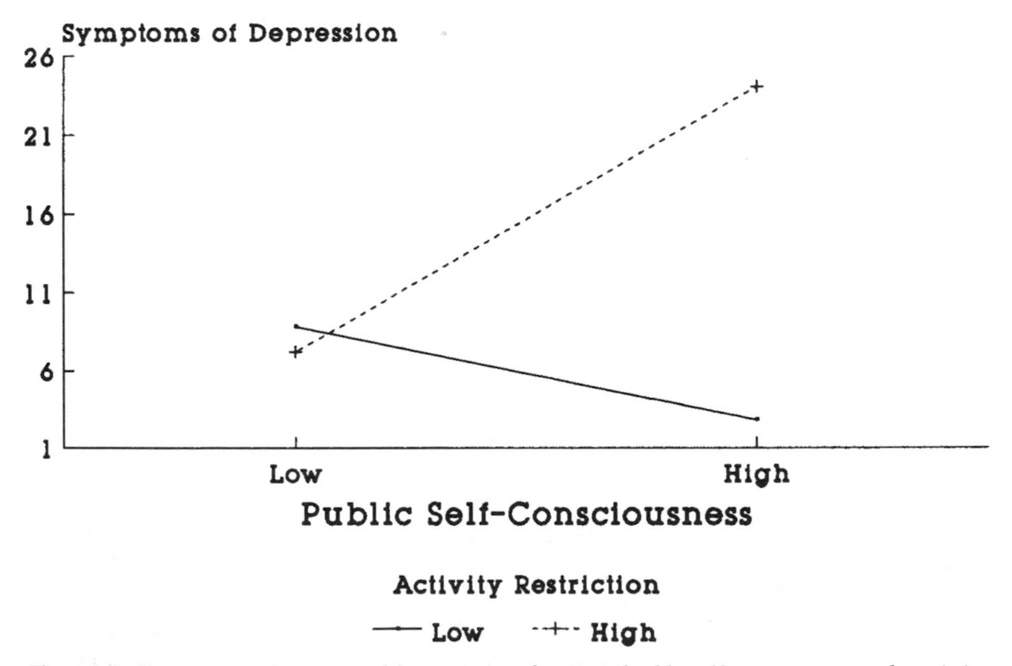

Figure 9.3. Breast cancer: Symptoms of depression as a function of public self-consciousness and restriction of normal activities.

women who were subject to a great deal of activity restriction (moderating effect). Moreover, high public self-consciousness appeared to result in more symptoms of depression, to the extent that being self-conscious led these women to forego normal activities (mediating effect).

Results also emerged for social support in the sample of breast cancer patients. Activity restriction partially mediated the impact of perceived social support on symptoms of depression; that is, more social support resources led to less negative affect, in part, to the extent that social support facilitated ability to conduct routine activities. These findings are consistent with those of our research on limb amputation (Williamson et al., 1994), in which satisfaction with social contacts was related to depressed affect in more than one way: (1) satisfaction with social contacts directly affected both symptoms of depression and activity restriction, (2) activity restriction was directly related to depressed affect, and (3) the effect of satisfactory social contacts on symptoms of depression was partially mediated by activity restriction. In other words, these patients were, in part, depressed because unsatisfactory levels of social support interfered with their ability to conduct normal activities.

Does activity restriction qualify the effects of psychosocial variables other than self-consciousness and social support? The answer is yes. In two studies, activity restriction intervened in the association between financial resources and depression. First, the relation between income adequacy and depressed affect among amputees was totally mediated by activity restriction (Williamson et al., 1994). Second, identical results emerged for highly communal relationships in our study of spousal caregivers of cancer patients (Williamson et al., 1998). Thus, income influences depressive symptomatology to the extent that ability to perform normal activities is restricted by inadequate financial resources.

Does Activity Restriction Predict Other Types of Affect?

Although not a major focus of our research (or this chapter), in making a case for the importance of activity restriction, we note that restriction of normal activities should predict affective reactions other than depression. Anxiety, anger, and resentment are likely candidates because they are often found among people undergoing stressful events (e.g., Cohen & Rodriguez, 1995). In our study of caregivers of spouses with recurrent cancer (Williamson et al., 1998), we found evidence to support this proposition, albeit evidence qualified by relationship quality. Activity restriction among caregivers in highly communal relationships was a strong predictor of depressed affect (but not resentment). However, activity restriction among those in less communal relationships predicted resentment toward the care recipient (but not depressed affect). As we noted earlier, depression and resentment were not correlated, leading to the conclusion that similarity in results for these two types of affect was not produced by shared variance.

Summary

We are satisfied that activity restriction is an important factor in adjustment to stressful life events. It independently accounts for significant portions of the variance in depression; it qualifies the effects of non-illness-related factors; and it predicts at least one kind (and we suspect, several kinds) of affect other than depression (e.g., resentment).

WHERE DO WE GO FROM HERE? IMPLICATIONS FOR FUTURE RESEARCH AND INTERVENTION

Our research indicates that depressed affect is partially a function of restricted normal activities. In terms of intervention strategies, the implication is that reducing activity restriction will result in less depression. Drawing on our results, a productive direction for future research is to identify the *processes* through which activities come to be restricted. Simply encouraging people to be more active is likely to be less efficacious than identifying *why* it is that people become less active and targeting interventions accordingly (Pate et al., 1995; Wolinsky, Stump, & Clark, 1995; but see Calfas et al., 1996; Marcus et al., 1997, for evidence suggesting that advice from primary care physicians *in conjunction with* repeated follow-up sessions can increase physical activity). In particular, we argue that identifying the (doubtless multiple) reasons that activities become restricted, beyond the obvious impact of decrements in health status, will provide clues relevant to numerous points of intervention (see Clark & Bennett, 1992; Cohen & Rodriguez, 1995, for similar arguments). Moreover, these efforts should aid in identifying individuals most at risk for poor adaptation (i.e., those who are most likely to benefit from early and appropriate intervention). In the following sections, we consider the results of our activity restriction studies both for what we have learned and for some clues they provide that warrant further investigation.

Is It Dollars or Is It Perceived Income Adequacy?

Our findings offer interesting insights into why research on the impact of income on adjustment to life stress often produces mixed results. Methodological issues seem to be a major player; that is, obtaining an effect of financial resources on activity restriction (and, by inference, on depression) may depend on how income is measured. First, our results indicate that there should be enough variability in income levels within the sample to allow for identifying differences attributable to absolute income. Second, the actual needs of the particular population should be taken into account. For example, are needs likely to be lifelong (e.g., as a result of limb amputation) or, relatively speaking, short term (e.g., as a result of breast cancer)? Finally, it may be most useful to assess perceptions of income adequacy rather than actual dollar amounts. We have repeatedly observed that individuals with low incomes do not necessarily perceive their financial resources as inadequate, and similarly, those with relatively high incomes do not uniformly report their financial resources as adequate. Thus, it seems that people who perceive their incomes as inadequate, regardless of absolute dollar amount, are likely to perceive their activities as more restricted by illness.

In terms of intervention, it may be most useful to focus on perceptions of income adequacy. Although it may be true that financial losses due to illness and disability prohibit participation in previously enjoyed activities, people can be pointed toward enjoyable but less costly substitute activities (see Zimmer, Hickey, & Searle, 1997, for a discussion of replacement activities). Social-psychological research yields clues about ways to increase contentment by helping people adjust their comparison standards and feel less relatively deprived (e.g., Dermer, Cohen, Jacobsen, & Anderson, 1979; Gibbons & Gerrard, 1989; Taylor, 1989). These findings would seem to be particularly relevant but are seldom considered in this context.

Is It Worthwhile to Study Personality?

By definition, personality traits are stable, and it should, therefore, be difficult to design interventions to alter these characterological dispositions. Does this mean that identifying traits that predispose people to poorer adjustment is a futile endeavor? We think not. Rather, we believe that individual differences in personality can help identify those who are likely to adapt poorly, and based on specific individual needs, target interventions where they are most likely to be effective.

Consider, for example, our two studies investigating the role of public self-consciousness in predicting activity restriction and depression. Among those who had experienced limb amputation, high public self-consciousness directly predicted feeling more vulnerable and less able to defend onself in public situations, and these feelings then predicted increased activity restriction (Williamson, 1995). Moreover, the activities of those who were highly self-conscious were significantly more restricted regardless of type of amputation, whereas the activities of their less self-conscious counterparts depended on the type of amputation they had undergone. These results indicate that among people less concerned about the impression they make on others, activities are restricted by actual functional limitations. Those who are more concerned about how they present themselves to others may relinquish usual activities as much out of concern about their public image as out of functional limitation. Our results also suggest points of intervention in this population. For example, interventions that increase the ability to defend oneself (or the perceptions thereof) could foster continuance of routine activities. We also found evidence in this study that among very self-conscious amputees, not only were activities that take place in the presence of other people (caring for others, shopping, visiting friends, and maintaining friendships) more restricted but also activities associated with self care and household tasks. Although we did not expect to find that activities primarily conducted in private would be restricted by public self-consciousness, we interpreted this finding as indicating that people who are especially concerned about the impression they make on others may more readily adopt a "sick role" (e.g., Waddell, 1987) than those less concerned with these matters. In other words, feeling insecure and uncomfortable about performing some types of activities may generalize such that those high in self-consciousness are more likely to expect to be functionally disabled in multiple ways (both public and private) as a normal part of their illness. Clearly, further investigation of these issues is warranted, including the extent to which "depression and physical disability may feed on themselves, setting off a spiraling decline in physical and psychological health" (p.176) (Bruce, Seeman, Merrill, & Blazer, 1994; also see Cohen & Rodriguez, 1995). In the meantime, these data point to the need for early intervention to prevent onset of the sick role process and "excess disability" (e.g., Femia, Zarit, & Johansson, 1997; Lawrence & Jette, 1996).

Additional support for the importance of public self-consciousness in adjustment to disfiguring health conditions comes from our work with breast cancer patients

(Williamson, in press) in which higher self-consciousness was directly related to greater activity restriction. Moreover, symptoms of depression were much higher among highly self-conscious women subject to a great deal of activity restriction, and high public self-consciousness appeared to result in more depression to the extent that being self-conscious led these women to forego normal activities. We also note as further evidence for the importance of this personality variable that among a wide variety of demographic, personality, and medical factors, only high public self-consciousness emerged as a predictor of greater likelihood of having reconstructive surgery following mastectomy (Williamson et al., 1999). Taken together, these results indicate that women who are dispositionally concerned about their physical image are likely to restrict their activities as a result of their cancer treatment and will undergo elective surgical treatment in order to regain a favorable body image. At this point, it remains unknown how satisfied highly self-conscious women are with the results of these efforts, and this is an important direction for future research. In terms of intervention, programs are widely available to help women maintain their physical appearance during treatment for breast cancer, and it may be that women who stand to benefit most from these programs are those dispositionally high in public self-consciousness.

Because of its hypothesized association with adjustment to disfiguring conditions, our activity restriction research has, to date, tended to focus on the personality variable of public self-consciousness. Other personality traits bear investigation as well. Among these are dispositional characteristics linked to coping and social support (see Hooker, Monahan, Bowman, Frazier, & Shifren, 1998, for an excellent discussion of these issues) such as optimism (e.g., Carver et al., 1993), neuroticism (e.g., McCrae & Costa, 1986), agency/mastery (e.g., Femia et al., 1997; Herzog et al., 1998), and communal orientation (e.g., Clark et al., 1987; Williamson & Clark, 1989; Williamson & Schulz, 1990). These are potentially fruitful, as yet untapped, directions for future research. For example, high optimism and its predisposition to adaptive coping should be related to less activity restriction following the onset of debilitating illness, while high neuroticism and its concomitant maladaptive responses to stress should foster more activity restriction. We further expect that individuals who are more agentically oriented and those with a strong sense of mastery find ways to continue some of their rewarding activities even when faced with severe illness and disability (see also Femia et al., 1997). In addition, people high in communal orientation most likely have developed better interpersonal relationships and social support systems than have those who do not dispositionally subscribe to the notion that people should help each other and, thereby, should have more resources to draw on in times of crisis. In fact, it is likely that many aspects of personality are related to the types of interpersonal relationships and social support people have (e.g., Hooker et al., 1998). We further propose that researchers can learn a great deal by attending more to interactions between personality and situational variables (e.g., social support) and less to simple main effects (for related discussions, see Hooker et al., 1998; Williamson & Silverman, 2000).

Consistent with this idea is evidence that social support buffers the impact of stressful life events (e.g., Cohen & Wills, 1985), and the substantial body of evidence that supports interpersonal relationships as a critical factor in well-being (e.g., Baron, Cutrona, Hicklin, Russell, & Lubaroff, 1990; Cohen, 1988; Jemmott & Magloire, 1988). We now turn to a discussion of what our research tells us about social support and interpersonal relationships, and some of the questions that remain unanswered.

Social Support, Interpersonal Relationships, and the Interplay Between Them

As noted previously, our research has tended to focus on functional, rather than structural, aspects of social support. Structural measures of social support (e.g., how many social relationships a person has and how frequently a person interacts with others) seem to be most likely to predict well-being in general (e.g., Cohen & Wills, 1985). Although "connectedness" with a social network has been related to experiencing fewer stressful life events (e.g., Berkman & Syme, 1979; Blazer, 1982; House, Robbins, & Metzner, 1982), structural measures may be virtually irrelevant for people who have already been subjected to major life stressors. Consequently, we questioned what happens when these events occur, as is often the case, despite being well connected to a social network. Among individuals who are trying to cope with a major life event (i.e., the participants in our studies), high levels of functional support appear to be more beneficial than frequent contact with many friends and family members (e.g., Cohen & Wills, 1985; Oxman & Hull, 1997). In addition, structural measures of support bypass the fact that social contact inherently encompasses not only positive but also negative exchanges (e.g., Cobb, 1976; Cohen & McKay, 1983; Kaplan et al., 1993; Schulz & Williamson, 1991; Suls, 1982 Taylor & Dakof, 1988; Wortman, 1984). Indeed, negative social interaction may have more detrimental effects on well-being than positive interactions have in promoting well-being (e.g., Cohen & McKay, 1983; Suls, 1982; Wortman, 1984; also see Kaplan et al., 1993, for a review).

We freely acknowledge that our data say little about the detrimental effects of negative social interactions on the adjustment of medically compromised individuals. However, they do indicate that less satisfaction with social contacts (Williamson et al., 1994) and lower perceptions of available support (Williamson, in press) contribute to more activity restriction. This is not direct evidence for the impact of negatively valenced interactions, but it also is not unlike data reported in the large social support literature in general. We think research specifically focused on negative social interaction is an extremely important direction for future research, not only for its potential to add to our understanding of the stress–coping process but also because negatively valenced social interactions and their associations with quality of interpersonal relationships are major targets for clinical intervention. However, we note that it is important to determine deficits in the exact components (e.g., structural vs. functional) of social support that are most distressing, since such distinctions have important implications for designing interventions (e.g., Kessler, Kendler, Heath, Neale, & Eaves, 1994; Krause, 1995; Oxman & Hull, 1997).

High levels of informal social support (services for which no pay is received–for example, care provided by family and friends) are almost exclusively the domain of communal relationships (ideally, those found among spouses, family members, and close friends). In fact, society *dictates* that spouses, family members, and (to a lesser extent) friends help each other in times of need and that such help should be given without resentment or the expectation of quid pro quo repayment. If the world (and the people in it) were perfect, this would always be the case. However, people may be disposed to follow society's rules because they feel duty-bound and/or because they fear societal disapproval. Still, they may not always be happy about following these rules (Williamson & Clark, 1989; Williamson & Schulz, 1990, 1995b; Williamson et al., 1998). We believe a situation of this nature has serious implications both for the person who provides care, and for the person who needs care and that interventions can (and should) be designed for both parties with these implications in mind.

In this regard, we particularly advocate research and interventions guided by the Theory of Communal Relationships (e.g., Clark & Mills, 1979, 1993; Mills & Clark, 1982). Probably because support for this theory has mostly been derived from experimental laboratory studies and published in mainstream social psychology journals (but see Williamson & Schulz, 1990, 1995b; Williamson et al., 1998, for exceptions), the theory has received little attention from practice-oriented psychologists (but see Clark & Bennett, 1992, for an exception). We feel this is unfortunate. To illustrate our point, consider the results of our work with family caregivers (Williamson & Schulz, 1990, 1995b; Williamson et al., 1998), showing that reactions to providing help vary dramatically according to quality of the preillness relationship between caregiver and care recipient. For example, among caregivers in historically high communal relationships, intimacy/affectional loss strongly predicted activity restriction, and activity restriction strongly predicted *depressed affect* but not resentment (Williamson et al., 1998). Among caregivers in historically low communal relationships, symptom severity predicted activity restriction which in turn predicted *resentment* toward the care recipient but not depressed affect. We think these results have a great deal to say about interventions most likely to benefit caregivers.

Caregivers in less communal relationships may profit most from interventions emphasizing *caregiving support*, that is, assistance provided by others with caregiving tasks. Support with the actual tasks of providing care should reduce the likelihood that caregiving will seriously restrict their idiosyncratic activities and, consequently, lessen their resentment of the caregiving duties they feel obligated to perform. This is not to say that caregivers in more communal relationships would not also benefit from caregiving support. However, the activity restriction that predicted their depressed affect was itself predicted by loss of intimate and affectional aspects of their relationship with the care recipient. This finding implies that these caregivers may profit most from *compensatory support*, that is, contacts with friends and family members that focus on activities that replace some positive aspects of the relationship that the recipient can no longer provide. Although compensatory support obviously cannot replace all valued aspects of a previously close and mutually supportive relationship, it should nonetheless moderate the direct impact of these losses and decrease vulnerability to depressed affect.

Going beyond their potential utility for caregivers, we also argue that our results, coupled with propositions of the Theory of Communal Relationships, are useful in predicting the quality of care frail elders are likely to receive. Most community-dwelling older people depend on a family member for care when they become ill or disabled (Select Committee on Aging, 1987). Perhaps most provide care of acceptable quality, but virtually no empirical evidence documents this assumption (Barer & Johnson, 1990). We are currently conducting longitudinal research to address this issue.

Although many caregivers adapt well to providing care, a substantial literature clearly shows that caregivers are at risk for increased depression, anger, and anxiety compared to their noncaregiving peers (see, e.g., reviews by Bookwala et al., Chapter 6, this volume; Schulz, O'Brien, Bookwala, & Fleissner, 1995; Schulz, Visintainer, & Williamson, 1990). Surprisingly, researchers have not taken what seems to be a next logical step: Does caregiver emotional distress influence the care provided, and if so, in what ways? Is a depressed (or angry, or anxious) caregiver able to provide high quality care? Are depressed caregivers more likely to neglect the elder's needs? Are angry caregivers more likely to be abusive? We have evidence that quality of the relationship between caregiver and care recipient before illness onset is directly associated with caregiver burden, activity restriction, and depression (Williamson & Schulz, 1990, 1995b; Williamson et al., 1998). It seems likely that caregivers who have not, for example, shared a mutually communal relationship with their care recipient in the past may resent having their activities restricted by caregiving, provide care that is less than adequate, and institutionalize the patient sooner. The results of our current research should begin to provide answers to these questions.

Our interests in conducting this project are very practical. We expect the results to lay concrete groundwork for research focused on improving caregiver mental health and preventing poor quality care. Our results should foster treatment intervention studies aimed at improving caregiver mental health using known methodologies for the treatment of depression, anxiety, and anger discontrol in elderly persons and assessing impact of such improvements on quality of care. In terms of prevention, our study will help determine caregiver mental health variables that mediate relations between known predisposing factors and subsequent poor quality care, allowing for identification of specific caregiver mental health problems, their antecedents (e.g., quality of prior relationship and activity restriction), and the magnitude necessary to impact quality of care negatively.

WHY IS ACTIVITY RESTRICTION SO IMPORTANT?

Why should activity restriction play such a central role in adjustment to stressful life events? Consider that illness conditions perceived as "a fate worse than death" are also those viewed as most interfering with valued life activities (Ditto, Druley, Moore, Danks, & Smucker, 1996). This sentiment is consistent with hypotheses derived from activity theory (Lemon et al., 1972; Longino & Kart, 1982; Reitzes et al., 1995) and underlying assumptions of the Activity Restriction Model. Being unable to continue performing activities related to personal independence and an enjoyable, meaningful lifestyle represents a threat to one's sense of self and loss of control over important aspects of one's life (e.g., Beck, Rush, Shaw, & Emery, 1979; Williamson & Schulz, 1992a; Williamson et al., 1994). The results of a few studies are consistent with these assumptions. For example, Heidrich, Forsthoff, and Ward (1994) found that decrements in functional status (i.e., activity restriction) were related to more discrepancy between the ideal and actual self and also to poorer adaptation. Merluzzi and Martinez Sanchez (1997) reported associations between maintenance of activity and independence, higher self-efficacy, and less psychological distress. More directly relevant are results by Herzog et al. (1998) indicating that frequent participation in leisure and productive activities predicts less depression, an effect partially mediated by agentic sense of self. In other words, as we have long expected, the impact of activity restriction on depressed affect may depend, at least in part, on the extent to which being unable to perform normal activities threatens one's sense of self.

LIMITATIONS AND CONCLUDING CAVEATS

We feel that our research makes a strong case for the central role of activity restriction in adaptation to major life stress. However, we also know we have just begun to scratch the surface with regard to factors that predict activity restriction, situations in which its impact is relevant, and aspects of adjustment that activity restriction has the potential to affect. Thus, some caveats seem warranted. First, although we have accounted for significant amounts of variance in activity restriction (up to 35%) and even more in depression (up to 59%), large portions of variance remain unexplained. Our list of psychosocial variables is by no means exhaustive, and in fact, we expect that other variables also are important in predicting activity restriction and depression (e.g., cognitive processes and other measures of personal and social resources). Second, the amount of variance explained by illness severity appears to depend on how comprehensively it is assessed and, we suspect, on the population under consideration. We have evidence that multiple health conditions play a role in activity restriction (Williamson & Schulz,1995a). Thus, we recommend comprehensive assessment of illness severity, both specific to the condition under study and in more general terms. This is particularly important given that multiple chronic illnesses are common among older adults (e.g., U.S. Senate Special Committee on Aging, 1991; Williamson & Schulz, 1995a). Third, we believe that the Activity Restriction Model applies to many types of life stress. Therefore, an interesting and potentially important direction for future research is to assess the generalizability of our model for predicting adjustment to stressors that are medically unrelated but nevertheless common among older adults (e.g., relocation and bereavement). Fourth, most of the figures we have cited for the amount of variance explained in activity restriction and depression pertain only to main effects and do not include variance accounted for by indirect and interactive effects of illness severity and psychosocial variables, mediational and moderational patterns that can explain significant additional variance beyond simple main effects.

To further complicate matters, some psychosocial variables may not have direct effects on activity restriction (and/or depression) at all. Instead, very different results may be obtained when samples are divided according to their high or low standing on these factors. For instance, we have repeatedly found no linear association between age and restricted activities (e.g., Williamson, 1998; Williamson et al., 1994; Williamson & Schulz, 1995a). However, differences appear when samples are split into younger and older age groups (Walters & Williamson, 1999; Williamson & Schulz, 1995a). In a similar vein, we found no linear association between past mutual communal behaviors and activity restriction among spouse caregivers of cancer patients, but very different patterns emerged when the sample was divided into low and high communal relationships (Williamson et al., 1998).

In summary, existing research is by no means definitive. On the other hand, as evidenced by the variety of approaches considered in this volume, important advances have been made since the 1991 NIH Consensus Statement in teasing apart the association between physical illness and depression. Severity of physical impairment is a clear contributor, but so are psychosocial factors. In addition, we endorse consideration of the extent to which restriction of usual activities and its antecedents contribute to psychological distress. More research is needed to identify the full range of factors that contribute to activity restriction and, subsequently, to depression. Meanwhile, interventions aimed at decreasing depressive symptomatology should attend to the predictors of activity restriction identified thus far. At this point, the most efficacious approach would seem to be interdisciplinary in nature (e.g., involving medical care providers, social workers, physical therapists, and psychologists) to fully assess and treat not only the symptoms of illness but also the psychosocial factors that promote relinquishing previously valued activities.

ACKNOWLEDGMENTS: Research summarized in this chapter was supported by grants from the Andrus Foundation, Grant No. MH41887 from the National Institute of Mental Health, Grant No. CA48635 from the National Cancer Institute (R. Schulz, principal investigator) and grants from the University of Georgia Faculty Research Grants Program and the Southeastern Center for Cognitive Aging (G. M. Williamson, principal investigator). Manuscript preparation was supported by funds to the first author from the Institute for Behavioral Research at the University of Georgia and Grant No. AG15321 from the National Institute on Aging (G. M. Williamson, principal investigator).

REFERENCES

Aldous, J. (1977). Family interaction patterns. *Annual Review of Sociology, 3,* 105–135.

Ashcroft, J. J., Leinster, S. J., & Slade, P. A. (1985). Breast cancer-patient choice of treatment: Preliminary communication. *Journal of the Royal Society of Medicine, 78,* 43–46.

Barer, B. M., & Johnson, C. L . (1990). A critique of the caregiving literature. *Gerontologist, 30,* 26–29.

Baron, R. S., Cutrona, C. E., Hicklin, D., Russell, D. W., & Lubaroff, D. M. (1990). Social support and immune function among spouses of cancer patients. *Journal of Personality and Social Psychology, 59,* 344–352.

Beck, A. T., Rush, A. J., Shaw, B. F., & Emery, G. (1979). *Cognitive therapy of depression.* New York: Guilford Press.

Berkman, L. F., & Syme, S. L. (1979). Social networks, host resistance, and mortality: A nine-year follow-up study of Alameda County residents. *American Journal of Epidemiology, 109,* 186–204.

Bernard, S. L., Kincade, J. E., Konrad, T. R., Arcury, T. A., Rabiner, D. J., Woomert, A., DeFriese, G. H., & Ory, M. G. (1997). Predicting mortality from community surveys of older adults: The importance of self-rated functional ability. *Journal of Gerontology, 52,*155–163.

Billings, A. G., & Moos, R. H. (1984). Coping, stress, and social resources among adults with unipolar depression. *Journal of Personality and Social Psychology, 46,* 877–891.

Blazer, D. G. (1982). Social support and mortality in an elderly community population. *American Journal of Epidemiology, 115,* 684–694.

Blichert-Toft, M. (1992). Breast-conserving therapy for mammary carcinoma: Psychosocial aspects, indications and limitations. *Annals of Medicine, 24,* 445–451.

Bradway, J. K., Malone, J. M., Racy, J., Leal, J. M., & Poole, J. (1984). Psychological adaptation to amputation: An overview. *Orthotics and Prosthetics, 38,* 46–50.

Bruce, M. L., Seeman, T. E., Merrill, S. S., & Blazer, D. G. (1994). The impact of depressive symptomatology on physical disability: MacArthur Studies of Successful Aging. *American Journal of Public Health, 84,* 1796–1799.

Buss, A. H. (1980). *Self-consciousness and social anxiety.* New York: Freeman.

Calfas, K. J., Long, B. J., Sallis, J. F., Wooten, W. J., Pratt, M., & Patrick, K. (1996). A controlled trial of physician counseling to promote the adoption of physical activity. *Preventive Medicine, 25,* 225–233.

Cantor, M. H. (1983). Strain among caregivers: A study of experience in the United States. *Gerontologist, 23,* 597–604.

Carver, C. S., Pozo, C., Harris, S. D., Noriega, V., Scheier, M .F., Robinson, D. S., Ketcham, A. S., Moffat, F. L., Jr., & Clark, K .C. (1993). How coping mediates the effect of optimism on distress: A study of women with early stage breast cancer. *Journal of Personality and Social Psychology, 65,* 375–390.

Clark, G. S., Blue, B., & Bearer, J. B. (1983). Rehabilitation of the elderly amputee. *Journal of the American Geriatrics Society, 31,* 439–448.

Clark, M. S., & Bennett, M. E. (1992). Research on relationships: Implications for mental health. In D. N. Ruble, P. R. Costanzo, & M. E. Oliveri (Eds.), *The social psychology of mental health: Basic mechanisms and application* (pp. 166–198). New York: Guilford Press.

Clark, M. S., & Mills, J. (1979). Interpersonal attraction in exchange and communal relationships. *Journal of Personality and Social Psychology, 37,* 12–24.

Clark, M. S., & Mills, J. (1993). The difference between communal and exchange relationships: What it is and is not. *Personality and Social Psychology Bulletin, 19,* 684–691.

Clark, M. S., Mills, J., & Corcoran, D. (1989). Keeping track of needs and inputs of friends and strangers. *Personality and Social Psychology Bulletin, 15,* 533–542.

Clark, M. S., Mills, J., & Powell, M. C. (1986). Keeping track of needs in communal and exchange relationships. *Journal of Personality and Social Psychology, 51,* 333–338.

Clark, M. S., Ouellette, R., Powell, M. C., & Milberg, S. (1987). Recipient's mood, relationship type, and helping. *Journal of Personality and Social Psychology, 53,* 94–103.

Clark, M. S., & Waddell, B. (1985). Perception of exploitation in communal and exchange relationships. *Journal of Social and Personal Relationships, 2,* 403–413.

Cobb, S. (1976). Social support as a moderator of life stress. *Psychosomatic Medicine, 38,* 300–314.

Cohen, S. (1988). Psychosocial models of social support in the etiology of physical disease. *Health Psychology, 7,* 269–297.

Cohen, S., & McKay, G. (1983). Interpersonal relationships as buffers of the impact of psychosocial stress on health. In A. Baum, S. E. Taylor, & J. E. Singer (Eds.), *Handbook of psychology and health* (Vol. 4, pp. 253–267). Hillsdale, NJ: Erlbaum.

Cohen, S., Mermelstein, R., Kamarck, T., & Hoberman, H. M. (1985). Measuring the functional components of social support. In I. G. Sarason & B. R. Sarason (Eds.), *Social support: Theory, research, and applications* (pp. 73–94). Boston: Martin Nijkoff.

Cohen, S., & Rodriguez, M. S. (1995). Pathways linking affective disturbances and physical disorders. *Health Psychology, 14,* 374–380.

Cohen, S., & Wills, T. A. (1985). Stress, social support, and the buffering hypothesis. *Psychological Bulletin, 98,* 310–357.

Darlington, R. B. (1990). *Regression and linear models.* New York: McGraw-Hill.

Dermer, M., Cohen, S. J., Jacobsen, E., & Anderson, E. A. (1979). Evaluative judgments of aspects of life as a function of vicarious exposure to hedonic extremes. *Journal of Personality and Social Psychology, 37,* 247–260.

Diagnosis and treatment of depression in late life. (1991). Reprinted from NIH Consensus Development Conference Consensus Statement, 1991. November 4–6:9(3).

Ditto, P. H., Druley, J. A., Moore, K. A., Danks, J. H., & Smucker, W. D. (1996). Fates worse than death: The role of valued life activities in health-state evaluations. *Health Psychology, 15,* 332–343.

Femia, E. E., Zarit, S. H., & Johansson, B. (1997). Predicting change in activities of daily living: A longitudinal study of the oldest old in Sweden. *Journal of Gerontology, 52,* 294–302.

Fenigstein, A. (1979). Self-consciousness, self-attention, and social interaction. *Journal of Personality and Social Psychology, 37,* 75–86.

Fenigstein, A., Scheier, M. F., & Buss, A. H. (1975). Public and private self-consciousness. *Journal of Consulting and Clinical Psychology, 43,* 522–527.

Frank, R. G., Kashani, J. H., Kashani, S. R., Wonderlich, S. A., Umlauf, R. L., & Ashkanazi, G. S. (1984). Psychological response to amputation as a function of age and time since amputation. *British Journal of Psychiatry, 144,* 493–497.

George, L. K., & Landerman, R. (1984). Health and subjective well-being: A replicated secondary data analysis. *International Journal of Aging and Human Development, 19,* 133–156.

Gibbons, H. B., & Gerrard, M. (1989). Effects of upward and downward social comparison on mood states. *Journal of Social and Clinical Psychology, 8,* 14–31.

Given, C. W., Collins, C. E., & Given, B. A. (1988). Sources of stress among families caring for relatives with Alzheimer's Disease. *Nursing Clinics of North America, 23,* 69–82.

Goldberg, R. T. (1984). New trends in the rehabilitation of lower extremity amputees. *Rehabilitation Literature, 45,* 2–11.

Heidrich, S. M., Forsthoff, C. A., & Ward, S. E. (1994). Psychological adjustment in adults with cancer: The self as mediator. *Health Psychology, 13,* 346–353.

Herzog, A. R., Franks, M. M., Markus, H. R., & Holmberg, D. (1998). Activities and well-being in older age: Effects of self-concept and educational attainment. *Psychology and Aging, 13,* 179–185.

Hooker, K., Monahan, D. J., Bowman, S. R., Frazier, L. D., & Shifren, K. (1998). Personality counts for a lot: Predictors of mental and physical health of spouse caregivers in two disease groups. *Journal of Gerontology, 53,* 73–83.

House, J. S., Landis, K. R., & Umberson, D. (1988). Social relationships and health. *Science, 241,* 540–545.

House, J. S., Robbins, C., & Metzner, H. L. (1982). The association of social relationships and activities with mortality: Prospective evidence from the Tecumseh Community Health Study. *American Journal of Epidemiology, 116,* 123–140.

Irvine, D., Brown, B., Crooks, D., Roberts, J., & Browne, G. (1991). Psychosocial adjustment in women with breast cancer. *Cancer, 67,* 1097–1117.

Jarrett, W. H. (1985). Caregiving within kinship systems: Is affection really necessary? *Gerontologist, 25,* 5–10.

Jemmott, J. B., III, & Magloire, K. (1988). Academic stress, social support, and secretory immunoglobulin A. *Journal of Personality and Social Psychology, 55,* 803–810.

Kaplan, G., Barell, V., & Lusky, A. (1988). Subjective state of health and survival in elderly adults. *Journal of Gerontology, 43,* 114–120.

Kaplan, R. M., Sallis, J. F., Jr., & Patterson, T. L. (1993). *Health and human behavior.* New York: McGraw-Hill.

Kaplan, R. M., & Toshima, M. T. (1990). Social relationships in chronic illness and disability. In B. R. Sarason, I. G., Sarason, & G. R. Pierce (Eds.), *Social support: An interactional view.* New York: Wiley.

Kessler, R. C., Kendler, K. S., Heath, A., Neale, M. C., & Eaves, L. J. (1994). Perceived support and adjustment to stress in a general population sample of female twins. *Psychological Medicine, 24,* 317–334.

Keyser, J. E. (1993). Diabetic wound healing and limb salvage in an outpatient wound care program. *Southern Medical Journal, 86,* 311–317.

Krause, N. (1995). Negative interaction and satisfaction with social support among older adults. *Journal of Gerontology, 50,* 59–73.

LaPlante, M. P. (1989). Disability risks of chronic illnesses and impairments. *Disability Statistics Report, 2,* 1–39.

Lawrence, R. H., & Jette, A. M. (1996). Disentangling the disablement process. *Journal of Gerontology, 51,* 173–182.

Lemon, B. W., Bengtson, V. L., & Peterson, J. A. (1972). An exploration of the activity theory of aging: Activity types and life satisfaction among in-movers to a retirement community. *Journal of Gerontology, 27,* 511–523.

Lewinsohn, P. M., Hoberman, H., Teri, L., & Hautzinger, M. (1985). An integrative theory of depression. In S. Reiss & R. R. Bootzin (Eds.), *Theoretical issues in behavior therapy* (pp. 331–359). San Diego, CA: Academic Press.

Lewinsohn, P. M., Rohde, P., Seeley, J. R., & Fischer, S. A. (1991). Age and depression: Unique and shared effects. *Psychology and Aging, 6,* 247–260.

Li, L. W., Seltzer, M. M., & Greenberg, J. S. (1997). Social support and depressive symptoms: Differential patterns in wife and daughter caregivers. *Journal of Gerontology, 52,* 200–211.

Liang, J. (1986). Self-reported physical health among aged adults. *Journal of Gerontology, 41,* 248–260.

Longino, C. F., Jr., & Kart, C. S. (1982). Explicating activity theory: A formal replication. *Journal of Gerontology, 37,* 713–722.

Marcus, B. H., Goldstein, M. G., Jette, A., Simkin-Silverman, L., Pinto, B. M., Milan, F., Washburn, R., Smith, K., Rakowski, W., & Dube, C. E. (1997). Training physicians to conduct physical activity counseling. *Preventive Medicine, 26,* 382–388.

Margolis, G., Goodman, R. L., & Rubin, A. (1990). Psychological effects of breast conserving cancer treatment and mastectomy. *Psychosomatics, 31,* 33–39.

McCrae, R. R., & Costa, P. T., Jr. (1986). Personality, coping, and coping effectiveness in an adult sample. *Journal of Personality, 54,* 385–405.

Merluzzi, T. V., & Martinez Sanchez, M. A. (1997). Assessment of self-efficacy and coping with cancer: Development and validation of the Cancer Behavior Inventory. *Health Psychology, 16,* 163–170.

Mills, J., & Clark, M. S. (1982). Communal and exchange relationships. In L. Wheeler (Ed.), *Review of personality and social psychology* (pp. 121–144). Beverly Hills, CA: Sage.

Moore, W. S., & Malone, J. M. (1989). *Lower extremity amputation.* Philadelphia: Saunders.

Moyer, A. (1997). Psychosocial outcomes of breast-conserving surgery versus mastectomy: A meta-analytic review. *Health Psychology, 16,* 284–298.
Mutran, E. J., Reitzes, D. C., Mossey, J., & Fernandez, M. E. (1995). Social support, depression, and recovery of walking ability following hip fracture surgery. *Journal of Gerontology, 50,* 354–361.
Oxman, T. E., & Hull, J. G. (1997). Social support, depression, and activities of daily living in older heart surgery patients. *Journal of Gerontology, 52,* 1–14.
Parmelee, P. A., Katz, I. R., & Lawton, M. P. (1991). The relation of pain to depression among institutionalized aged. *Journal of Gerontology, 46,* 15–21.
Pate, R. R., Pratt, M., Blair, S. N., Haskell, W. L., Macera, C. A., Bouchard, C., Buchner, D., Ettinger, W., Heath, G. W., King, A. C., Kriska, A., Leon, A. S., Marcus, B. H., Morris, J., Paffenberger, R. S., Patrick, K., Pollock, M. L., Rippe, J. M., Sallis, J., & Wilmore, J. H. (1995). Physical activity and public health: A recommendation from the Centers for Disease Control and the American College of Sports Medicine. *Journal of the American Medical Association, 273,* 402–407.
Pell, J. P., Donnan, P. T., Fowkes, F. G. R., & Ruckley, C. V. (1993). Quality of life following lower limb amputation for peripheral arterial disease. *European Journal of Vascular Surgery, 7,* 448–451.
Petrisek, A. C., Laliberte, L. L., Allen, S. M., & Mor, V. (1997). The treatment decision-making process: Age differences in a sample of women recently diagnosed with nonrecurrent, early-stage breast cancer. *Gerontologist, 37,* 598–608.
Pinzur, M. S., Graham, G., & Osterman, H. (1988). Psychologic testing in amputation rehabilitation. *Clinical Orthopaedics, 229,* 236–240.
Radloff, L. (1977). The CES-D Scale: A self-report depression scale for research in the general population. *Applied Psychological Measurement, 1,* 385–401.
Reiber, G. E., Pecoraro, R. E., & Koepsell, T. D. (1992). Risk factors for amputation in patients with diabetes mellitus. *Annals of Internal Medicine, 117,* 97–105.
Reitzes, D. C., Mutran, E. J., & Verrill, L. A. (1995). Activities and self-esteem. *Research on Aging, 17,* 260–277.
Russell, D. W., & Cutrona, C. E. (1991). Social support, stress, and depressive symptoms among the elderly: Test of a process model. *Psychology and Aging, 6,* 190–201.
Rybarczyk, B. D., Nyenhuis, D. L., Nicholas, J. J., Schulz, R., Alioto, R. J., & Blair, C. (1992). Social discomfort and depression in a sample of adults with leg amputations. *Archives of Physical Medicine and Rehabilitation, 73,* 1169–1173.
Ryckman, R. M., Robbins, M. A., Thornton, B., Kaczor, L. M., Gayton, S. L., & Anderson, C. V. (1991). Public self-consciousness and physique stereotyping. *Personality and Social Psychology Bulletin, 17,* 400–405.
Sarason, B. R., Sarason, I. G., & Pierce, G. R. (1990). Traditional views of social support and their impact on assessment. In B. R. Sarason, I. G., Sarason, & G. R. Pierce (Eds.), *Social support: An interactional view* (pp. 9–25). New York: Wiley.
Schain, W. S. (1991). Psychosocial factors in mastectomy and reconstruction. In R. B. Noone (Ed.), *Plastic and reconstructive surgery of the breast* (pp. 327–343). Philadelphia: Decker.
Scheier, M. F., & Carver, C. S. (1985). The self-consciousness scale: A revised version for use with general populations. *Journal of Applied Social Psychology, 15,* 687–699.
Schulz, R., O'Brien, A. T., Bookwala, J., & Fleissner, K. (1995). Psychiatric and physical morbidity effects of Alzheimer's disease caregiving: Prevalence, correlates, and causes. *Gerontologist, 35,* 771–791.
Schulz, R., Visintainer, P., & Williamson, G. M. (1990). Psychiatric and physical morbidity effects of caregiving. *Journal of Gerontology, 45,* 181–191.
Schulz, R., & Williamson, G. M. (1991). A 2-year longitudinal study of depression among Alzheimer's caregivers. *Psychology and Aging, 6,* 569–578.
Select Committee on Aging, U.S. House of Representatives. (1987). *Exploding the myths: Caregivers in America* (Committee Publication No. 99-611). Washington, DC: U.S. Government Printing Office.
Sneeuw, K. C. A., Aaronson, N. K., Yarnold, J. R., Broderick, M., Regan, J., Ross, G., & Goddard, A. (1992). Cosmetic and functional outcomes of breast conserving treatment for early stage breast cancer: 2. Relationship with psychosocial functioning. *Radiotherapy and Oncology, 25,* 160–166.
Stanton, A. L., & Snider, P. R. (1993). Coping with a breast cancer diagnosis: A prospective study. *Health Psychology, 12,* 16–23.
Stewart, C. P. U., Jain, A. S., & Ogston, S. A. (1992). Lower limb amputee survival. *Prosthetics and Orthotics International, 16,* 11–18.

Stump, T. E., Clark, D. O., Johnson, R. J., & Wolinsky, F. D. (1997). The structure of health status among Hispanic, African American, and white older adults. *Journal of Gerontology, 52,* 49–60.

Suls, J. (1982). Social support, interpersonal relations, and health: Benefits and liabilities. In G. S. Saunders & J. Suls (Eds.), *Social psychology of health and illness* (pp. 255–277). Hillsdale, NJ: Erlbaum.

Taylor, S. E. (1989). *Positive illusions: Creative self-deception and the healthy mind.* New York: Basic Books.

Taylor, S. E., & Dakof, G. A. (1988). Social support and the cancer patient. In S. Spacapan & S. Oskamp (Eds.), *The social psychology of health: Claremont Symposium on Applied Social Psychology* (pp. 95–116). Newbury Park, CA: Sage.

Thornhill, H. L., Jones, G. D., Brodzka, W., & VanBockstaele, P. (1986). Bilateral below-knee amputations: Experience with 80 patients. *Archives of Physical Medicine and Rehabilitation, 67,* 159–163.

U.S. Senate Special Committee on Aging. (1991). *Aging America: Trends and projections, 1991 edition.* Washington, DC: Department of Health and Human Services.

Waddell, G. (1987). A new clinical model for the treatment of low-back pain. *Spine, 12,* 632–644.

Walters, A. S., & Williamson, G. M. (1999). The role of activity restriction in the association between pain and depressed affect: A study of pediatric patients with chronic pain. *Children's Health Care, 28,* 33–50.

Wethington, E., & Kessler, R. C. (1986). Perceived support, received support, and adjustment to stressful life events. *Journal of Health and Social Behavior, 27,* 78–89.

Williamson, G. M. (1995). Restriction of normal activities among older adult amputees: The role of public self-consciousness. *Journal of Clinical Geropsychology, 1,* 229–242.

Williamson, G. M. (1996, August). *The centrality of restricted routine activities in psychosocial adjustment to illness and disability: Advances in the identification of contributing factors.* Invited symposium presentation at the 104th Convention of the American Psychological Association, Toronto, Ontario, Canada.

Williamson, G. M. (in press). Extending the Activity Restriction Model of Depressed Affect: Evidence from a sample of breast cancer patients. *Health Psychology.*

Williamson, G .M. (1998). The central role of restricted normal activities in adjustment to illness and disability: A model of depressed affect. *Rehabilitation Psychology, 43,* 327–347.

Williamson, G. M., & Clark, M. S. (1989). The communal/exchange distinction and some implications for understanding justice in families. *Social Justice Research, 3,* 77–103.

Williamson, G. M., & Clark, M. S. (1992). Impact of desired relationship type on affective reactions to choosing and being required to help. *Personality and Social Psychology Bulletin, 18,* 10–18.

Williamson, G. M., Clark, M. S., Pegalis, L., & Behan, A. (1996). Affective consequences of refusing to help in communal and exchange relationships. *Personality and Social Psychology Bulletin, 22,* 34–47.

Williamson, G. M., Jones, D. J., & Ingram, L. A. (1999). Medical and psychosocial predictors of breast cancer treatment decisions. In D. C. Park, R. W. Morrell, & K. Shifren (Eds.), *Processing of medical information in aging patients: Cognitive and human factors perspectives* (pp. 69–91). Mahwah, NJ: Erlbaum.

Williamson, G. M., & Schulz, R. (1990). Relationship orientation, quality of prior relationship, and distress among caregivers of Alzheimer's patients. *Psychology and Aging, 5,* 502–509.

Williamson, G. M., & Schulz, R. (1992a). Pain, activity restriction, and symptoms of depression among community-residing elderly. *Journal of Gerontology, 47,* 367–372.

Williamson, G. M., & Schulz, R. (1992b). Physical illness and symptoms of depression among elderly outpatients. *Psychology and Aging, 7,* 343–351.

Williamson, G. M., & Schulz, R. (1995a). Activity restriction mediates the association between pain and depressed affect: A study of younger and older adult cancer patients. *Psychology and Aging, 10,* 369–378.

Williamson, G. M., & Schulz, R. (1995b). Caring for a family member with cancer: Past communal behavior and affective reactions. *Journal of Applied Social Psychology, 25,* 93–116.

Williamson, G. M., Schulz, R., Bridges, M., & Behan, A. (1994). Social and psychological factors in adjustment to limb amputation. *Journal of Social Behavior and Personality, 9,* 249–268.

Williamson, G. M., & Shaffer, D. R. (1998). Implications of communal relationships theory for understanding interpersonal loss among family caregivers. In J. H. Harvey (Ed.), *Perspectives on personal and interpersonal loss: A sourcebook* (pp. 173–187). Philadelphia: Taylor & Francis.

Williamson, G. M., Shaffer, D. R., & Schulz, R. (1998). Activity restriction and prior relationship history as contributors to mental health outcomes among middle-aged and older caregivers. *Health Psychology, 17,* 152–162.

Williamson, G. M., & Silverman, J. G. (2000). Violence against a female partner: Direct and interactive effects of family history, communal orientation, and peer-related variables. Manuscript submitted for publication.

Wolberg, W. H., Tanner, M. A., Romsaas, E. P., Trump, D. L., & Malec, J. F. (1987). Factors influencing options in primary breast cancer treatment. *Journal of Clinical Oncology, 5,* 68–74.

Wolinsky, F. D., Stump, T. E., & Clark, D. O. (1995). Antecedents and consequences of physical activity and exercise among older adults. *Gerontologist, 35,* 451–462.

Wortman, C. B. (1984). Social support and the cancer patient. *Cancer, 53,* 2339–2360.

Zeiss, A. M., Lewinsohn, P. M., Rohde, P., & Seeley, J. R. (1996). Relationship of physical disease and functional impairment to depression in older people. *Psychology and Aging, 11,* 572–581.

Zeiss, A. M., Lewinsohn, P. M., & Rohde, P. (1996). Functional impairment, physical disease, and depression in older adults. In P. M. Kato & T. Mann (Eds.), *Handbook of diversity issues in health psychology* (pp. 161–184). New York: Plenum Press.

Zimmer, Z., Hickey, T., & Searle, M. S. (1997). The pattern of change in leisure activity behavior among older adults with arthritis. *Gerontologist, 37,* 384–392.

10

We *Should* Measure Change—and Here's How

CHARLES E. LANCE, ADAM W. MEADE, and GAIL M. WILLIAMSON

Thirty years ago, Cronbach and Furby's "How Should We Measure 'Change'—Or Should We?" (1970) critically reviewed various approaches to the measurement of change on psychological variables over time. Focusing specifically on the measurement of change across (only) two measurement waves, they concluded that (1) simple change (difference) scores should *not* be used to assess change over time; (2) in addition to simply indexing changes in level over time, assessments of change must also consider study variables' covariance structure; and (3) analysis of change should be based on estimated true (vs. observed) scores (Cronbach, 1992). As we point out later, analytic methods that accomplish these objectives (especially for three or more waves of measures) have been available for some time but have only recently become more widely accessible to gerontological researchers. As Cronbach (1992) points out, the Cronbach and Furby (1970) paper has often been miscited. However, citation data indicating that it is one of the five most often cited *Psychological Bulletin* articles (Sternberg, 1992) are testimony to the fact that the measurement of change has been a long-standing and controversial topic in the psychological literature (Burr & Nesselroade, 1990; Collins, 1996; Rogosa, 1988).

"Change" is an important theoretical construct in many areas of psychology. But the need for appropriate methodological approaches to the measurement of change in longitudinal research is perhaps more integral and critical for developmental psychology than any other of psychology's subdisciplines. Naturally, this is just as true of developmental research on populations of older adults as it is for other populations, and of research on relationships between depression and physical illness and disability in older adults more specifically. One need for the study of longitudinal change

---

CHARLES E. LANCE, ADAM W. MEADE, and GAIL M. WILLIAMSON • Department of Psychology, University of Georgia, Athens, Georgia 30602.

Physical Illness and Depression in Older Adults: A Handbook of Theory, Research, and Practice, edited by Gail M. Williamson, David R. Shaffer, and Patricia A. Parmelee. Kluwer Academic/Plenum Publishers, New York, 2000.

stems from the possibility that some developmental processes just may not be (as) detectable in cross-sectional research designs. For example, substantial research indicates that more severe health problems are associated with higher levels of depression (e.g., Mathew, Weinman, & Mirabi, 1981; Parmelee, Katz, & Lawton, 1991; Williamson & Schulz, 1992). Thus, it would seem that since older adults do tend to experience more illness conditions than younger adults, they should also experience more depression. However, a growing body of literature suggests that this is not the case (Blazer, Hughes, & George, 1987; Schulz, 1985; Williamson & Schulz, 1995). Rather, all types of affective responses, including depression, appear to decrease with age. Yet these seemingly contradictory findings are based largely on cross-sectional research, leaving open questions about possible effects such as cohort differences, history effects, and other "selection effects" (Meredith, 1964a, 1964b, 1993; Mulaik, 1972) that threaten the construct validity of comparisons among intact groups (Nesselroade & Thompson, 1995). Consequently, aging researchers have increasingly begun to conduct longitudinal (and, in some cases, large-scale population-based) studies in attempts to specify more clearly age-related relationships, including the nature of the link between physical illness and depression in older adults. These more sophisticated designs call for commensurately sophisticated analytic techniques, in particular, techniques that directly investigate longitudinal change over time in the critical variables.

This points to the three main questions that we intend to address in this chapter. First, what is the "state of the practice" in the assessment of change in aging research? Rather than focusing on research on physical illness and depression specifically, we framed this question more broadly in order to gain perspective on methods currently used in aging research in general. To answer this question, we undertook a comprehensive review of the empirical literature in three prominent journals in gerontology (*The Gerontologist*, the *Journal of Gerontology*, and *Psychology and Aging*), with the aim of summarizing the approaches that researchers have used to analyze longitudinal data in empirical articles published over the last 5 years. Second, how does the "state of the practice" compare to the "state of the art" for analyzing longitudinal data? In conjunction with our review of current practices of analyzing longitudinal data, we evaluate the various analytic approaches in terms of the types of inferences they support, their various strengths and limitations, and their relationships to what would be desired of "state of the art" approaches to the measurement of longitudinal change. Finally, how *should* aging researchers measure longitudinal change given present "state of the art" analytic tools? Here, we describe an application of linear structural equation modeling to latent growth models (LGMs) of longitudinal change that have seen increasing application in some areas of social science research but have not yet been widely applied in the aging literature. In this section, we also present an empirical example of an application of LGM. In a concluding section, we discuss some extensions to LGM as they are appropriate for "nonstandard" longitudinal designs.

MEASUREMENT OF LONGITUDINAL CHANGE: THE "STATE OF THE PRACTICE"

The measurement of change has been a long-standing and controversial topic in psychology, and the literature offers little consensus on what constitutes the best methods for assessing change (Burr & Nesselroade, 1990; Hertzog, 1996). In light of this, we sought to determine what methods aging researchers are currently using to assess change; that is, what is the "state of the practice" in measuring longitudinal change in the aging literature?

To answer this question, we conducted a manual search of all articles published in *The Gerontologist*, *Psychology and Aging*, and the *Journal of Gerontology* between January 1993 and June 1998, inclusive. In the process of our review, we first determined whether each article reported results of an empirical study (vs., e.g., editorials, theoretical or review articles, articles discussing practical applications of research findings, commentaries, book reviews) and restricted our review to empirical articles. Second, we further restricted our review to empirical articles that reported results of longitudinal, nonexperimental research, in which there were at least two measurement waves spanning at least 6 months. Examples of studies that were excluded from the review based on this criterion were (1) case studies, (2) studies that only reported descriptive data cross-sectionally, (3) cross-sectional studies (e.g., correlational studies, experimental studies, or comparisons of intact groups using cross-sectional data), and (4) experimental or quasi-experimental studies that included only short-term follow-up posttests. In the final step, we coded each longitudinal study as to the (1) number of measurement waves included in the study, (2) total time frame (in years) covered by the study, (3) length of the interval spanning measurement waves, and (4) primary analytic procedure used. The first set of results is summarized in Table 10.1.

The top portion of Table 10.1 shows the number of empirical articles published, by year, aggregated across the three journals that we reviewed.[1] The first row shows that there was little fluctuation in the number of empirical articles published per year—approximately 200. The second row indicates that a minority of these tended to be longitudinal studies, from a low of 6.2% in 1993, to a high of 15.3% in 1996. The correlation between study year and the percentage of studies published that were longitudinal ($r = .78$, $N = 6$, $p = .07$) indicates an increasing trend toward publishing proportionally more longitudinal studies. This likely reflects recent trends for funding agencies to give priority to longitudinal studies and data from these studies becoming more widely accessible to researchers.

The next portion of Table 10.1 summarizes the number of measurement waves included in longitudinal studies by year of publication. Overall, the mean number of measurement waves was 3.63, and there is no apparent trend across years (the correlation between year and mean number of measurement waves was $r = -.32$, $N = 6$, $p = .54$). Note, however, that the SD and range suggest that there is considerable skewness in the number of measurement waves. In fact, skewness for the data aggregated across publication year was 4.07, $p < .001$. Half of the studies included only two measurement occasions, and 90% included six or fewer. The remaining studies reported seven or more measurement occasions, with one study reporting as many as 25. Noteworthy is the fact that the *modal* number of measurement occasions is two in each year. These data indicate a positive skew toward a relatively small number of measurement occasions and that the modal longitudinal study (with two measurement occasions) is only minimally "longitudinal" (Rogosa, 1988).

Next, Table 10.1 presents data on the total duration of the longitudinal studies

[1] Results tabulated for each journal separately are available from Charles E. Lance.

Table 10.1. Design Characteristics of Longitudinal Studies in *The Gerontologist, Psychology and Aging,* and the *Journal of Gerontology,* 1993–1998

	Publication year						
	1993	1994	1995	1996	1997	1998	Total
Total empirical articles	211	207	204	163	202	100	1087
Longitudinal studies	13	18	24	25	24	13	117
Number of measurement waves							
Mean	3.18	4.53	2.86	4.17	3.96	2.46	3.63
S.D.	2.09	3.21	1.21	2.97	4.75	0.88	3.06
Mode	2	2	2	2	2	2	2
Range	7	10	5	13	23	2	23
Total study duration (years):							
Mean	5.96	5.52	11.95	8.84	7.40	4.73	7.91
S.D.	5.46	5.38	16.21	10.19	9.28	2.82	10.27
Mode	1	2	3	2	3	3	3
Range	17.58	23.10	69.25	40.50	39.50	10.50	69.58
Measurement interval duration							
Mean	3.88	2.17	7.73	3.30	5.12	3.46	4.44
S.D.	4.00	1.70	10.88	3.00	8.55	2.15	6.71
Mode	2	2	3	2	3	1	2
Range	14.60	6.10	39.40	11.75	39.99	6.00	39.99

reviewed. Overall, the mean total study duration was 7.91 years, and this remained constant across publication year (the correlation between publication year and mean study duration was $r = -.07$, $N = 6$, $p = .89$). Overall, the modal study duration was 3 years. Again, there is tremendous variability and skew (skewness = 3.16, $p < .001$) with a small number of studies reporting data collected over long time frames. Half of the studies' durations were 4 years or shorter, but one study reported a 70-year data-collection span. However, Table 10.1 does indicate a trend for the modal study length to be longer in later publication years.

Finally, Table 10.1 summarizes data relating to the duration of the interval separating measurement waves. Overall, the mean measurement interval was 4.44 years, but, again, a relatively small number of studies reported relatively lengthy intervals, skewing the distribution (skewness = 3.73, $p < .001$). Over half (52%) of the studies reported measurement intervals of 2 years or less, and one study reported a measurement interval as long as 40 years. The modal measurement interval was 2 years overall, with no apparent trend for it to increase or decrease across publication date.

In summary, the modal longitudinal study in the literature reviewed here included two measurement occasions over a time span of about 3 years. However, a minority of studies sometimes reported many more waves of measurement (as many as 25), over long time spans (as long as 70 years), and sometimes with widely spaced measurement intervals (as long as 40 years). More importantly for this review, however, are the methods researchers employed to analyze their data in these studies.

Table 10.2 summarizes the primary analytic approaches used to analyze longitudinal data by year of publication. Analytic strategies were categorized as:

- *Descriptive statistics only.* These studies typically reported only descriptive statistics that were compared narratively across measurement waves. For example,

Burkhauser, Duncan, and Hauser (1994) reported descriptive economic (income-related) data tracked over a 5-year period in the 1980s separately by age group, gender, and, for the United States and Germany, to assess differential patterns of economic growth and well-being. Trends in the data over time and across nations were noted, but no inferential statistics were reported.

- *Change scores as DVs.* These studies used change (i.e., difference) scores as dependent variables, measured by substracting subjects' Time 1 scores from scores obtained at (a) later time(s). For example, in a study of age-related changes in visual memory and verbal intelligence, Giambra, Arenberg, Zonderman, Kawas, and Costa (1995) measured change in visual memory (verbal ability) by subtracting Time 1 scores on the Benton Visual Retention Test (Vocabulary subscale of the Wechsler Adult Intelligence Scale) from scores obtained 6.8, 13.1, 19.0 and 25.1 years later. Change scores were then regressed on factors thought to be predictive of change, such as age at first testing, first test score, and year of testing.
- t *tests.* These studies used t tests to measure changes in dependent variables across time. For example, Manton, Stallard, and Corder (1995) assessed changes in disease prevalence rates by conducting t tests of differences in raw proportions of subjects reporting the presence of various diseases over three measurement waves.
- *ANOVA or MANOVA.* These studies used (typically, repeated measures) analysis of variance (ANOVA) or multivariate ANOVA (MANOVA) to assess changes on group means longitudinally. For example, Small, Basun, and Backman (1998) used MANOVAs and follow-up ANOVAs to assess differential change in cognitive ability, memory, visuospatial, and verbal performance across two measurement waves over a 3-year span for very old adults as a function of presence or absence of apolipoprotein E genotype.
- *Lagged regression.* Although not necessarily concerned with assessing change

Table 10.2. Methods Used to Measure Longitudinal Change in *The Gerontologist*, *Psychology and Aging*, and the *Journal of Gerontology*, 1993–1998

	Publication year						
	1993	1994	1995	1996	1997	1998	Total
Number of studies	13	18	24	25	24	13	117
Analytic Approach							
1. Descriptive statistics only	1	2	0	0	0	0	3
2. Change scores as DV	0	1	1[1]	0	0	0	2
3. t tests	0	1[1]	1	0	0	0	2
4. ANOVA or MANOVA	4[1]	5	8[2]	3[1]	4[3]	3	27
5. Lagged regression	4	7	10	12	14	5	52
6. Lagged regression, Y_{t-i} covariate	2	3[1]	3[1]	8[2]	8[3]	1	25
7. Hazard or survival analysis	2	0	3	1[1]	1	2	9
8. Time series	1[1]	0	0	0	0	0	1
9. Longitudinal factor analysis	1	0	0	0	1	1	3
10. Latent growth/curve analysis	0	0	0	2	0	0	2

Note. In some cases, more than one primary analytic strategy was used. These cases are indicated here by a superscript denoting the number of studies which used a given technique, along with one or more others, as the primary analytic approach.

per se, these studies modeled longitudinal relationships by regressing dependent measures collected at some later time on predictors measured (concurrently and) in some previous measurement wave(s). This category also includes logistic regression analysis in which subject status (e.g., mortality) is predicted from data obtained at some earlier measurement occasion(s). For example, in a study of women's employment status and assumption of caregiving responsibilities, Pavalko and Artis (1997) predicted Time 2 caregiver status from various demographic and employment status indicators measured 3 years earlier.

- *Lagged regression, Y_{t-i} covariate.* These studies typically modeled change by regressing dependent measures obtained at some later time period on predictors measured earlier or contemporaneously, but with some prior measure of the dependent variable entered as a covariate. For example, in a study of parents' and adult children's exchange patterns and older adults' psychological wellbeing, Davey and Eggebeen (1998) predicted Time 2 depression and overall life satisfaction from various demographic factors and relationship variables measured both concurrently and 6 years earlier. Time 1 values of depression and life satisfaction were entered in the first stage of a hierarchical regression procedure "to control for the stability" (p. 90) of the scores. Additional predictors were entered in subsequent steps of the regression model.
- *Hazard or Survival Analysis.* Typically, these studies seek to predict attrition or mortality (or alternately, survival) over some measurement window based on factors measured earlier, usually at the beginning of the measurement window. For example, Tucker, Friedman, Tsai, and Martin (1995) analyzed data from the longitudinal study begun by Terman in 1921 to investigate whether playing with pets (as assessed in 1977) was associated with mortality risk in a sample of 343 men and 300 women over a 13-year time frame.
- *Time series.* No studies reported fitting autoregressive moving average (ARMA-type) time series models to longitudinal data. The one study that used some form of a time series approach was Ekerdt and DeViney's (1993) study of factors involved in older workers' retirement process, in which they reported a pooled (cross-sectional) time series analysis of 4-wave panel data collected over a 9-year period.
- *Longitudinal factor analysis.* These studies report factor analyses conducted repeatedly across measurement waves. Issues addressed were the stability of a theoretical factor structure and the assessment of longitudinal measurement equivalence. Ranzijn, Keeves, Luszcz, and Feather (1998) addressed the first issue using confirmatory factor analysis to compare alternative theoretical factor structures underlying a self-esteem measure and tested the stability of the best-fitting models identified from Time 1 data, using Time 2 data, obtained from the same sample 2 years later. As an example of the latter, Schaie, Maitland, Willis, and Intrieri (1998) studied the measurement equivalence of cognitive ability measures across age groups.
- *Latent growth/curve analysis.* These studies use structural equation modeling techniques to estimate true initial status and change on variables measured longitudinally and to model predictors of individual differences in growth trajectories. For example, Jones and Meredith (1996) modeled individual differences in

stability/change in six personality factors for 211 subjects over a 30- or 40-year time span. As a second example, Walker, Acock, Bowman, and Li (1996) used latent growth modeling techniques to model coresidence status and duration of caregiving as exogenous predictors of initial status and straight-line change in the amount of care given and caregiver satisfaction with the caregiving experience over four time periods spanning 2 years.

Several things are apparent from Table 10.2. First, during the 5-year period encompassed by our review, very few studies used the most elementary techniques as their primary approach to analyzing longitudinal relationships: Only three studies just reported descriptive statistics for variables measured over time, and only one study each assessed change using difference scores and conducted t tests as the *sole* analytic approach to measuring change over time. This is the good news. Second, the lion's share of studies used some variant of the general linear (regression) model (GLM) to assess change and/or over-time relationships: 89% of the studies reviewed (correcting for use of multiple applications) used some form of ANOVA or regression-based approach. Overall, there was a trend toward increased use of GLM-based techniques across publication year ($r = .73$, $N = 6$, $p = .10$), but this was due to a trend toward the use of lagged regression techniques in more recent years ($r = .80$, $N = 6$, $p = .06$). Frequency of using ANOVA procedures and lagged regression with previous values of the dependent variable as a covariate did not vary systematically with publication year. Third, more complex analytic approaches such as longitudinal factor analysis and LGM were rarely used. As we point out later, some of these techniques offer analytic and inferential advantages over more typically used GLM techniques. Thus, their underutilization is perhaps the bad news. Overall, the picture painted by this review of approaches to longitudinal data analysis is that researchers on aging, like researchers in other subfields of psychology (e.g., Stone-Romero, Weaver, & Glenar, 1995), tend to rely upon the family of GLM-based analytic techniques that form the core of many traditional postgraduate methodology and statistics course sequences, sequences that tend "to be rooted in past traditions, rather than exemplifying current advancements in the area of quantitative methods" (Byrne, 1996, p. 76).

Summary. So, what is the "state of the analytic practice" in longitudinal studies on aging? The typical data set is multivariate, and consists of two (or a few) waves of measurement, collected over about 3 years. Overall, researchers tended most often to choose one or more traditional analytic tools from the family of GLM procedures (e.g., ANOVA, MANOVA, some form of regression analysis) to the exclusion of more recently developed and perhaps more powerful approaches. Among these procedures, lagged regression was most often the choice. Finally, although many of the longitudinal studies that we reviewed *did* have the measurement of change as a primary goal, many others did not. Studies that did have as an explicit goal the modeling of change most often used LGM, ANOVA, MANOVA, or regression approaches with earlier values of some dependent variable(s) as covariates. In other studies, however, the time-related aspect of the data was exploited to (1) test lagged or predictive relations over time, or (2) cross-validate or test the stability of cross-sectional relationships across multiple time periods. In the next section, we turn our attention to the former set of studies and evaluate the current practice of change measurement relative to what may be desired of change measures and the current state of the art.

MEASUREMENT OF LONGITUDINAL CHANGE: THE STATE OF THE PRACTICE VERSUS THE STATE OF THE ART

We begin this section by posing the question: "What properties would characterize an 'idealized' approach to the measurement of change?" The list we propose is nonexhaustive, but includes issues such as the following:

1. The ability to model change at the individual level of analysis, as well as mean change at the aggregate (sub)group level, that is, the ability to track individuals' patterns of change as well as mean change patterns averaged across subjects.
2. Assessment of the extent of individual differences in initial status and rate of change over time and in relation to group averages, that is, assessment of the degree of homogeneity or heterogeneity in individuals' change patterns.
3. Measurement of change on (estimated) true scores rather than on fallible scores containing measurement error.
4. Estimation of various patterns of change (e.g., linear, quadratic, cubic, or some "optimal" change trajectory).
5. Estimation of concomitant change on multiple critical variables simultaneously.
6. Prediction of initial status and change on critical study variables on the basis of other factors, that is, modeling aspects of individuals' change trajectories as functions of some set of explanatory variables.

This is an idealized list but not necessarily an unrealistic one. To what extent do the present "state of the practice" analytic approaches to change measurement satisfy this list of desired characteristics of an idealized approach to change measurement? Table 10.3 lists the major analytic approaches from Table 10.2 that, in our literature review, were used with the expressed intent of assessing change on study variables, along with our assessment of the extent to which they meet each of the seven desired qualities of an approach to change measurement.

The first question posed in Table 10.3 is whether change can be (or is typically) measured at the *individual level of analysis* as well as at the aggregate (mean) level. Modeling change at the individual level of analysis allows for the possibility that there are meaningful individual differences in change trajectories, that is, that variability about some aggregate (e.g., mean) change function may not solely reflect measurement error or random shocks (Burchinall & Applebaum, 1987). As applied in the literature we reviewed, descriptive statistics (e.g., sample means) used as indicators of change were always calculated across individuals; no individual-level change indicators were reported. Individual-level change scores were also never reported but must be calculated for analysis (i.e., the change score for *n*th individual on the *p*th measure calculated as its value on the *k*th time period as compared to some earlier time period $k-i$, $D_{Pn} = P_{nk} - P_{nk-i}$), typically as dependent measures. Yet change scores only reflect change with respect to two measurement occasions, and their psychometric properties have a long history of unfavorable reviews (see, e.g., Bedeian, Day, Edwards, Tisak, & Smith, 1994; Burr & Nesselroade, 1990; Collins, 1996; Cronbach, 1992; Cronbach & Furby, 1970; Edwards, 1994; Rogosa, 1988; Williams & Zimmerman, 1996). Thus, while change scores do index individual-level change, their utility is limited. Applications of *t* tests, ANOVA, or MANOVA assess change only at the aggregate level; within-cell

Table 10.3. Comparisons among Analytic Approaches against Desirable Criteria for Change Assessment

Modeling	Descriptive statistics	Change scores	t tests	ANOVA	MANOVA	Lagged regression, Y_{t-i} covariate	Longitud. factor analysis	Latent growth
Individual- and group-level change?	No	Yes	No	No	No	Yes	Limited	Yes
Individual differences in change?	No	Yes	Limited	Limited	Limited	Yes	Limited	Yes
Change at true score level?	No	No	No	No	No	No	Yes	Yes
Various forms of change?	No	No	No	Yes	Yes	No	No	Yes
Concomitant change?	Yes	Yes	No	No	Yes	No	Yes	Yes
Prediction of change?	No	Yes	No	No	No	Yes	Limited	Yes

variability contributes to the "error term" (Keppel, 1982), so that potentially meaningful individual differences in change trajectories may be misconstrued as lack of a coherent overall or mean pattern of change. Individual-level change scores are estimable using lagged regression as residualized gain scores (e.g., Y_2 as residualized from its regression on Y_1), which do account for the interrelationships between variables measured over time (vs. unit weights applied to simple difference scores). However, they share most of the undesirable properties of simple change scores (Burr & Nesselroade, 1990). Individual-level change patterns are estimable in longitudinal factor analysis from estimated factor scores (however, see McDonald & Mulaik, 1979, and Mulaik, 1972, on indeterminacy of factor scores) or from mean and covariance structure analysis (e.g., Little, 1997; Millsap & Everson, 1991), but this is rarely done. Of the techniques used in the studies we reviewed, only LGM has as an explicit goal the modeling of individual growth trajectories as well as overall mean growth functions (Duncan & Duncan, 1995; Duncan, Duncan, & Stoolmiller, 1994; McArdle, 1988; McArdle & Aber, 1990; McArdle & Anderson, 1990; Willett & Sayer, 1994).

The second question posed in Table 10.3 is whether *individual differences in change* trajectories are quantifiable. Simple descriptive statistics do not support this, of course. Variability in change scores or residualized change scores (from lagged regression) is estimable, but acknowledged limitations to these indices limit inferences about their variability as well. Individual differences are apparent as within-time period variability in t tests, ANOVA, and MANOVA designs, but this is uninformative about individual differences in change trajectories over time. Following from its explicit emphasis on modeling individual-level change trajectories, LGM is the only one of these analytic approaches that facilitates direct assessment of individual differences (i.e., variability) in longitudinal change trajectories.

The third characteristic, *estimation of change at the true score level,* concerns whether assessment of change is conducted in the absence of measurement error. Although in principle, corrections for attenuation due to unreliability are often possible in GLM applications, in practice, this is almost never done (Schmidt & Hunter, 1996). Table 10.3 reflects this for all approaches to change measurement except longitudinal factor analysis and LGM. If the goal of longitudinal factor analysis is to model change over time (Tisak & Meredith, 1990), and the focus of change is at the factor level (i.e., the level of the latent variable), then assessments of change are disattenuated for measurement error because latent variables are, at least theoretically, perfectly reliable (James, Mulaik, & Brett, 1982). Similarly, LGM seeks to model individual growth trajectories in terms of latent variables representing subjects' true initial status and change over time (see, e.g., Duncan et al., 1994; McArdle, 1988; Willett & Sayer, 1994). Thus, longitudinal factor analysis and LGM are the only analytic methods of those listed in Table 10.3 that estimate change at the true score level, at least, as they are typically implemented.

The fourth characteristic listed in Table 10.3 concerns whether *various forms of change* are estimable. Only linear change is estimable in two-wave longitudinal designs (this is one reason for Rogosa's [1988] limited acceptance of a two-wave design as constituting a "longitudinal study"), to which change scores, *t* tests, and lagged regression analyses are typically applied to evaluate change. Alternative patterns of change are testable in ANOVA and MANOVA designs, for example, by choosing alternative patterns of orthogonal constrasts (e.g., linear, quadratic; see Keppel, 1982), although this is rarely done. Similarly, alternative patterns of change are testable in an LGM framework by selecting alternative (fixed) factor loadings for the latent change variables that represent linear, quadratic, and soon, growth trajectories (see Willett & Sayer, 1994). As we show later, this is a flexible feature of LGM that allows a priori tests of alternative theoretically plausible growth trajectories.

The fifth characteristic listed in Table 10.3 is whether *concomitant change* can be assessed on several variables simultaneously. Descriptive statistics on a number of variables can be evaluated simultaneously, but only in a narrative sense. Change scores can be treated multivariately (e.g., intercorrelations among change scores), but acknowledged limitations of change scores militate against the utility of doing so. Univariate *t* tests and ANOVAs assess change only on single variables, but MANOVA applications are designed to assess change simultaneously on a number of dependent variables. One way in which longitudinal factor analysis can be applied to evaluate concomitant change is through McArdle's (1988, p. 584ff.) curve-of-factor-scores (CUFFS) growth model, in which second-order growth factors are fit to first-order (occasion-specific) factors that summarize interrelationships among a number of study variables, each of which is measured at Time *k*. However, none of the studies we reviewed applied this model. Finally, although early developments of LGM were restricted to modeling individuals' growth trajectories on single variables, more recent work has shown how this general approach can be extended to modeling change on a number of variables simultaneously (e.g., Duncan & Duncan, 1994, 1996; Raykov, 1994; Willett & Sayer, 1996).

The last row of Table 10.3 lists the *prediction of change* or whether, in addition to assessing change in some way, each approach is concerned with modeling aspects of individuals' change trajectories as functions of some set of explanatory variables. This is the explicit concern in studies using change scores and lagged regression techniques, but shortcomings mentioned earlier limit their utility. Individual-level change is not measured in studies using *t* tests, ANOVA, or MANOVA approaches, so this precludes the prediction of individual differences in change. But as Willett and Sayer (1996) and others have shown, the prediction of individual differences in initial status and change is a direct extension of LGM approaches to change measurement.

Summary. It is clear that the utility of using descriptive statistics alone, change scores, and t-tests to evaluate change is quite low as compared to our criteria for an idealized approach to the measurement of change, and it is good to know from Table 10.2 that aging researchers adopt these techniques infrequently. What is less comforting is that (1) aging researchers' mainstay analytic approaches to change measurement (i.e., ANOVA, MANOVA, lagged regression) are also quite limited when evaluated against our "idealized" criteria, and (2) techniques that fare much better against these criteria are seldom applied. These trends support Byrne's (1996) claim that advances in psychologists' substantive areas often seem to outpace applications of similar advances in statistical, measurement, and research methods. This gap may reflect a disparity that Hertzog (1996) suggested between researchers' knowledge of advanced analytic techniques and their behavior in incorporating them into their research. On the other hand, this gap may reflect lack of exposure to some of the more advanced analytic approaches in their graduate training (Byrne, 1996) and/or subsequent experience. It is our goal in the remainder of this chapter to outline and illustrate one of these more advanced approaches.

SO, HOW *SHOULD* WE MEASURE CHANGE?

In this section, we describe and illustrate one approach to change measurement (i.e., latent growth modeling, LGM) that is gaining more widespread attention and application in psychology as well as in other disciplines. Our presentation is intended to be introductory, although we assume some familiarity with regression and LISREL-based models. To simplify our presentation, we generally assume that a total of N subjects each are observed on P variables on each of T (≥ 3) equally spaced time intervals.

LGM as an Approach to Change Measurement

LGM is not new, but recent advances in structural equation modeling (SEM) and widespread availability of SEM software have made its application much more accessible. For example, McArdle (1988) traces LGM's roots back to time-based factor analysis work reported by Cattell (1966), Rao (1958), and Tucker (1958). McArdle and his colleagues (e.g., McArdle, 1988; McArdle & Anderson, 1990; McArdle & Hamagami, 1992) have written extensively on LGM, but more recent work by Willett and Sayer (1994), Duncan and Duncan (1995), and Duncan et al. (1994) have made the logic, procedures, and application of LGM more widely accessible to applied researchers. We consider only some of the more basic aspects of LGMs here. More advanced treatments, particularly as they relate to specific design considerations can be found elsewhere (e.g., Duncan & Duncan, 1994; Duncan, Duncan, & Hops, 1996; Duncan, Duncan, Alpert, Hops, Stoolmiller, & Muthén, 1997; McArdle, 1988; McArdle & Anderson, 1990; McArdle & Hamagami, 1992; McArdle, Hamagami, Elias, & Robbins, 1991; Raykov, 1994; Tisak & Meredith, 1990).

Conceptually, LGM is a two-stage process. In the first stage, individual-level growth trajectories are fit to individuals' scores as they vary over time. This is the "within-individual" stage that is designed to model intraindividual change. In the second stage, additional variables are introduced into the model as predictors of individual differences in subjects' growth trajectories. This represents the "between-individual" stage that seeks to model individual difference determinants of intraindividual change.

The Within-Individual Model

Assuming for the moment that we are interested only in linear change, the first stage of the model may be represented as:

$$Y_{it} = a_i + b_i t_{it} + e_{i,} \tag{1}$$

where Y_{it} represents the ith individual's score on some variable Y at the tth measurement occasion, t_{it} is some time-related variable (e.g., subject age, or occasion of measurement), and a_i and b_i are estimated parameters that link the individual's scores on Y to the time-related variable. Thus, the parameters a_i and b_i define the linear growth trajectory that describes the ith individual's change on Y over time. The intercept a_i represents the subject's estimated value on Y at $t_{it} = 0$, and the slope b_i reflects the subject's (positive or negative) rate of change over the T measurement occasions on Y as a function of a unit change on t_{it}. Assume that we have a simple (and minimal) longitudinal design in which N individuals are measured on Y on $T = 3$ equally spaced occasions. Then the ith individual's linear growth trajectory can be shown in matrix terms as being estimated from:

$$\begin{bmatrix} Y_{i1} \\ Y_{i2} \\ Y_{i3} \end{bmatrix} = \begin{bmatrix} 1 & t_{i1} \\ 1 & t_{i2} \\ 1 & t_{i3} \end{bmatrix} \begin{bmatrix} a_i \\ b_i \end{bmatrix} + \begin{bmatrix} e_{i1} \\ e_{i2} \\ e_{i3} \end{bmatrix} \tag{2}$$

This equation has the familiar form of the ordinary least squares (OLS) regression equation $y = Xb + e$. Note however, that while OLS-type models are most often used to estimate relationships between variables across persons, in this case, it is invoked to model over-time relations (i.e., change) on Y for a single individual. Note also that OLS-type models typically are applied with the intent of drawing inferences from sample data to a larger population with some probability. Here, however, the application is a purely descriptive one; that is, the purpose of the linear regression equations shown in Equations 1 and 2 is to fit some estimated linear growth function for the ith individual in a descriptive sense. Now, assume that we choose to code the measurement occasions $t_{i1} = 0$ through $t_{i3} = 2$ as follows:

$$\begin{bmatrix} Y_{i1} \\ Y_{i2} \\ Y_{i3} \end{bmatrix} = \begin{bmatrix} 1 & 0 \\ 1 & 1 \\ 1 & 2 \end{bmatrix} \begin{bmatrix} a_i \\ b_i \end{bmatrix} + \begin{bmatrix} e_{i1} \\ e_{i2} \\ e_{i3} \end{bmatrix} \tag{3}$$

Using this coding scheme, a_i represents the ith individual's initial status on Y at the first measurement occasion (i.e., measurement occasion 1 is coded "0"), and b_i reflects the rate of the ith individual's linear change across measurement occasions. Thus, this approach to the measurement of change satisfies one of the desirable properties that Cronbach and Furby (1970) mentioned 30 years ago, namely, that change be assessed directly without invoking some form of difference score. Willett and Sayer (1994) show that this intraindividual change model (shown in Equations 1 through 3) can also be written as a measurement model in LISREL notation, so that

$$\begin{bmatrix} Y_{i1} \\ Y_{i2} \\ Y_{i3} \end{bmatrix} = \begin{bmatrix} 0 \\ 0 \\ 0 \end{bmatrix} + \begin{bmatrix} 1 & t_{i1} \\ 1 & t_{i2} \\ 1 & t_{i3} \end{bmatrix} \begin{bmatrix} a_i \\ b_i \end{bmatrix} + \begin{bmatrix} e_{i1} \\ e_{i2} \\ e_{i3} \end{bmatrix} \tag{4}$$

corresponds to the LISREL measurement (i.e., confirmatory factor) model:

$$y = \tau_y + \Lambda_y \eta + \varepsilon \tag{5}$$

with $y' = [Y_{i1}, Y_{i2}, Y_{i3}]$, $\eta' = [a_i, b_i]$, and $\varepsilon' = [e_{i1}, e_{i2}, e_{i3}]$. However, one unusual feature of the models shown in Equations 2 through 4 (as compared to typical applications of LISREL for covariance structure modeling) is that the coefficient matrices τ_y and Λ_y contain entirely fixed parameters. As Willett and Sayer (1994) note, the function of these fixed parameter values is to pass the critical individual-level growth parameters a_i and b_i to the latent η vector with corresponding elements η_1 and η_2, respectively. In particular, (1) fixing elements in the t_y vector to zero permits estimation of latent variable means, that is, mean initial status (i.e., α_1, the mean of η_1) and mean change (i.e., α_2, the mean of η_2) across subjects; (2) fixing variables' factor loadings on η_1 to unit constants defines η_1 as the intercept of the estimated growth trajectory; and (3) fixing variables' factor loadings on η_2 to values that reflect some linear function defines η_2 as the slope of the estimated growth trajectory.

Figure 10.1 shows a path model that corresponds to these measurement relations.

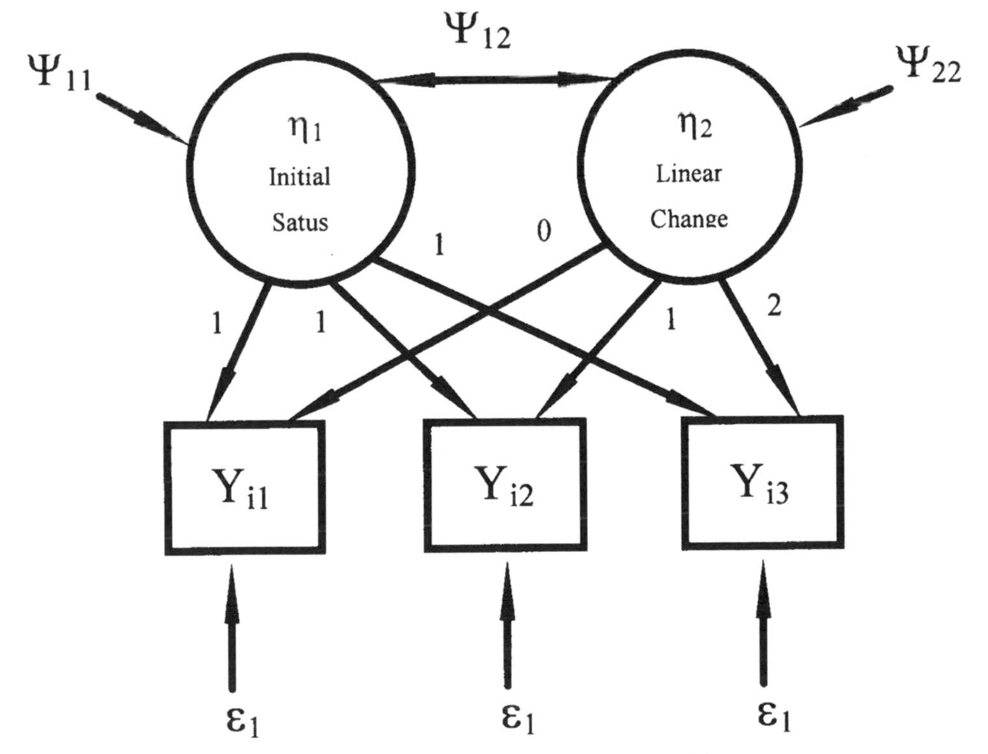

Figure 10.1. Latent growth model of longitudinal change.

With LGM parameters coded as in Equation 3, η_1 in Figure 10.1 corresponds to a_i in Equation 3 and represents individual initial status on Y; η_2 corresponds to b_i, and represents linear change on Y. Note that individual initial status and change are now operationalized at the latent variable level, so that η_1 and η_2 are said to reflect true initial status and true change, respectively (Willett & Sayer, 1994), with measurement errors reflected in the ε_i. In this sense, the LGM approach to change measurement satisfies another of Cronbach and Furby's (1970) desirable properties of a change measurement scheme, namely, that change be assessed at the level of (estimated) true scores. Figure 10.1 also is useful in illustrating some aspects of LGM relating to the covariance model.

The covariance equation following from Equation 5 (assuming that $E[\eta,\varepsilon'] = 0$) is:

$$\Sigma_{yy'} = \Lambda_y \Psi \Lambda'_y + \Theta_{\varepsilon}, \tag{6}$$

where $\Sigma_{yy'}$ is the $T \times T$ matrix of covariances among the Y variables as measured on T occasions, Λ_y contains the fixed factor pattern coefficients (as in Equation 3) linking the Ys to the initial status and change latent variables (η_1 and η_2, respectively), Ψ contains variances and covariances of the ηs, and Θ_ε contains variances and covariances among the residuals ε_i. The Λ_y matrix has been discussed previously, but elements in Ψ and Θ_ε warrant further comment.

Referring back to Figure 10.1, elements ψ_{11} and ψ_{22} of Ψ represent the variances of the latent initial status (η_1) and change (η_2) variables, respectively, and indicate the degree of variability of individuals' initial status and change growth trajectory parameters in the sample. Thus ψ_{11} and ψ_{22} indicate the extent to which initial status and change is relatively homogeneous or heterogeneous in the sample and, with some probability, in the larger population; that is, statistically significant estimates for ψ_{11} and ψ_{22} indicate significant heterogeneity in individuals' initial status and growth parameters, respectively. Values of ψ_{11} and ψ_{22} are also useful for plotting mean growth trajectories. For example, aggregate (mean) growth functions can be computed and plotted from $\alpha_1 \pm 2^*\psi^{1/2}_{11}$ and $a_2 \pm 2^*\psi^{1/2}_{22}$. The element ψ_{21} in the Ψ matrix represents the covariance (the correlation, in standardized units of η_1 and η_2) between initial status and change. For example, a significantly positive ψ_{21} could be indicative of a fan-spread pattern of change across subjects—those whose initial status (η_1) is higher (lower) have higher rates of positive (negative) change (η_2) over the course of the study. On the other hand, a significantly negative ψ_{21} could be indicative of a ceiling or floor effect in which those whose initial status is higher (lower) have little room to change toward the asymptotic level on Y toward which most subjects eventually tend.

Elements in Θ_ε also deserve mention. In typical measurement model applications of confirmatory factor analysis, Θ_ε is considered as a diagonal matrix whose elements are free to be estimated. This reflects typical, common factor-analytic assumptions of tau-equivalent measurement properties (Lord & Novick, 1968) and that the ε_is contain specific and nonsystematic error components (Mulaik, 1972). Referring back to Figure 10.1, this parameterization of Θ_ε would be one in which the residual variances are assumed to be heterogeneous and uncorrelated across measurement occasions. Assuming that σ^2_y is homogeneous over time, significantly heterogeneous residual variances would suggest differential reliability in the Y measure over time. However, other forms of Θ_ε also are plausible, including one in which the residual variances are homogeneous across measurement occasions (Willett & Sayer, 1994). This model can be effected by constraining estimates of the diagonal elements in Θ_ε to be equal and represents a reasonable and testable assumption about the LGM's error structure. Finally, a first-order autocorrelational structure may need to be imposed to account for the Ys' covariances, in addition to modeling their common dependency upon the latent initial status and growth variables (see Willett & Sayer, 1994). This possibility too can be tested for designs with sufficient numbers of measurement occasions (for identification purposes) by estimating elements of the form $\Theta_{\varepsilon t, \varepsilon t-1}$ in the off-diagonal portion of Θ_{ε}.

Some Simple Extensions

The previous section introduced, at a conceptual level, the basic single-variable, linear, individual-level portion of the LGM. In this section, we discuss two extensions to this basic model: (1) modeling nonlinear growth trajectories, and (2) simultaneous estimation of LGMs for multiple variables measured longitudinally.

First, assume now that we have measured Y on four equally spaced occasions for N subjects and that there is some reason to expect nonlinearity in at least some subjects' growth trajectories. For example, subjects' growth trajectories may be expected to increase relatively rapidly early on but eventually reach some asymptotic "ceiling," or alternately, may decrease relatively rapidly toward some asymptotic "floor." Both of these scenarios reflect negatively accelerated change over time, or a quadratic trend. To estimate this trend, we may augment Equation 1 with a quadratic term:

$$Y_{it} = a_i + b_{1i}t_{it} + b_{2i}t^2_{it} + e_{i,} \tag{7}$$

so that longitudinal change in Y is modeled as a quadratic function of the time-related factor, in addition to some aspect of change in Y that is linear. Then the ith individual's growth trajectory can be shown in matrix terms as being estimated linear and quadratic components from:

$$\begin{bmatrix} Y_{i1} \\ Y_{i2} \\ Y_{i3} \\ Y_{i4} \end{bmatrix} = \begin{bmatrix} 1 & t_{i1} & t^2_{i1} \\ 1 & t_{i2} & t^2_{i2} \\ 1 & t_{i3} & t^2_{i3} \\ 1 & t_{i4} & t^2_{i4} \end{bmatrix} \begin{bmatrix} a_i \\ b_{1i} \\ b_{2i} \end{bmatrix} + \begin{bmatrix} e_{i1} \\ e_{i2.} \\ e_{i3} \\ e_{i4} \end{bmatrix} \tag{8}$$

Adopting a comparable coding scheme for measurement occasion as we did previously (i.e., $t_{i1} = 0$ through $t_{i4} = 3$), coefficients for the intercept and linear and quadratic change components would appear in the individual-level change measurement model as:

$$\begin{bmatrix} Y_{i1} \\ Y_{i2} \\ Y_{i3} \\ Y_{i4} \end{bmatrix} = \begin{bmatrix} 1 & 0 & 0 \\ 1 & 1 & 1 \\ 1 & 2 & 4 \\ 1 & 3 & 9 \end{bmatrix} \begin{bmatrix} a_i \\ b_{1i} \\ b_{2i} \end{bmatrix} + \begin{bmatrix} e_{i1} \\ e_{i2} \\ e_{i3} \\ e_{i3} \end{bmatrix} \tag{9}$$

Implementation of a quadratic change component as a latent variable in a LISREL-type measurement model would be effected similarly with the addition of a third factor (i.e., η_3) linked to the Y variables measured longitudinally with fixed factor loadings corresponding to the third column in Equation 9. Thus, this model would appear as Figure 10.1 with the addition of (1) a fourth measurement occasion (i.e., Y_{i4}), (2) the latent variable η_3 that reflects the quadratic function, and (3) a series of four factor loadings linking η_3 to the Ys that are fixed equal to 0, 1, 4, and 9 for the first through the fourth measurement occasions, respectively. Similarly, the individual-level LGM could be augmented with additional, higher-order growth trajectory functions (e.g., cubic, quartic) for designs that include additional measurement occasions (for identification purposes), but these higher-order functions rarely contribute additional explanatory value. As a practical matter, it is useful to evaluate alternative growth functions serially (e.g., linear only, linear plus quadratic) using nested model comparisons (Medsker & Williams, 1994) to determine the incremental contribution (e.g., in terms of the reduction in the overall model χ^2 statistic) of higher-order functions in describing individuals' growth trajectories.

A second simple extension to the basic within-individual LGM involves simultaneous estimation of growth trajectories on multiple variables measured longitudinally. For example, assume that we are interested in modeling linear change on both Y and Z measured longitudinally on N individuals for $T = 3$ equally spaced measurement occasions. Thus, we are interested in simultaneous estimation of two equations of the form of Equation 1, or

$$Y_{it} = a_{yi} + b_{yi} t_{yit} + e_{yi} \tag{10a}$$

$$Z_{it} = a_{zi} + b_{zi} t_{zit} + e_{zi} \tag{10b}$$

Continuing with the same coding scheme that we adopted earlier for measurement occasions, we can represent the simultaneous estimation of initial status and linear change on Y and Z in matrix terms as

$$\begin{bmatrix} Y_{i1} \\ Y_{i2} \\ Y_{i3} \\ Z_{i1} \\ Z_{i2} \\ Z_{i3} \end{bmatrix} = \begin{bmatrix} 1 & 0 & 0 & 0 \\ 1 & 1 & 0 & 0 \\ 1 & 2 & 0 & 0 \\ 0 & 0 & 1 & 0 \\ 0 & 0 & 1 & 1 \\ 0 & 0 & 1 & 2 \end{bmatrix} \begin{bmatrix} a_{yi} \\ \mathrm{b}_{yi} \\ a_{zi} \\ b_{zi} \end{bmatrix} + \begin{bmatrix} e_{yi1} \\ e_{yi2} \\ e_{yi3} \\ e_{zi1} \\ e_{zi2} \\ e_{zi3} \end{bmatrix} \tag{11}$$

Note that initial status and linear change on Y (a_{yi} and b_{yi}, respectively) and Z (a_{zi} and b_{zi}, respectively) are estimated separately. However (as will be shown shortly), relationships among these parameter estimates are themselves estimable. As before, Equation 11 can also be represented as a LISREL-type measurement model of the form of Equation 5 (i.e., $y = \tau_y + \Lambda_y \eta + \varepsilon$), but partitioned into those portions that refer to the Y and Z variables specifically:

$$\begin{bmatrix} \mathrm{y} \\ \hline \mathrm{z} \end{bmatrix} = \begin{bmatrix} 0 \\ \hline 0 \end{bmatrix} + \left[\begin{array}{c|c} \Lambda_{\mathrm{y}} & 0 \\ \hline 0 & \Lambda_{\mathrm{z}} \end{array}\right] \begin{bmatrix} \eta_{\mathrm{y}} \\ \hline \eta_{\mathrm{z}} \end{bmatrix} + \begin{bmatrix} \varepsilon_{\mathrm{y}} \\ \hline \varepsilon_{\mathrm{z}} \end{bmatrix} \tag{12}$$

where Λ_y and Λ_z refer to the upper-left and lower-right portions of the coefficient matrix in Equation 11 that pattern initial status and linear change on the Y and Z variables, respectively, η_y contains elements η_{Y1} and η_{Y2} that express latent initial status and change on Y measured longitudinally, η_z contains elements η_{Z1} and η_{Z2} that express latent initial status and change on Z measured longitudinally, and ε_y and ε_y contain the vectors of residuals for the Y and Z variables, respectively. More interesting, however, is the covariance equation that follows from Equation 12:

$$\begin{bmatrix} \Sigma_{yy'} & \Sigma_{yz} \\ \Sigma_{zy} & \Sigma_{zz'} \end{bmatrix} = \begin{bmatrix} \Lambda_y & 0 \\ 0 & \Lambda_z \end{bmatrix} \begin{bmatrix} \Psi_{\eta(y)} & \Psi_{\eta(y),\eta(z)} \\ \Psi_{\eta(z),\eta(y)} & \Psi_{\eta(z)} \end{bmatrix} \begin{bmatrix} \Lambda'_y & 0 \\ 0 & \Lambda'_z \end{bmatrix} + \begin{bmatrix} \Theta_{\varepsilon(y)} & \Theta_{\varepsilon(y),e(z)} \\ \Theta_{\varepsilon(z),\varepsilon(y)} & \Theta_{\varepsilon(z)} \end{bmatrix} \quad (13)$$

Equation 13 is of the same form as Equation 6 (i.e., $\Sigma = \Lambda\Psi\Lambda' + \Theta_\varepsilon$), but elements in the matrices in Equation 13 are partitioned to reflect their distinctions as relating to the Y variables, the Z variables, or to relationships between them. For example, $\Sigma_{yy'}$ and $\Sigma_{zz'}$ refer to portions of the Σ matrix that contain covariances *among* the Y and Z variables, respectively, and $\Sigma_{yz} = \Sigma'_{zy}$ contains covariances *between* the Y and Z variables. As before, $\Psi_{\eta(y)}$ contains the variances of and the covariance between the latent initial status (η_{1y}) and change (η_{2y}) variables underlying the measurement of Y; $\Psi_{\eta(z)}$ contains corresponding quantities for the measurement of Z. $\Psi_{\eta(y),\eta(z)} = \Psi'_{\eta(z),\eta(y)}$ contains covariances (with latent variables in standard score form, correlations) *between* η_{1y} and η_{2y} and η_{1z} and η_{2y}. Thus, $\Psi_{\eta(y),\eta(z)}$ contains information pertaining to concomitant change on Y and Z, that is, (1) whether initial status on Y covaries with initial status on Z across subjects (i.e., $\Psi_{\eta1(y),\eta1(z)}$), (2) whether initial status on one variable is related to linear change on the other (i.e., $\Psi_{\eta1(y),\eta2(z)}$, or $\Psi_{\eta1(z),\eta2(y)}$), and whether linear change on one variable is concomitantly related to linear change on the other (i.e., $\Psi_{\eta2(y),\eta2(z)}$). Also, as before, $\Theta_{\varepsilon(y)}$ contains variances and covariances among the e_{yi}s; correspondingly, $\Theta_{\varepsilon(z)}$ contains variances and covariances among the e_{zi}s. And, as before, alternative error structures may be tested within the $\Theta_{\varepsilon(y)}$ and $\Theta_{\varepsilon(z)}$ portions of Θ_ε. $\Theta_{\varepsilon(y),\varepsilon(z)} = \Theta'_{\varepsilon(z),\varepsilon(y)}$ contains covariances between the Y residuals (ε_{yi}s) and the Z residuals (ε_{zi}s). Under usual factor analysis assumptions, these are zero, but in the present case, this may not be a reasonable assumption. Specifically, occasion-specific measurement factors may cause additional covariance between Y and Z measured at the same time, beyond that which is accounted for by Y's and Z's modeled change trajectories over time. Thus, it may be reasonable to consider a model which, in addition to modeling the error structure *within* variables, also includes covariances between residuals for variables measured at the same time.

In summary, the within-individual portion of the LGM seeks to model (some function of) initial status and change at the level of the latent variable in a measurement (confirmatory factor) model LISREL-based parameterization. Simple extensions to the basic one-variable linear-change LGM include the estimation of other forms of change trajectories and simultaneous modeling of concomitant change on several variables. We now turn to the between-individuals portion of LGM.

The Between-Individual Model

The second portion of the LGM seeks to model intraindividual differences in change trajectories as a function of some set of explanatory variables; that is, given that we can model individual initial status and change on variables measured longitudinally, what additional variables allow us to explain these change patterns? In a sense, this problem is analogous to commonly employed regression problems in which some criterion variable Y is regressed on some set of putatively explanatory variables X_k:

$$Y = a + \Sigma b_k X_k + e. \tag{14}$$

Here however, the criterion (or endogenous) variables are the latent initial status and change variables from the within-individual phase of LGM; that is, the η_1 and η_2 variables shown in Figure 10.1 become endogenous variables to be predicted on the basis of some other factors. Take the simple case that linear change is modeled on one variable measured longitudinally, and change on the longitudinal variable is predicted on the basis of two individual difference variables. This relationship can be written as

$$\eta = \alpha + \Gamma\xi + \zeta \tag{15}$$

or more explicitly as

$$\begin{bmatrix} \eta_1 \\ \eta_2 \end{bmatrix} = \begin{bmatrix} \alpha_1 \\ \alpha_2 \end{bmatrix} + \begin{bmatrix} \gamma_{11} & \gamma_{12} \\ \gamma_{21} & \gamma_{22} \end{bmatrix} \begin{bmatrix} \xi_1 \\ \xi_2 \end{bmatrix} + \begin{bmatrix} \zeta_1 \\ \zeta_2 \end{bmatrix} \tag{16}$$

where ζ_1 and ζ_2 are the explanatory individual difference variables, αs (now, rather than representing the means on the respective ηs) represent the ηs' intercepts from their regressions on the ξs, γs represent regression parameters linking the ηs to the ξs, and ζs represent regression residuals. Modeled relationships between η_1 and the ξs are tantamount to estimated cross-sectional relationships, as η_1 reflects (some function of) individuals' initial status on the variable measured longitudinally. On the other hand, modeled relationships between η_2 and the ξs implicitly represent hypothesized moderated relationships; that is, (1) since η_2 reflects change on some measured variable (say, Y) as it relates to time, and (2) if some ξ bears a nonzero relationship with η_2, then (3) the relationship between ξ and η_2 indicates that Y's relationship with time varies as a function of values on ξ. The covariance equation also is informative:

$$\text{COV}(\eta,\eta') = \Gamma\Phi\Gamma' + \Psi. \tag{17}$$

Here, Φ contains variances and covariances among the ξs but Ψ now contains the residual variances and covariances from the ηs' regressions on the ξs (where earlier, in the individual-level portion of LGM, Ψ contained the variances and covariances among the ηs). Since one is now interested in modeling covariances among the ηs, the Ψ matrix is often constrained to be a diagonal matrix at this point, thus containing only ηs' residual variances.

One extension to the between-subjects' portion of the LGM includes the modeling of latent initial status and change endogenous variables as functions, not only of exogenous predictors, but also of other initial status and change latent variables. This may be the case if, for example, some set of variables X_k are thought to predict change on variables Y and Z, both measured longitudinally, but variable Y measured longitudinally is also thought to predict Z measured longitudinally. In this case, the structural Equation 16 is augmented to include endogenous predictors (possibly including latent initial status and change predictors) of other latent initial status and change variables:

$$\eta = B\eta + \Gamma\xi + \zeta \tag{18}$$

where B contains parameters linking latent initial status and change variables on one variable measured longitudinally to comparable parameters on some other variable measured longitudinally, and with covariance equation:

$$\text{COV}(\eta,\eta') = (\text{I} - \text{B})^{-1}(\Gamma\Phi\Gamma' + \Psi)(\text{I} - \text{B})'^{-1} \tag{19}$$

Although seemingly more complicated than Equation 17, Equation 19 is of the same form but takes into account specified directional relationships among some hs in the *B* matrix.

An Example

Data. Data reported here were collected at three points in time by professional interviewers trained to work with physically ill respondents, as part of a project focused primarily on home care of patients with advanced cancer.[2] The second measurement wave (T2) occurred 4 months after the first interview (T1), and the third wave of data (T3) was collected 4 months after the second interview. With the exception of stable demographic characteristics, all variables assessed at T1 were also measured at T2 and T3.

Sample and Recruitment. The T1 sample consisted of 268 outpatients with various types of cancer who were recruited from five hospitals in the Pittsburgh, Pennsylvania area. To be included in the study, patients had to be (1) diagnosed as having cancer, (2) receiving outpatient radiation therapy for palliative purposes such as alleviation of pain or recurrence of cancer, and (3) residing in the community (institutionalized patients and patients in hospice programs were excluded). Personnel (nurses and technicians) at the referring hospitals notified project staff of potential participants. A letter describing the project and inviting participation was mailed to individuals meeting study criteria. Approximately 1 week after letters were mailed, an interviewer contacted each potential participant (usually by telephone, but occasionally at the treatment site). For those who agreed to participate, an interview was scheduled at the patient's convenience. These interviews took about 90 minutes to complete and were usually conducted in respondents' homes.

Of the 575 individuals referred by hospital staff, 268 (46.6%) agreed to participate in the study. There were no gender differences between the group who participated (49.3% male) and those who did not (49.5%). When a reason was given for not participating, patients most commonly said they were either too ill (30.9%) or too busy (28.7%). Women comprised 50.7% of the sample. Most respondents were white (90.7%); the remaining were black (9.3%). Mean age was 62.7 years ($SD = 11.1$). A substantial percentage (39.9%, $n = 107$) of the participants did not complete all three interviews. Of these, 26 (9.7%) withdrew from the study or could not be located, 11 (4.1%) became too ill to continue in the study, and 70 (26.1%) died. Thus, data reported here are on the remaining 161 participants who completed all three interviews.

Measures. The data we report here were collected on two variables measured longitudinally (Illness Severity and Pain) and two other variables whose measures we include only from the first measurement occasion (Chronic Diseases and Social Support). *Illness Severity* (SEVERE) was measured as a count of the number of 15 symptoms that respondents reported experiencing due to their cancer (e.g., poor appetite,

[2] Data used for example analyses were collected in research supported by Grant No. CA48625 from the National Cancer Institute (R. Schulz, principal investigator). We thank Dr. Schulz for his permission to use these data.

weight loss, nausea, fever, shortness of breath; coefficient alpha = .73 at T1). *Pain* (PAIN) was measured as the sum of three self-report items assessing the respondent's current level of experienced pain (e.g., "My pain right now is . . . " 1 = Mild to 5 = Excruciating; $\alpha = .91$ at T1). *Chronic diseases* (DISEASE) was a dichotomous variable (0 = no, 1 = yes) indicating whether respondents experienced any chronic health problems (e.g., diabetes, heart condition, digestive problems, stroke) in addition to their cancer. Finally, *social support* (SUPPORT) was measured with a 6-item version of the Interpersonal Support Evaluation List (ISEL; Cohen, Mermelstein, Kamarck, & Hoberman, 1985; Williamson & Schulz, 1992), which assesses the extent to which respondents felt they could call upon others for support if needed (e.g., "If I were too sick to do my daily chores, I could easily find someone to help me"; 1 = Definitely false to 4 = Definitely true, a = .70).

Traditional Analyses

Descriptive statistics and correlations among study variables are shown in Table 10.4 for the 161 patients who completed all three measurement waves. Both PAIN and SEVERE means declined over the three measurement occasions, suggesting that, on the average, participants reported less pain and fewer illness symptoms over time. One-way ANOVAs confirmed that these mean differences were statistically significant for both PAIN [$F(2,480) = 6.54$, $p < .01$] and SEVERE [$F(2,480) = 20.16$, $p < .01$]. Across-time correlations exhibited a simplex-like pattern for both PAIN and SEVERE, with correlations between adjacent measurement occasions being larger than T1–T3 correlations. However, larger over-time correlations indicated somewhat greater stability for SEVERE than for PAIN. Also, contemporaneous correlations between PAIN and SEVERE generally were larger than were lagged correlations. Finally, neither putative predictor (SOCIAL SUPPORT nor DISEASE) measured at T1 correlated significantly with PAIN or SEVERE at any measurement wave.

As is often done, we also used lagged regression to effect covariance adjustments (see Maris, 1998) in investigating possible predictors of change on PAIN and SEVERE, and considering SEVERE as a possible determinant of PAIN. We began by modeling T1 cross-sectional relationships. As suggested by the zero-order correlations in Table 10.4, T1 SEVERE was a significant predictor of T1 PAIN, but T1 DISEASE was not related to either T1 PAIN or T1 SEVERE. To assess possible predictors of T1–T2 and T2–T3 change in PAIN, we entered values of PAIN from the previous measurement

Table 10. 4. Descriptive Statistics and Correlations among Study Variables ($n = 161$)

Variable	Mean	S. D.	1.	2.	3.	4.	5.	6.	7.
1. PAIN–Time 1	4.81	2.83	–						
2. PAIN–Time 2	4.24	2.66	.42**	–					
3. PAIN–Time 3	3.85	2.64	.33**	.53**	–				
4. SEVERE–Time 1	24.33	6.61	.53**	.34**	.26**	–			
5. SEVERE–Time 2	21.43	5.94	.42**	.44**	.29**	.62**	–		
6. SEVERE–Time 3	20.81	5.95	.36**	.51**	.53**	.49**	.66**	–	
7. SOCIAL SUPPORT–Time 1	22.23	2.59	–.01	–.01	.02	.01	-.06	-.08	–
8. DISEASE–Time 1	1.32	0.65	.02	–.04	.02	–.01	.11	-.09	.04

**$p < .01$

Table 10.5. Lagged Regression Results Predicting PAIN and SEVERE Measured Longitudinally

Criterion	Predictor(s)	R^2	ΔR^2
PAIN–Time 1	.53** × SEVERE–Time 1 + .02 × DISEASE–Time 1	.28**	–
PAIN–Time 2			
Step 1	.42** × PAIN–Time 1	.18**	–
Step 2	.19** × SEVERE–Time 1 + .02 × SOCIAL SUPPORT–Time 1	.21**	.03**
PAIN–Time 3			
Step 1	.53** × PAIN–Time 2	.28**	-
Step 2	.10 × SEVERE-Time 1 + .02 × SOCIAL SUPPORT–Time 1	.29**	.01
SEVERE–Time 1	–.01 × DISEASE–Time 1	.00	–
SEVERE–Time 2			
Step 1	.62** × SEVERE–Time 1	.38**	–
Step 2	–.03 × SOCIAL SUPPORT–Time 1	.39**	.01
SEVERE–Time 3			
Step 1	.66** × SEVERE–Time 2	.44**	–
Step 2	–.05 × SOCIAL SUPPORT–Time 1	.44**	.00

**$p < .01$

wave (to control for temporal stability) in Step 1 of the regression model, and putative predictors of change (T1 SEVERE and T1 SOCIAL SUPPORT) in Step 2. Comparable procedures were followed for the assessment of change in SEVERE. As Table 10.5 shows, T1 SEVERE was instrumental in the prediction of change in PAIN from T1 to T2, but not from T2 to T3 (SOCIAL SUPPORT was not a significant predictor of change in PAIN across either measurement interval). Thus, results were equivocal as to whether T1 SEVERE was a significant predictor of change in PAIN. On the other hand, T1 SOCIAL SUPPORT showed no evidence of being predictive of change in SEVERE.

LGM Analyses

The Within-Individual Model. Since we were interested in modeling linear change only on SEVERE and PAIN concomitantly, we structured the within-individual measurement model using the LISREL program (Jöreskog & Sörbom, 1993) as described earlier in Equations 10–13. Specifically, (1) the Y and Z variables were SEVERE and PAIN, respectively, measured longitudinally over the three data collection waves, (2) we specified fixed factor loadings in the Λ_y matrix as patterned in Equation 11; and (3) we estimated alternative models for the error structure in the Θ_ε matrix. Concerning this last point, we estimated alternative models with heterogeneous residual variances, homogeneous residual variances, and homogeneous residual variances with covariances estimated between residuals for variables measured concurrently.

Table 10.6 shows overall goodness-of-fit indices for these models, where df refers to model degrees of freedom, χ^2 refers to the overall model chi-squared statistic, RMSEA is Browne and Cudeck's (1993) root mean squared error of approximation (p-close refers to the associated probability of model close fit), NFI refers to Bentler and Bonett's (1980) normed fit index, TLI refers to the Tucker–Lewis index (see Marsh, Balla, & McDonald, 1988), and CFI refers to Bentler's (1990) comparative fit index. The overall fit of each of the models tested was nearly equivalent: The overall χ^2 indicated that each model could be rejected on purely a statistical basis, but the RMSEA and associ-

Table 10.6. Goodness-of-Fit for Within-Individual Models

Model	*df*	χ^2	RMSEA	*p*-close	NFI	TLI	CFI
Heterogeneous residual variances	7	21.34*	.11	.027	.94	.91	.96
Homogeneous residual variances	11	30.13*	.10	.032	.92	.93	.95
Correlated residuals	8	24.47*	.11	.027	.93	.91	.95

Note. *df* = degrees of freedom, χ^2 = overall chi-squared statistic, RMSEA = (root mean sqared error of approximation, *p*-close = probability of close fit based on RMSEA, NFI = normed fit index, TLI = Tucker–Lewis fit index, CFI = comparative fit index.
* $p < .01$

ated *p*-close indices indicated that the models provided a reasonable approximation to the data, and the NFI, TLI, and CFI indices all were within ranges that are commonly interpreted as indicating acceptable model fit. Therefore, we undertook more specific model comparisons. The difference χ^2 between the models specifying heterogeneous versus homogeneous error structures ($\Delta\chi^2[4] = 8.79$, $p > .05$), indicated that the specification of a homogeneous error structure was reasonable. However, this model contained one improper solution: The correlation between the SEVERE-Change and PAIN-Change latent variables was estimated to be 1.09 (which, of course, represents an improper solution). This suggested (as discussed earlier) that occasion-specific factors might be responsible for covariance among measured variables beyond that which is structured by modeling linear change. Thus, we estimated a third model with homogeneous error structures, but with covariances estimated between variables measured contemporaneously. Although this model failed to provide a significant reduction in the overall χ^2 statistic ($\Delta\chi^2[3] = 5.66$, $p > .05$), this model did result in admissible parameter estimates. The input LISREL runstream for this model is included as Appendix A.

Table 10.7 shows parameter estimates obtained for this model. The right-hand portion shows the matrix of variances and covariances among Initial Status and Linear Change latent variables (i.e., the Ψ matrix referred to earlier in Equation 13). The diagonal of this matrix indicates that each variable's variance is significantly different from zero; that is, there are significant individual differences in change trajectories, both in terms of initial status and linear change, for both variables measured longitudinally. This inference is not supported by more traditional analyses of longitudinal data; that is, a single, group-level, estimated change trajectory (as might be plotted

Table 10.7. LGM Within-Individual Model Parameter Estimates—Correlated Residuals Model

Latent variable	Mean	Covariance matrix (Ψ)			
SEVERE–Initial status	23.95**	30.66**			
SEVERE–Linear change	−1.76**	−5.58*	4.22**		
		(−.49)			
PAIN–Initial status	4.76**	7.93**	−0.61	4.22**	
		(.70)	(−.14)		
PAIN–Linear Change	−0.50**	−1.91*	0.84	−0.82*	0.72*
		(−.41)	(.48)	(−.47)	

Note. Standardized values are shown in parentheses.
*$p < .05$
**$p < .01$

from repeated measures ANOVA means) would not be representative of the collection of change trajectories that describe the present sample. Looking at the covariances in the off-diagonals, the rate of linear change is negatively related to initial status for both SEVERE and PAIN. This pattern, too, is not recoverable from more traditional ANOVA-based analysis of longitudinal data, and reflects the tendency for individuals whose initial status is high to either increase less rapidly or decrease over time, compared to those whose initial status is low. Table 10.7 also shows that Initial Status on SEVERE is highly related to Initial Status on PAIN, another result that is obtained using LGM but not ANOVA; that is, those individuals that reported more illness symptoms at T1 also tended to experience more pain. However, the two variables' Linear Change parameters are not significantly related. This indicates that patterns of subsequent change in Illness Severity are not significantly related to concomitant change in Pain. Note that this pattern is only estimable at the group-mean level of analysis in ANOVA applications. Thus, these results show that the extent of individual variability in change trajectories is directly estimable using LGM, as are elements of concomitant change.

The column of means reported in Table 10.7 shows the estimated means on the SEVERE and PAIN Initial Status and Change latent variables (reported in the LISREL **a** vector discussed earlier). All estimated means are statistically significant. For the initial status variables, this indicates that mean levels of Illness Severity and Pain differed from zero at T1. More interestingly, statistically significant latent change means indicate that there is significant linear change over time at the aggregate, or group, level. This is one inference that *is* supported by the ANOVA results reported earlier. However, the present results are with respect to *latent* mean differences, whereas the test of ANOVA mean differences are with respect to fallible measured variables. Note that latent mean values (just as the ANOVA means) can be used to plot aggregate change over time. For example, the aggregate change trajectory for SEVERE starts out at a value of 23.95 at T1 (i.e., Initial Status) and declines to 20.43 at T3 (i.e., 23.95 – 2*1.76). Mean change is plotted in Figure 10.2 for SEVERE and PAIN, along with two other representative change trajectories. These were chosen to reflect the negative relationship between initial status and change for both SEVERE and PAIN. Consequently, the trajectory labeled "+1SD" represents initial status at 1 *SD* unit above the mean and change at –1 *SD* below the mean change. On the other hand, the trajectory

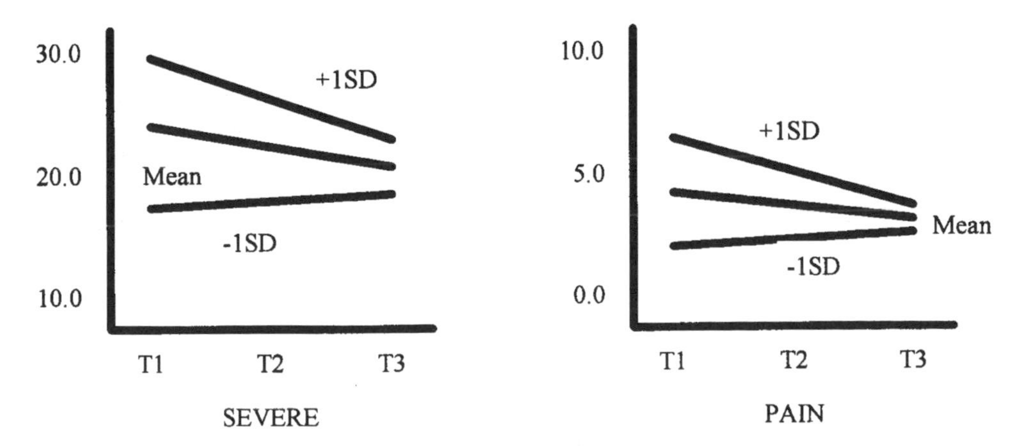

Figure 10.2. Representative longitudinal change trajectories for SEVERE and PAIN.

Table 10.8. LGM Between-Individual Model Parameter Estimates

Endogenous variable	Intercept	Predictors: Endogenous	Exogenous		Residual variance
		SEVERE–IS	DISEASE	SUPPORT	
SEVERE–IS	23.95	–	−0.03 (.01)	–	22.20**
SEVERE–LC	−1.76	–	–	−.01 (−.30)	0.66
PAIN–IS	4.77	0.32** (.81)	0.07 (.02)	–	1.23**
PAIN–LC	−0.51	−0.05* (−.40)	– (.04)	0.01	0.25

Note. IS = Initial status; LC = Linear change; standardized values are shown in parentheses.
*$p < .05$
**$p < .01$

labeled "−1SD" represents initial status at −1 *SD* unit and change at +1 *SD* above the change mean. Of course, other representative change trajectories could be plotted as well, including estimated individual-level trajectories. The plots in Figure 10.2 have similar forms, a general convergence of scores over time that reflect the negative association between initial status and linear longitudinal change.

Table 10.8 shows results of estimated between-individual associations. Here (as in the previous "traditional" analyses of change), we viewed SEVERE as antecedent to PAIN, and both DISEASE and SUPPORT as antecedent to SEVERE and PAIN. Thus, the form of the model that we tested here corresponds to the model that we presented in Equations 18 and 19, in which there are both endogenous and exogenous predictors of latent initial status and growth parameters. For example, the column labeled "Endogenous–SEVERE– IS" in Table 10.8 represents hypothesized predictive relationships from initial status on SEVERE to initial status and linear change on PAIN (parameters in the *B* matrix discussed earlier), and both of these were estimated to be statistically significant. Thus, Initial Status on SEVERE bore a significant cross-sectional relationship with PAIN at T1 (consistent with the cross-sectional results reported in Table 10.4), and served to moderate the relationship between Pain as it varied over time. This latter result is counter to the "equivocal" results predicting change on PAIN from SEVERE found in more traditional analyses of change, and points to the value of assessing change directly as a latent variable using LGM. The columns labeled "Exogenous" in Table 10.8 represent hypothesized relationships between DISEASE and SUPPORT variables measured at T1 and SEVERE and PAIN initial status and linear change variables. Thus, these represent elements contained in the Γ matrix discussed earlier in Equations 18 and 19. Contrary to hypotheses, however, and consistent with results reported in Table 10.4, none of these exogenous effects were statistically significant.

Comparison of Results

What information was gained by conducting LGM analyses? We conclude this section by comparing findings from more traditional analyses of the data just discussed (Tables

10.4 and 10.5) to the LGM analyses (Tables 10.6 though 10.8) using the desirable criteria for a change assessment approach (see Table 10.3) as a comparative framework:

1. *Individual- versus group-level change*. ANOVA provided information about change over time but only at the aggregate level. Thus, while ANOVA results correctly identified mean declines in pain and other cancer-related symptoms over the course of the study for the sample as a whole, they yielded no information about change at the individual level. Based solely on these results, it might be concluded that the symptoms of recurrent cancer patients who are receiving palliative radiation therapy improve over an 8-month period, with a concomitant decrease in depression (see Williamson & Schulz, 1995, for data indicating that decreases in physical symptoms are related to decreases in depression in this sample). Such a conclusion would seem to be overly optimistic given that during this period, almost one-third of the T1 participants either died or became too ill to complete all three waves of data collection. It might then be concluded that, at least among those who were able to complete the study, pain and other symptoms (and, by inference, depression) uniformly declined. Most likely, this conclusion is also erroneous or, at the very least, overly simplistic given substantial variation about the means (see Table 10.4).

Lagged regression analyses were somewhat more informative in this regard by identifying predictors of changes in pain and symptom severity across adjacent time periods. Essentially, these analyses indicated that the strongest predictor of pain (and symptom severity) at T2 or T3 was the analogous measure at the immediately preceding measurement point (with the possible exception of the impact of T1 symptom severity on T2 pain). Still, these analyses did not model individual-level change trajectories directly. On the other hand, LGM analyses explicitly modeled individual-level trajectories over the entire data collection time frame, and supported inferences regarding (a) the sample's mean change trajectory (consistent with ANOVA findings), (b) degree of variability in individuals' change trajectories over time, and (c) shapes of specific change trajectories within the range of those exhibited by research participants.

The latter data are particularly helpful for determining participants that show improvements in pain and other symptoms (and declines in depression) consistent with conclusions drawn from ANOVA results for aggregate data. They are also useful for pointing to individuals who show different trajectories that may be influenced by factors the researcher has not considered.

2. *Individual differences in change.* ANOVA results were uninformative on this point, since deviations from mean change over time in pain and symptom severity would be regarded as error variance. Regression analyses predicted change in these variables by modeling residualized change across adjacent time periods as functions of the same variables at previous measurement points. However, these results (a) proved to be ambiguous (in that symptom severity appeared to predict change in pain between T1 and T2, but not between T2 and T3), and (b) were limited to change assessment only between two time periods and not over the duration of the study. LGM analyses identified significant variability in individuals' change trajectories in both pain and symptom severity across the duration of the study and permitted tests of exogenous (T1) variables in predicting individual differences in change trajectories. Neither the presence of other chronic diseases nor perceived social support predicted change in pain

or symptom severity, clarifying the ambiguous results obtained from regression analyses; that is, symptom severity was, in fact, a significant predictor of true change in pain across the study's duration.

3. *Assessment of change at the true score level.* ANOVA and regression analyses are conducted on observed scores; and thus, assessments of change are subject to influences of measurement error. By contrast, LGM operationalizes aspects of change as latent variables which are, at least theoretically, perfectly reliable and contain no measurement error (James et al., 1982). Consequently, Willett and Sayer (1994) and others have referred to η_1 and η_2 as "true initial status" and "true change," respectively. This clear advantage of the LGM approach allows for a more accurate representation of participant status, both initially and over time.

4. *Assessment of various forms of change.* In our example, ANOVAs were not restricted to any particular form of change, although linear, quadratic, and higher-order orthogonal contrasts could be conducted to test specific hypotheses regarding the shape of the mean trajectory over time. This is a strength of the ANOVA approach, but one that must be tempered with the reminder that the only information available in these types of analysis relates to change at the aggregate, not individual, level. By necessity, regression analyses were restricted to assessing linear change since higher-order functions cannot be identified when change is assessed across only two time periods (e.g., T1–T2, T2–T3). LGM analyses were restricted to the assessment of linear change in this example in the interest of parsimony and because linear change functions fit the data well. However, with sufficient (i.e., at least three) measurement occasions, other, higher-order change functions could be fit as well, including optimal change functions (Duncan et al., 1997; McArdle, 1988; Willett & Sayer, 1994). This, too, is a strength of the LGM approach, although one that was not utilized in our example.

5. *Assessment of concomitant change.* The ANOVA and regression results related only to the assessment of change in single variables, as is almost always the case. Of course, multivariate extensions of these approaches (MANOVA and multivariate regression) have been available for some time, but these strategies are seldom applied (see our earlier analysis and discussion of the recent literature). As we have noted, one of the extensions to the most basic latent growth models is simultaneous assessment of change on multiple variables measured longitudinally. An example of this extension was presented in Table 10.7. This example showed that (a) aspects of change can be assessed on multiple variables simultaneously, and (b) interrelationships among initial status and change parameters associated with different variables measured longitudinally are directly estimable. This is particularly useful in determining whether change in one variable measured longitudinally is positively, negatively, or unrelated to concomitant change in some other variable. For example, although severity of illness symptoms and pain were strongly and positively correlated at T1, linear change in illness severity was negatively related to linear change in pain over time. These results suggest that smaller decreases in illness severity (excluding pain) were related to more (rather than less) alleviation of pain. Most likely, these results reflect progression of the cancer despite palliative radiation therapy with a concomitant increase in efforts to control progressive pain. These relationships would likely have gone undetected in more traditional analyses in which change per se is assessed (a) only indirectly, and (b) separately (i.e., not concomitantly) for different variables that are tracked longitudinally.

6. *Prediction of change.* ANOVAs are useful for determining whether there is significant aggregate-level change in some variable(s) measured longitudinally, but do not support analyses of the factors that might be responsible for measured change. Of course, predicting change is the goal of regression analyses of the form presented here, but as our literature review indicated, change is usually assessed across only two time periods (recall that the LGM approach requires at least three waves of measurement). On the other hand, the second basic extension to LGM that we discussed involves direct tests of hypotheses that individual differences in change trajectories are predictable on the basis of individual differences on other measured variables. As we mentioned earlier, these represent hypothesized moderator effects such that (a) a given variable, (b) as it varies across time, (c) is contingent upon values of a third variable. Thus, we speculated that individual differences in linear change trajectories for pain and symptom severity would vary as functions of the presence of social support and other disease conditions (in addition to cancer) measured at T1, a speculation that was not supported in this data set.

DISCUSSION

The purposes of this chapter were to (1) review the current "state-of-the-practice" in the aging literature with respect to the analysis of longitudinal data, and the measurement of change in particular; (2) assess the current "state-of-the-practice" in view of an idealized set of criteria for the measurement of change, or the desired "state-of-the-art"; and (3) to present an introduction to one relatively recent approach to change measurement that meets many of our criteria for an optimal approach to change measurement.

Generally speaking, the current state-of-the-practice in aging research can be characterized as consisting of a minority of longitudinal studies (although the proportion of empirical studies that report longitudinal research appears to be increasing), many of which are minimally longitudinal (Rogosa, 1988), in which data are analyzed using some variant of traditional GLM-based techniques. The analytic techniques in use generally are appropriate but tend to fall short of the criteria that we proposed for an "idealized" approach to change measurement. One analytic approach that more fully satisfies these criteria is LGM, but its application to change measurement in the aging literature is still infrequent. Thus, our goals were to provide a basic introduction to the logic and implementation of LGM, and point to some of its advantages relative to other approaches to the measurement of change. Some of the advantages that the LGM approach to change measurement present include (1) measurement of change at the individual level of analysis as well as assessment of overall mean change trajectories, (2) the ability to explore alternative functional forms of change over time, (3) assessment of concomitant change on a number of variables simultaneously, and (4) the ability to model individual-difference predictors of intraindividual change.

We used an existing data set to illustrate how LGM can provide additional information and clarify results obtained with traditional analytic methods. From LGM analyses, beyond the information gleaned from more commonly used ANOVA and regression approaches, we demonstrated that change assessed at the aggregate level may present an erroneous picture of change at the individual level. These differences in results have important implications for both research and intervention, especially in terms of the degree of variability in, and the nature of, individuals' trajectories of change over time in response to treatment. They also point out that factors other than those investigated in a particular study may be responsible for such differences. For example, neither social support nor other illness conditions at T1 predicted changes in symptom severity and pain. However, it is likely that other variables (e.g., the development of new illness conditions and other major life events) that we did not include in these analyses contribute to subsequent increases in symptoms and pain.

In addition, LGM allows (and, in fact, requires) us to look at changes that are not restricted to comparisons between only two measurement points. Advantages include more clearly specifying the trajectory of change over longer periods of time as well as clarifying potentially ambiguous results (e.g., when two variables are related between two particular points in time but not between two other points in time). Returning to our example analyses, ANOVA and traditional regression results indicated that changes in symptom severity predicted changes in pain from T1 to T2 but not from T2 to T3. A literal way to interpret this finding is that pain due to recurrent cancer is lessened by treatment for other cancer-related symptoms in the early, but not the later, phases of treatment. This conclusion is most likely incorrect, since LGM analyses revealed that severity of symptoms (other than pain) significantly predicted true change in pain across the entire study period—results that have implications in terms of interventions aimed at alleviating symptoms of illness, including pain.

We do not, however, endorse LGM as *the* method of choice for the analysis of longitudinal data. The choice of method must follow from the nature of the research questions that are being addressed and design characteristics underlying the collection of study data. Nevertheless, we do see LGM as an attractive data-analytic option for the measurement of change in studies in which participants are tracked longitudinally over at least three measurement occasions. For these studies, LGM seems to address Cronbach and Furby's (1970) challenges that measures of change should (1) not use difference scores—LGM models individual change trajectories in the within-individual phase of analysis, (2) model the covariance structure among longitudinal data—LGM permits the structuring of alternative change functions on multiple variables simultaneously, and (3) measure change using estimated true scores—LGM models change at the level of latent initial status and change variables. But many other methods are available for research questions that LGM cannot, and is not designed to, address, including (1) relative survival rates of subjects as a function of some (set of) initial characteristics, as is appropriately addressed using, for example, proportional hazards models; (2) simpler cross-sectional relationships that might be more easily studied using experimental or correlational implementations of GLM-based methods; and (3) questions of measurement equivalency, which are appropriately addressed using confirmatory longitudinal factor analysis. Thus, the research question and the data must drive the analytic method, not the reverse.

Our presentation of LGM was introductory, as intended. Consequently, there were issues, some more advanced, that we did not consider. One of these is the question of *measurement equivalence*, or "whether or not, under different conditions of observing and studying phenomena, measurement operations yield measures of the same attribute" (p.117) (Horn & McArdle, 1992). The key to measurement equivalence is comparable operationalization, in measurement terms, of common underlying constructs across (sub)populations, different situations, or over time. Measurement equivalence represents a critical, yet rarely tested, assumption for longitudinal analysis and the measurement of change in particular. If the assumption of measurement equivalence is not met, and if observed measures actually reflect changing or different constructs, then comparisons of measures collected across time may be tantamount to comparing apples, bicycles, and gubernatorial elections. The issue of measurement equivalence has been discussed extensively in literature on factorial invariance (e.g., Marsh, 1994; Meredith, 1993), and more recently in the aging literature (Horn & McArdle, 1992; Schaie & Hertzog, 1985). Yet it is seldom acknowledged as an aspect of construct validation that is a logical prerequisite to empirical tests of substantive hypotheses (Bagozzi & Edwards, 1998; Byrne, 1989; McArdle & Prescott, 1992). Nesselroade and Thompson (1995) present an excellent example of considering measurement equivalence issues prior to considering more substantive research questions.

There are also a number of issues pertaining to LGM that we did not consider here. For example, McArdle (1988) provides an excellent discussion of alternative models for time structured data, including his "curves-of-factors" (CUFFS) and "factor-of-curves" (FOCUS) second-order factor parameterizations, and Cumsille and Sayer (1998), Duncan and Duncan (1996), Lance and Vandenberg (in press), and Sayer (1998) provide excellent examples. Duncan and Duncan (1994) discuss modeling longitudinal change using LGM with incomplete longitudinal data; McArdle and Hamagami (1992) discuss how this approach can be extended to estimation of the complete longitudinal sequence from incomplete subsets of the entire sequence; and McArdle et al. (1991) extend this even further to combinations of cross-sectional and longitudinal series (see also, McArdle & Hamagami, 1998; McArdle, Prescott, Hamagami, & Horn, 1998). Other papers offer excellent treatments of alternative designs such as cohort sequential designs (e.g., Duncan et al., 1996; Duncan & Duncan, 1995; McArdle & Anderson, 1990; Tisak & Meredith, 1990). Also, Duncan et al. (1997) present models for simultaneous analysis of multilevel and longitudinal data, and Duncan and Duncan (1996) discuss higher-order LGMs, treating first-order intercept (initial status) and slope (change) parameters for multiple variables as multiple indicators for second-order factors. Finally, there are excellent introductions to LGMs by Duncan et al. (1994), McArdle and Aber (1990), McArdle and Anderson (1990), Muthén (1991), Raykov (1992, 1993, 1994), and Willett and Sayer (1994, 1996).

In conclusion, we *should* measure change. LGM, along with various extensions that have come forward the last 10 years, present a number of attractive features that compel researchers in the field of aging to focus on longitudinal research and analyze their data using state-of-the-art methods.

ACKNOWLEDGMENTS: Preparation of this manuscript was facilitated by Grant No. AG15321-02 from the National Institutes of Health (G. M. Williamson, principal investigator). We thank David R. Shaffer for his helpful comments on earlier versions of this chapter.

REFERENCES

Bagozzi, R. P., & Edwards, J. R. (1998). A general approach for representing constructs in organizational research. *Organizational Research Methods, 1*, 45–87.

Bedeian, A. G., Day, D. V., Edwards, J. R., Tisak, J. T., & Smith, C. S. (1994). Difference scores: Rationale, formulation, and interpretation. *Journal of Management, 20*, 673–674.

Bentler, P. M. (1990). Comparative fit indexes structural models. *Psychological Bulletin, 107*, 238–246.

Bentler, P. M., & Bonett, D. G. (1980). Significance tests and goodness-of-fit in the analysis of covariance structures. *Psychological Bulletin, 88*, 588–600.

Blazer, D., Hughes, D. C., & George, L. K. (1987). The epidemiology of depression in an elderly community population. *Gerontologist, 27*, 281–287.

Browne, M. W., & Cudeck, R. (1993). Alternative ways of assessing fit. In K. A. Bollen & J. S. Long (Eds.), *Testing structural equation models* (pp. 136–162). Newbury Park, CA: Sage.

Burchinall, M., & Appelbaum, M. I. (1987). Estimating individual developmental functions: Methods and their assumptions. *Child Development*, 62, 23–43.

Burkhauser, R. V., Duncan, G. J., & Hauser, R. (1994). Sharing prosperity across the age distribution: A comparison of the United States and Germany in the 1980s. *Gerontologist, 34*, 150–160.

Burr, J. A., & Nesselroade, J. R. (1990). Change measurement. In A. von Eye (Ed.), *Statistical methods in longitudinal research* (Vol. 1, pp. 3–34). Boston: Academic Press.

Byrne, B. M. (1989). Multigroup comparisons and the assumption of equivalent construct validity across groups: Methodological and substantive issues. *Multivariate Behavioral Research, 24*, 503–523.

Byrne, B. M. (1996). The status and role of quantitative methods in psychology: Past, present and future perspectives. *Canadian Psychology, 37*, 76–80.

Cattell, R. B. (1966). *Handbook of multivariate experimental psychology*. Chicago: Rand McNally.

Cohen, S., Mermelstein, R., Karmack, T., & Hoberman, H. M. (1985). Measuring the functional components of social support. In I. G. Sarason & B. R. Sarason (Eds.), *Social support: Theory, research, and applications* (pp. 73–94). Boston: Martin Nijkoff.

Collins, L. M. (1996). Measurement of change in research on aging: Old and new issues from an individual growth perspective. In J. E. Birren & K. W. Schaie (Eds.), *Handbook of the psychology of aging* (4th ed., pp. 38–56). San Diego: Academic Press.

Cronbach, L. J. (1992). Four *Psychological Bulletin* articles in perspective. *Psychological Bulletin*, 112, 389–392.

Cronbach, L. J., & Furby, L. (1970). How should we measure "change"—or should we? *Psychological Bulletin*, 74, 68–80.

Cumsille, P. E., & Sayer, A. G. (1998, October). *A comparison of first- and second-order latent growth modeling of alcohol expectancies in adolescence*. Paper presented at the conference New Method for the Analysis of Change, State College, PA.

Davey, A., & Eggebeen, D. J. (1998). Patterns of intergenerational exchange and mental health. *Journal of Gerontology: Psychological Sciences, 53B*, P86–P95.

Duncan, S. C., & Duncan, T. E. (1994). Modeling incomplete longitudinal substance use data using latent variable growth curve methodology. *Multivariate Behavioral Research, 29*, 313–338.

Duncan, S. C., & Duncan, T. E. (1996). A multivariate latent growth curve analysis of adolescent substance use. *Structural Equation Modeling, 3*, 323–347.

Duncan, S. C., Duncan, T. E., & Hops, H. (1996). Analysis of longitudinal data within accelerated longitudinal designs. Psychological Methods, 1, 236–248.

Duncan, T. E., & Duncan, S. C. (1995). Modeling the processes of development via latent variable growth curve methodology. *Structural Equation Modeling, 2*, 178–213.

Duncan, T. E., Duncan, S. C., Alpert, A., Hop, H., Stoolmiller, M., & Muthén, B. (1997). Latent variable modeling of longitudinal and multilevel substance use data. *Multivariate Behavioral Research, 32*, 275–318.

Duncan, T. E., Duncan, S. C., & Stoolmiller, M. (1994). Modeling developmental processes using latent growth structural equation modeling. *Applied Psychological Measurement, 18*, 343–354.

Edwards, J. R. (1994). The study of congruence in organizational behavior research: Critique and a proposed alternative. *Organizational Behavior and Human Decision Processes, 58*, 141–155.

Ekerdt, D. J., & DeViney, S. (1993). Evidence for a preretirement process among older male workers. *Journal of Gerontology: Social Sciences, 2*, S35–S43.

Giambra, L. M., Arenberg, D., Zonderman, A. B., Kawas, C., & Costa, P. T., Jr. (1995). Adult life span changes in immediate visual memory and verbal intelligence. *Psychology and Aging, 10*, 123–139.

Hertzog, C. (1996). Research design in studies of aging and cognition. In J. E. Birren & K. W. Schaie (Eds.), *Handbook of psychology of aging* (4th ed., pp. 24–37). San Diego: Academic Press.

Horn J. L., & McArdle, J. J. (1992). A practical and theoretical guide to measurement invariance in aging research. *Experimental Aging Research, 18,* 117–144.

James, L. R., Mulaik, S. A., & Brett, J. M. (1982). *Causal analysis: Models, assumptions, and data.* Beverly Hills, CA: Sage.

Jones, C. J., & Meredith, W. (1966). Patterns of personality change across the life span. *Psychology and Aging, 11,* 57–65.

Jöreskog, K. G., & Sörbom, D. (1993). *LISREL 8 user's guide.* Chicago: Scientific Software.

Keppel, G. (1982). *Design and analysis: A researcher's handbook* (2nd ed.). Englewood Cliffs, NJ: Prentice-Hall.

Lance, C. E., & Vandenberg, R. J. (in press). Latent growth models of individual change: The case of newcomer adjustment. *Organizational Behavior and Human Processes.*

Little, T. D. (1997). Mean and covariance structures (MACS) analysis of cross-cultural data: Practical and theoretical issues. *Multivariate Behavioral Research, 32,* 53–76.

Lord, F. M., & Novick, M. R. (1968). *Statistical theories of mental test scores.* Reading, MA: Addison-Wesley.

Manton, K. G., Stallard, E., & Corder, L. (1995). Changes ion morbidity and chronic disability in the U.S. elderly population: Evidence from the 1982, 1984, and 1989 national long term care surveys. *Journal of Gerontology: Social Sciences, 4,* S194–S204.

Maris, E. (1998). Covariance adjustment versus gain scores–revisited. *Psychological Methods, 3,* 309–327.

Marsh, H. W. (1994). Confirmatory factor analysis models of factorial invariance: A multifaceted approach. *Structural Equation Modeling, 1,* 5–34.

Marsh, H. W., Balla, J. R., & McDonald, R. P. (1988). Goodness-of-fit indexes in confirmatory factor analysis: The effect of sample size. *Psychological Bulletin, 103,* 391–410.

Mathew, R., Weinman, M., & Mirabi, M. (1981). Physical symptoms of depression. *British Journal of Psychiatry, 139,* 293–296.

McArdle, J. J. (1988). Dynamic but structural equation modeling of repeated measures data. In J. R. Nesselroade & R. B. Cattell (Eds.), *Handbook of multivariate experimental psychology* (2nd ed., pp. 561–614). New York: Plenum Press.

McArdle, J. J., & Aber, M. S. (1990). Patterns of change within latent variable structural equation models. In A. von Eye (Ed.), *Statistical methods in longitudinal research* (Vol. 1, pp.151–224). Boston: Academic Press.

McArdle, J. J., & Anderson, E. (1990). Latent growth models for research on aging. In J. E. Birren & K. W. Schaie (Eds.), *Handbook of the psychology of aging* (3rd ed., pp. 21–44) San Diego: Academic Press.

McArdle, J. J., & Hamagami, F. (1992). Modeling incomplete longitudinal and cross- sectional data using latent growth structural models. *Experimental Aging Research, 18,* 145–166.

McArdle, J. J., & Hamagami, F. (1998, October). *Linear dynamic analyses with incomplete longitudinal data using raw data structural equation modeling techniques.* Paper presented at the conference New Method for the Analysis of Change, State College, PA.

McArdle, J. J., Hamagami, F., Elias, M. F., & Robbins, M. A. (1991). Structural modeling of mixed longitudinal and cross-sectional data. *Experimental Aging Research, 17,* 29–52.

McArdle, J. J., & Prescott, C. A. (1992). Age-based construct validation using structural equation modeling. *Experimental Aging Research, 18,* 87–115.

McArdle, J. J., Prescott, C. A., Hamagami, F., & Horn, J. L. (1998). A contemporary method for developmental-genetic analyses of age changes in intellectual abilities. *Developmental Neuropsychology, 14,* 69–114.

McDonald, R. P., & Mulaik, S. A. (1979). Determinacy of common factors: A nontechnical review. *Psychological Bulletin, 86,* 297–306.

Medsker, G. J., & Williams, L. J. (1994). A review of current practices for evaluating causal models in organizational behavior and human resources management research. *Journal of Management, 20,* 439–464.

Meredith, W. (1964a). Notes on factorial invariance. *Psychometrika, 29,* 177–185.

Meredith, W. (1964b). Rotation to achieve factorial invariance. *Psychometrika, 29,* 187–206.

Meredith, W. (1993). Measurement invariance, factor analysis, and factorial invariance. *Psychometrika, 58,* 525–543.

Millsap, R. E., & Everson, H. (1991). Confirmatory measurement models using latent means. *Multivariate Behavioral Research, 26,* 479–497.

Mulaik, S. A. (1972). *Foundations of factor analysis.* New York: McGraw-Hill.

Muthén, B. (1991). Analysis of longitudinal data using latent variable models with varying parameters. In L. Collins & J. Horn (Eds.), *Best methods for the analysis of change: Recent advances, unanswered ques-*

tions, future directions (pp. 1–17). Washington, DC: American Psychological Association.

Nesselroade, J. R., & Thompson, W. W. (1995). Selection and related threats to group comparisons: An example comparing factorial structures of higher and lower ability groups of adult twins. *Psychological Bulletin, 117,* 271–284.

Parmelee, P. A., Katz, I. R., & Lawton, M. P. (1991). The relation of pain to depression among institutionalized aged. *Journal of Gerontology, 46,* 15–21.

Pavalko, E. K., & Artis, J. E. (1997). Women's caregiving and paid work: Causal relationships in late midlife. *Journal of Gerontologv: Social Sciences, 52B,* S170–S197

Ranzijn, R., Keeves, J., Luszcz, M., & Feather, N. T. (1998). The role of self-perceived usefulness and competence in the self-esteem of elderly adults: Confirmatory factor analyses of the Bachman revision of Rosenberg's self-esteem scale. *Journal of Gerontology: Psychological Sciences, 53B,* P96–P104.

Rao, C. R. (1958). Some statistical methods for the comparison of growth curves. *Biometrics, 14,* 1–17.

Raykov, T. (1992). Structural models for studying correlates and predictors of change. *Australian Journal of Psychology, 44,* 101–112.

Raykov, T. (1993). A structural equation model for measuring residualized change and discerning patterns of growth or decline. *Applied Psychological Measurement, 17,* 53–71.

Raykov, T. (1994). Studying correlates and predictors of longitudinal change using structural equation modeling. *Applied Psychological Measurement, 18,* 63–77.

Rogosa, D. (1988). Myths about longitudinal research. In K. W. Schaie, R. T. Campbell, W. Meredith, & S. C. Rawlings (Eds.), *Methodological issues in aging research* (pp. 171–209). New York: Springer.

Sayer, A. G. (1998, October). *Second-order latent growth models.* Paper presented at the New Methods for the Analysis of Change Conference, State College, PA.

Schaie, K. W., & Hertzog, C. (1985). Measurement in the psychology of adulthood and aging. In J. E. Birren & K. W. Schaie (Eds.), *Handbook of the psychology of aging* (2nd ed., pp. 61–92). New York: Van Nostrand Reinhold.

Schaie, K. W., Maitland, S. B., Willis, S. L., & Intrieri, R. C. (1998). Longitudinal invariance of adult psychometric ability factor structures across 7 years. *Psychology and Aging, 13,* 8–20.

Schmidt, F. L., & Hunter, J. E. (1996). Measurement error in psychological research: Lessons from 26 research scenarios. *Psychological Methods, 1,* 199–223.

Schulz, R. (1985). Emotions and affect. In J. E. Birren & K. W. Schaie (Eds.), *Handbook of the psychology of aging* (2nd ed., pp. 531–543). New York: Van Nostrand Reinhold.

Small, B. J., Basun, H., & Backman, L. (1998). Three-year changes in cognitive performance as a function of apolipoprotein E genotype: Evidence from very old adults without dementia. *Psychology and Aging, 13,* 80–87.

Sternberg, R. J. (1992). Psychological Bulletin's top 10 "Hit Parade." *Psychological Bulletin, 112,* 387–388.

Stone-Romero, E. F., Weaver, A. E., & Glenar, J. L. (1995). Trends in research design and data analytic strategies in organizational research. *Journal of Management, 21,* 141–157.

Tisak, J., & Meredith, W. (1990). Descriptive and associative developmental models. In A. von Eye (Ed.), *Statistical methods in longitudinal research* (Vol. 2, pp. 387–406). Boston: Academic Press.

Tucker, J. S., Friedman, H. S., Tsai, C. M., & Martin, L. R. (1995). Playing with pets and longevity among older people. *Psychology and Aging, 10,* 3–7.

Tucker, L. R. (1958). Determination of parameters of a functional relation by factor analysis. *Psychometrika, 23,* 19–23.

Walker, A. J. Jr., Acock, A. C., Bowman, S. R., & Li, F. (1996). Amount of care given and caregiving satisfaction: A latent growth curve analysis. *Journal of Gerontology: Psychological Sciences, 3,* P130–P142.

Willett, J. B., & Sayer, A. G. (1994). Using covariance structure analysis to detect correlates and predictors of individual change over time. *Psychological Bulletin. 116,* 363–381.

Willett, J. B., & Sayer, A. G. (1996). Cross-domain analyses of change over time: Combining growth modeling and covariance structure analysis. In G. A. Marcoulides & R. E. Schumacker (Eds.), *Advanced structural equation modeling: Issues and techniques* (pp. 125–157). Mahwah, NJ: Erlbaum.

Williams, R. H., & Zimmerman, D. W. (1996). Are simple gain scores obsolete? *Applied Psychological Measurement, 20,* 59–69.

Williamson, G. M., & Schulz, R. (1992). Physical illness and symptoms of depression among elderly outpatients. *Psychology and Aging, 7,* 343–351.

Williamson, G. M., & Schulz, R. (1995). Activity restriction mediates the association between pain and depressed affect: A study of younger and older cancer patients. *Psychology and Aging, 10,* 369–378.

APPENDIX A: LISREL CODE FOR WITHIN-INDIVIDUAL MODEL

```
ILLNESS SEVERITY & PAIN MODEL - 1 ERRORS & CORR'D RESIDUALS
DA NI = 8 NO = 161 MA = CM
KM
*
1.000
 .016 1.000
-.037 .416 1.000
 .017 .329 .528 1.000
 .042 -.013 -.013 .021 1.000
-.008 .528 .335 .255 .013 1.000
 .105 .421 .436 .285 -.055 .616 1.000
-.093 .357 .510 .526 -.077 .490 .661 1.000
SD
*
.65 2.83 2.66 2.64 2.59 6.61 5.94 5.94
ME
*
1.32 4.81 4.24 3.85 22.23 24.33 21.43 20.81
LA
*
DISEASE1 PAIN1 PAIN2 PAIN3 TOTSUP1 SEVERE1 SEVERE2 SEVERE3
SE
*
6 7 8 2 3 4 /
MO NY = 6 NE = 4 PS = SY, FR BE = ZE TE = SY, FR TY = Z E AL = FR
LE
*
ISSEV CHSEV ISPAIN CHPAIN
VA 1.0 LY(1,1) LY(2,1) LY(3,1)
VA 1.0 LY(4,3) LY(5,3) LY(6,3)
VA 0.0 LY(1,2) LY(4,4)
VA 1.0 LY(2,2) LY(5,4)
VA 2.0 LY(3,2) LY(6,4)
PA TE
*
1
0 1
0 0 1
1 0 0 1
0 1 0 0 1
0 0 1 0 0 1
EQ TE(1,1) TE(2,2) TE(3,3)
EQ TE(4,4) TE(5,5) TE(6,6)
OU SE TV MI SS SC AD = OFF
```

APPENDIX B: LISREL CODE FOR BETWEEN-INDIVIDUALS MODEL

```
ILLNESS SEVERITY & PAIN BETWEEN-INDIVIDUALS PREDICTIVE MODEL
DA NI = 8 NO = 161 MA = CM
KM
*
1.000
 .016 1.000
-.037 .416 1.000
 .017 .329 .528 1.000
 .042 -.013 -.013 .021 1.000
-.008 .528 .335 .255 .013 1.000
 .105 .421 .436 .285 -.055 .616 1.000
-.093 .357 .510 .526 -.077 .490 .661 1.000
SD
*
.65 2.83 2.66 2.64 2.59 6.61 5.94 5.94
ME
*
1.32 4.81 4.24 3.85 22.23 24.33 21.43 20.81
LA
*
DISEASE1 PAIN1 PAIN2 PAIN3 TOTSUP1 SEVERE1 SEVERE2 SEVERE3
SE
*
6 7 8 2 3 4 1 5 /
MO NY = 6 NX = 2 NK = 2 BE = ZE GA = FU, FR LX =I D TD = ZE NE = 4 PS = DI, FR BE = FU,
FI TX = ZE KA = FR TE = SY, FR TY = ZE AL = FR
LE
*
ISSEV CHSEV ISPAIN CHPAIN
PA BE
*
0 0 0 0
0 0 0 0
1 0 0 0
1 0 0 0
PA GA
*
1 0
0 1
1 0
0 1
VA 1.0 LY(1,1) LY(2,1) LY(3,1)
VA 1.0 LY(4,3) LY(5,3) LY(6,3)
VA 0.0 LY(1,2) LY(4,4)
VA 1.0 LY(2,2) LY(5,4)
VA 2.0 LY(3,2) LY(6,4)
PA TE
*
```

```
1
0 1
0 0 1
1 0 0 1
0 1 0 0 1
0 0 1 0 0 1
EQ TE(1,1) TE(2,2) TE(3,3)
EQ TE(4,4) TE(5,5) TE(6,6)
OU SE TV MI SS SC AD = OFF
```

III

Diagnosis and Treatment

11

Depression and Physical Illness in Older Primary Care Patients

Diagnostic and Treatment Issues

HERBERT C. SCHULBERG, RICHARD SCHULZ, MARK D. MILLER, and BRUCE ROLLMAN

The interaction between depression and physical illness, and the diverse symptoms and disorders associated with this interaction are of major concern to both theoreticians and clinicians (Dew, 1998; Lyness et al., 1996; Schulberg, McClelland, & Burns, 1987). Perhaps nowhere are concerns about this interaction experienced more strongly than in the primary medical care sector, which serves for most older persons as the portal of entry to the health care delivery system. By 1996, 13% of older Americans were enrolled in HMO-type medical programs, with the rate twice as high in urban areas (Medicare Payment Advisory Commission, 1998). Medicare-financed inducements for older persons to join managed care programs are positioning the primary care physician ever more strongly as this age group's critical gatekeeper to health care.

The primary care physician's growing gatekeeping role augurs well for coordinating and optimizing the manner in which late-life depression is diagnosed and treated. However, this development also has troublesome features, since many primary care physicians do not perceive their clinical skills as adequate to these psychiatric tasks (Banazak, 1996; Callahan, Nienaber, Hendrie, & Tierney, 1992). Moreover, the clinical outcomes of depressed older adults may not improve even when their primary care physicians are provided patient-specific recommendations for treating the mood disorder (Callahan et al., 1994).

In light of these realities, this chapter seeks to improve the quality of care provided older primary care patients experiencing depression by focusing on the following:

HERBERT C. SCHULBERG • Department of Psychiatry, Weill Medical College of Cornell University, White Plains, New York 10605. **RICHARD SCHULZ, MARK D. MILLER, and BRUCE ROLLMAN** • Department of Psychiatry, University of Pittsburgh School of Medicine, Pittsburgh, Pennsylvania 15213.

Physical Illness and Depression in Older Adults: A Handbook of Theory, Research, and Practice, edited by Gail M. Williamson, David R. Shaffer, and Patricia A. Parmelee. Kluwer Academic/Plenum Publishers, New York, 2000.

1. A theoretic model for understanding the relationship between medical illness and depression.
2. Assessment strategies pertinent to a differential diagnosis when physical signs and symptoms co-occur with mood disturbances.
3. Treatments suited to the clinical complexities resulting from the associated physical and psychiatric disorders.

HEALTH DECLINE, CONTROL, AND DEPRESSION IN OLDER PERSONS

Geriatric depression is caused by the complex interplay of psychosocial and biological factors acting over the lifetime of the individual. The substrate for the emergence of depression is provided by genetically influenced characteristics, early developmental experiences, and later-life exposures that shape the biological, psychological, and behavioral resources of the individual. Individuals who are biologically robust and have developed effective psychological and behavioral coping strategies are more likely to withstand the challenges of late-life than those who are biologically and psychologically compromised. As noted in subsequent sections of this chapter, this complex interaction of multiple factors poses profound challenges for both diagnosis and treatment of depression in late-life.

In general, the impact of late-life challenges such as physical illness and disability are thought to influence depression through two major pathways. The psychosocial pathway focuses on the loss of independence and control over one's life as well as the psychological and behavioral coping strategies individuals have effectively used to deal with these stressors. The second pathway focuses on how organic disease may compromise brain function, which may directly lead to depression and/or reduce the individual's cognitive-behavioral coping capacity. Since subsequent sections of this chapter examine the relation between physical illness and brain disease, we focus our attention here on the psychosocial pathway to depression in late-life.

Our conception of the relationship between physical illness and depression in older persons is derived from a life-span model of development (Heckhausen & Schulz, 1995; Schulz & Heckhausen, 1996). The model emphasizes the role of four types of control strategies that moderate the older person's affective response to physical illness and disability: (1) selective primary control, which involves the investment of internal resources such as effort, time, and ability in order to attain important goals; (2) compensatory primary control, which involves the use of external resources such as getting help from others or using technical aids in order to facilitate goal attainment; (3) selective secondary control, which is a cognitive strategy designed to increase the motivation for obtaining a derived goal by enhancing its value or by devaluing competing goals; and (4) compensatory secondary control, which is aimed at facilitating coping with failure or loss through cognitive processes such as goal disengagement, and making self-protective attributions and social comparisons. The first three strategies are aimed at attaining desired goals or responding to threatened losses in primary control. The fourth strategy, compensatory secondary control, is aimed at cognitively compensating for losses or failures. It protects the individual's motivational resources and, therefore, supports primary control striving.

When confronted with major physical illnesses and disability, most individuals effectively use a wide array of available compensation and selection strategies (e.g., disengagement from unattainable goals, self-protective causal attributions, strategic social and intraindividual comparisons, and use of external resources and technical aids). These strategies seek to maintain and expand existing levels of control in important life domains. When these strategies fail to redress the threatened loss of control, the individual experiences negative affect (Schulz, Heckhausen, & O'Brien, 1994) and associated disturbances in cognitive behavior, motivation, and somatic functioning.

Figure 11.1 graphically illustrates a highly simplified sequential process linking late-life physical illness and disability to clinical depression and associated cognitive-behavioral, motivational, affective, and somatic deficits. Initially, individuals respond to declines or losses in physical health through selection and compensation mechanisms designed to protect and enhance their primary control potential. When these strategies fail, negative affect is likely to result, the individual's cognitive and behavioral competencies are undermined, and his or her motivational and somatic resources are compromised. Physical illness may also directly contribute to clinical depression through neuroanatomical or biochemical changes linked to the illness. It is important to note that this model has many important feedback loops, only two of which are illustrated in Figure 11.1 with a dotted line.

Although the proposed model is primarily unidirectional, we recognize that the emergence and impact of clinical depression in late-life is a complex, dynamic process wherein responses at one stage of the model may subsequently feed back to earlier stages. The relationship between depression, disability, and somatic functioning represents one example of this process. Depression may deplete somatic resources, which in turn aggravate existing physical illnesses and age-related developmental declines. This might happen, for example, when depression leads to poor sleep, which in turn leads to falls, ultimately exacerbating existing medical conditions and disabilities.

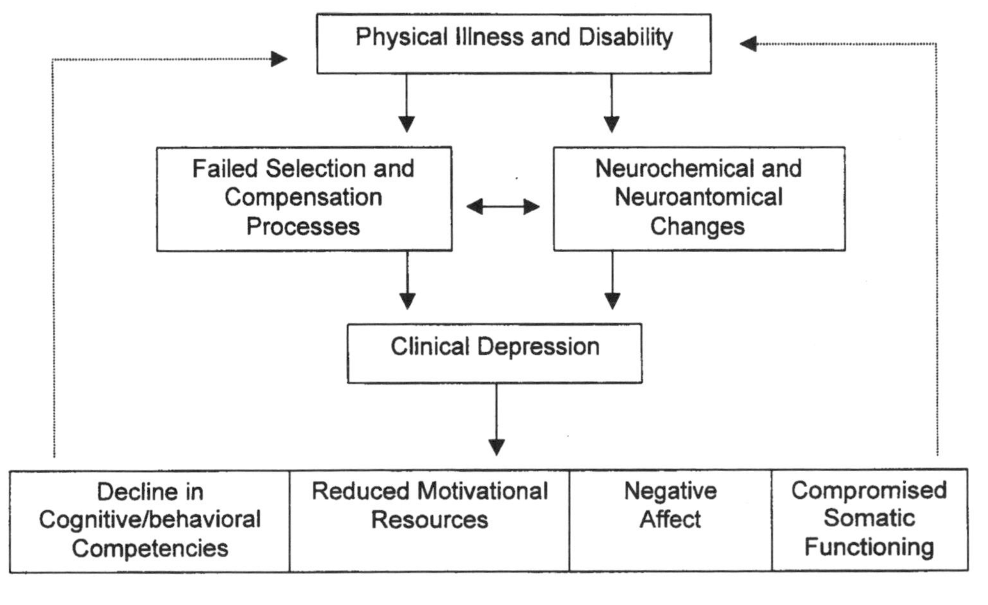

Figure 11.1. Psychosocial and biological pathways linking physical illness/disability to clinical depression.

PHYSICAL ILLNESS, CONTROL, AND DEPRESSION IN OLDER PERSONS

Among older individuals, the empirical literature has focused primarily on the relationship between physical illness, disability, and depression. Although the prevalence of clinical depression among older individuals residing in the community is generally thought to be low (Blazer, Hughes, & George, 1987), the prevalence of major depression among older primary care patients is considerably higher, ranging from 4% to 30% (Caine, Lyness, & Conwell, 1996; Turrima et al., 1994; Williams, Kerber, Mulrow, Medina, & Aguilar, 1995). Our own research suggests that the point prevalence of major depression among older primary care patients is approximately 6–8% (Schulberg et al., 1998). The prevalence of major depression is considerably higher among older individuals with specific medical conditions such as rheumatoid arthritis or osteoarthritis (Creed & Ash, 1992; Dexter & Brandt, 1994), Parkinson's disease (Timberlake et al., 1997), or stroke (Spencer, Tompkins, & Schulz, 1997). Studies that focus on the functional consequences of physical illness, assessed in terms of either limitations in instrumental or basic activities of daily living or in terms of restrictions in normal activities, also report high levels of depression (Alexopoulos et al., 1996; Kennedy, Kelman, & Thomas, 1990; Parmelee, Katz, & Lawton, 1992a; Williamson & Schulz, 1992a, 1992b). Finally, several studies show that increases in depression are associated with changes in functional status in older adults (Armenian, Pratt, Gallo, & Eaton, 1998; Parmelee, Katz, & Lawton, 1992b; Turner & Noh, 1988), in individuals with Parkinson's disease (Cummings, 1992), and in those with arthritis (Creed & Ash, 1992). Little is known, however, about the extent to which chronic medical conditions differ in degree of psychological distress (particularly depression) that they inflict upon affected older persons. A study of psychological distress in a community sample found that it was highest among older persons with hearing impairment, neurological disease, visual impairment, and arthritis (Ormel et al., 1997).

ASSESSMENT STRATEGIES

As indicated previously, the relationship between physical illness, disability, and depression is dynamic and multidimensional rather than stable and unidirectional. Consequently, the interactions between physical illness, disability, and depression may be manifested in any of the following clinical patterns (Deitch & Zetin, 1983):

1. *A functional depression may mimic a recognized physical illness.* A major depressive disorder (MDD) is a complex psychological syndrome involving mood disturbance, neurovegetative signs, and/or cognitive schema. From the physiological perspective, MDD involves multilevel impairments of brain neurotransmitters, hormonal systems, circadian rhythms, and rapid eye movement (REM) sleep. However, MDD can also be a severe reaction to overwhelming stressful events and losses. With regard to why patients express depression somatically rather than dysphorically, this symptom pattern is thought to be a more personally and culturally acceptable idiom of distress for older persons (Schulberg et al., 1987). Sociocultural factors can also influence and reinforce somatic expressions of a mood disorder when the culture lacks proper terms for describing emotional states or poses strong sanctions against perceiving depression as an emotional state.

Since physically symptomatic older persons with an underlying depression typically seek help from a primary care physician, do these persons present distinctive signs that can aid the physician in arriving at a diagnosis? Chronic pain and/or fatigue in the absence of physical pathology have long been considered such markers, but they still receive inadequate attention in the diagnostic process. Not surprisingly, therefore, older primary care patients with depressive symptoms utilize medical services more extensively and incur greater total diagnostic test charges compared to those without such symptoms (Callahan, Kesterson, & Tierney, 1997; Koenig & Kuchibhatla, 1999; Unutzer et al., 1997).

2. *A recognized physical illness may mimic a functional depression.* Just as depressive disorders can mimic physical illness, so can various organic conditions mimic depression. The clinician, therefore, is challenged to determine whether an ostensible affective disorder is truly functional or whether it belies organic pathology. It is important to note in this regard that while there has been much concern expressed about the primary care physician's difficulty in properly diagnosing late-life depression (Williams-Russo, 1996), equally troublesome is the possibility that underlying medical morbidity can be misdiagnosed by the mental health specialist as an affective disorder (Morrison, 1997).

What biological and psychological processes lead organic illnesses to resemble or generate depression? Earlier investigators had considered this form of somatopsychic illness to result from pathological nervous system monoamine changes stimulated by the physical illness (Katon, 1982), and/or as a response to the stress of coping with such illness (as described previously). More recently, structural brain imaging studies have found depression to be associated with enlargement of lateral ventricles; cortical atrophy; more frequent periventricular hyperintensities, deep white-matter lesions, and basal ganglia lesions; and smaller caudate and putamen (Krishnan & Gadde, 1996). These findings are consistent with what Alexopoulos et al. (1997) have described as "vascular depression." Katz (1996) has offered a unifying hypothesis, which proposes that physical or neurological disorders interact with depression. The mechanisms producing such interaction are associated with pathology of the brain structure (e.g., atrophy, stroke, and subclinical cerebrovascular disease) as well as disease-related disturbances of the cerebral environment related to activation of certain cytokine systems. Within this perspective, Katz suggests that the decision to initiate treatment should be more influenced by the scope of "sickness behavior" stimulated in the patient by the symptoms rather than by their physical versus functional etiology.

Given the uncertain manner in which physical illnesses produce depressive symptoms, are there particular illnesses clinically manifesting themselves in primary care practice as dysphoric and related mood distress in older ambulatory medical patients? Among the most commonly cited organic illnesses thought to cause depressive signs are hypothyroidism, diabetes mellitus, cancer, cardiovascular disease, multiple sclerosis, and nutritional deficiencies. Additionally, various medications such as antihypertensives (particularly beta-blockers), corticosterioids, and sedative-hypnotics are thought to produce the same dysphoric affect (Shua-Haim, Shua-Haim, Comsti, & Ross, 1998). These diverse clinical patterns had previously led many to suggest that primary care physicians should distinguish depression-mimicking physical illness from a functional disorder as they seek to improve the clinical course of a patient's presenting symptoms (Schulberg et al., 1987). Presently, however, treatment plans are more typically formulated within the perspective suggested by Katz (1996); that is, treat the depressive symptoms regardless of their etiology.

3. *Depressive reactions to physical illness and disability.* In addition to the diagnostic complexities that are produced when the symptoms of physical and depressive illnesses mimic each other, the primary care physician also is challenged by (a) depressive symptoms that constitute a functional reaction to the disability induced by physical illness; and (b) depression that occurs simultaneously with an unrelated physical illness (Stoudemire, 1995). Much of the clinician's concern about these further types of comorbidity pertains to their ambiguous causality; for example, does chronic pain mask an underlying depression, or is the severity of chronic pain sufficient to induce a depressive reaction? In the common event wherein patients attribute declining functional abilities to a worsening chronic illness and are unaware of the mood component of their distress, physicians may pursue unneeded tests and prescribe inappropriate medications (Callahan et al., 1998; Katon, 1996). While, as noted previously, the quest for a definitive causal model was central to many earlier analyses of this issue (Schulberg et al., 1987), we again note that the emphasis has shifted to effective modes of intervention regardless of the symptoms' chronological sequence. For an expanded analysis of this issue, see Gurland, Katz, and Pine (Chapter 15, this volume).

Diagnostic Systems

Given these diverse patterns of physical and depressive illnesses, and their varying implications for treatment, it had been thought that the assessment process requires a diagnostic system that reliably accounts for the etiology and course of such common primary care complaints as fatigue and difficulty concentrating. As Koenig, Pappas, Holsinger, and Bachar (1995) observed, DSM-III followed a strictly etiological approach in that it required the clinician to judge whether particular symptoms were attributable to psychiatric or other causes. Only symptoms of the former type were to be counted toward a diagnosis of major depression or another mood disorder.

The difficulties experienced by even skilled clinicians in distinguishing psychiatric versus organic etiologies for various presenting symptoms have led to at least three alternative strategies for diagnosing major depression in older patients with physical illness (Cohen-Cole & Stoudemire, 1987). The first alternative is the inclusive approach, which counts all symptoms toward the diagnosis of depression, regardless of perceived etiology. Although highly sensitive and reliable, the specificity of the inclusive approach is imperfect, since it likely assigns the diagnosis of depression to individuals experiencing a physical rather than psychiatric disorder. The risk of unnecessarily treating an older person for depression is minimized, however, when the physician prescribes a selective serotonin reuptake inhibitor (SSRI) medication given its relative safety and favorable side-effect profile. A second diagnostic strategy is the exclusive approach, which eliminates symptoms of ambiguous etiology (e.g., fatigue), from the list of depression criteria. The resulting diagnoses are likely to be highly specific but they are achieved at the cost of diminished sensitivity. The third diagnostic strategy is the substitutive approach (Endicott, 1984). It eliminates depressive symptoms likely confounded by physical illness (e.g., loss of energy and weight loss) and replaces them with more psychologically based items (e.g., irritability and tearfulness). This third strategy minimizes etiologic ambiguities but permits few comparisons to other reports of late-life depression.

DSM-IV is more inclusive than DSM-III, but it still requires that clinicians make an etiological determination of whether a particular symptom reflects the physiological effects of an organic illness, or whether it is a manifestation of a mood disorder. Given the benefits and limitations of the standard DSM system and its alternatives, Koenig et al. (1995) investigated the ability of mental health clinicians to assess major depression within these diverse diagnostic systems. They found that these specialists could be trained to make reliable etiological judgments and that the alternative classificatory strategies produced only marginal, if any, advantage relative to the DSM-IV approach. It appears justified, therefore, to maintain DSM-IV parameters for the differential diagnosis of depression versus physical illness. However, this approach can only be reliable in primary care practice when generalist physicians are skilled in distinguishing the subtleties of a symptom's presentation and employ a proper threshold to determine whether the symptom is physically or psychiatrically founded (Koenig, George, Peterson, & Pieper, 1997). For example, patients with a chronic depression perceive their physical illnesses as much more serious and incapacitating than would an objective assessment of these symptoms (Schrader, 1997). Since depressed mood and perception of physical illness can be mutually reinforcing, the primary care physician must be careful to avoid attributing undue etiological significance to physical symptoms on the basis of their sheer reported severity.

Depression Case-Finding

Accepting that DSM-IV, or its DSM-IV-PC variation (American Psychiatric Association, 1995), will constitute the diagnostic framework within which depression is distinguished from physical illness, procedures are needed to identify older primary care patients at high risk of experiencing a major depression. The value of efficiently constructing what Katon et al. (1997) have described as an "epidemiological map" of mood morbidity is evident in findings such as those by Wells et al. (1989) that depressed patients perceive their general health and ability to perform expected roles as poor, and by Covinsky, Fortinsky, Palmer, Kresevic, and Landefeld (1997) that the physical health of medical patients with depressive symptoms is less likely to improve and more likely to deteriorate.

As an initial step in the differential diagnosis of physical illness and depression, or to clarify whether they are co-occurring, a thorough physical exam should be performed to rule out any organic problems that might be risk factors in the onset or exacerbation of depression. Particular attention should be paid to the cardiovascular, cerebrovascular, and endocrine systems, as various dysfunctions in these systems have been associated with depression (see Lyness & Caine, Chapter 3, this volume). Clinical signs and symptoms suggestive of possible neurodegenerative illnesses associated with depression such as Parkinson's disease, Huntington's disease, Alzheimer's disease or normal pressure hydrocephalus should be carefully screened out (Shua-Haim et al., 1998).

The number of screening tests to be performed following a thorough somatic anamnesis and physical exam is subject to quite diverse thinking (Mookhoeg, Sterrenburg, & Nieuwegiessen, 1998). Most clinicians would agree, however, that screening blood work should include an assessment of complete blood count (CBC) and differential to rule out anemia or occult signs of infection; a thyroid profile to rule out hyper- or hypothyroidism; and serum electrolytes, blood, urea, nitrogen (BUN), creatinine and calcium to rule out occult metabolic/hormonal abnormalities. Serum B12 levels are frequently low in older persons and should be checked, since a deficiency state can lead to depression as well as cognitive dysfunction, paranoid psychosis, anemia, and neurological symptoms (Blazer, Busse, Craighead, & Evans, 1996). More recently, attention has focused on the role of sex hormones in mood disorders and to the replacement of deficiency states in late-life for both men and women (Lavretsky, 1998; Seidman & Walsh, 1999).

The dexamethasone suppression test (DST) was once recommended as a diagnostic blood test for depression-induced dysregulation of the adrenal–pituitary axis. However, many other stressful conditions such as physical illness or certain drugs can impair the DST, causing it to remain too nonspecific to be useful as a biological screen for depression. A blunted response of thyroid-stimulating hormone to an infusion of thyrotropin-releasing hormone has been reported to occur in depression, but it is also seen with advancing age, again rendering a blunted response useless as a screening tool. Sleep parameter changes measured by electroencephalography (EEG), such as increased sleep discontinuity, reduced slow-wave sleep (Stages 3 and 4), and decreased REM latency (i.e., the time from sleep onset to the first REM period) have been shown to be reliable markers for depression. Sleep EEG studies are expensive, however, and not available in many medical centers, thus limiting their use to the differentiation of complex or ambiguous cases rather than as a case-finding tool in routine practice.

Depression has long been recognized as a common complication of stroke. Recent research has broadened and refined this clinical phenomenon to include the presence of microstrokes evident on magnetic resonance imaging (MRI) as white matter hyperintensities (WMHs). The strength of co-occurring hypertension, coronary artery disease, and vascular dementia with depressive syndromes has led to the designation of vascular depression as a subtype of mood disorders. The hypothesized mechanism of this subtype requires vascular disruption through ischemic lesions of the prefrontal circuits necessary for the maintenance of mood, or cumulative ischemic events exceeding a threshold of damage that disregulates mood (Alexopoulos et al., 1997; Krishnan, 1993). While an MRI of the brain is not an efficient or effective screening tool for late-life depression, it can help the clinician in the differential diagnosis of neurological conditions. This is particularly true when the patient presents with focal, soft neurological signs.

In addition to the physical exam and laboratory tests pertinent to distinguishing physical and affective illnesses, various psychiatric rating scales are available for case-finding and diagnostic purposes. The first type of such scales follows the previously described exclusive diagnostic approach; that is, it is devoid of items that potentially overlap both depression and physical illness or dementia (e.g. Alexopoulos, Abrams, Young, & Shamoian, 1988a, 1988b). A second type of rating scale utilizes informant responses as well as patient self-reports to enhance the validity and reliability of the gathered data. The third type of scale is administered to patients alone. A number of psychometrically sound instruments are available for this purpose. Perhaps that with the longest history of use in the primary care sector is the Center For Epidemiologic Studies—Depression Scale (CES-D; Radloff, 1977), which consists of 20 items assessing the patient's functioning in the past week. A study by Schulberg et al. (1998), in which the CES-D was administered to primary care patients age 60 and older, found that with a point prevalence of 9% for major depression in this clinical population, the

CES-D had a positive predictive power (PPP) of 34.6% when the cutscore was set at 11. As would be expected, the PPP of the CES-D to identify major depression increased to 51.6% when the cutscore was elevated to ≥ 26.

The other instrument commonly administered to older primary care patients for case identification purposes is the Geriatric Depression Scale (GDS; Lyness et al., 1997; Yesavage, Brink, Rose, & Adey, 1983). It consists of 30 items assessing the patient's functioning during the past week. Garrard et al. (1998) found that over a 2-year period, approximately half of the older patients scoring ≥ 11 on this instrument were identified by their primary care physicians as possibly depressed. In keeping with similar findings from research conducted with midlife primary care patients (Schwenk, Coyne, & Fechner-Bates, 1996), Garrard et al. found primary care physicians seven times more likely to recognize feelings of depression among patients scoring 26–30 versus those scoring only 6–10. While multi-item case-finding instruments typically are more psychometrically acceptable than brief questionnaires, investigators have found that 10, four-, and even single-item versions of the GDS are valid and reliable when utilized to identify older depressed primary care patients (Chochinov, Wilson, Enns, & Lander, 1997; Mahoney et al., 1994; van Marwijk et al., 1995).

Yet another self-report instrument is the Symptoms Checklist (SCL-90), from which a Depression Factor Score can be extracted. Magni, Schifano, and deLeo (1986) found this score to have high sensitivity and specificity with an older medical population. However, the SCL-90 is more difficult to complete compared to the GDS since each item has five choices compared to only two choices on the GDS. Finally, the Primary Care Evaluation of Mental Diseases–MD (PRIME–MD), which includes two items pertaining to depression in its patient-completed questionnaire, can help primary care physicians identify cases of psychiatric illness. Surprisingly, however, case identification with the PRIME-MD did not lead providers to increase the frequency with which they intervened in the mood or other psychiatric disorder of older patients (Valenstein et al., 1998).

Instruments that use objective raters to assess the severity of a major depression are also applicable to the assessment of older primary care patients who may be experiencing co-occurring physical illnesses. Of particular utility is the widely used Hamilton Rating Scale for Depression (HRSD), whose content has been analyzed with regard to whether the total score is confounded when an older patient is experiencing cognitive or physical impairment. Mulsant et al. (1994) found this not to be the case since medical burden accounted for less than 6% of the variance in baseline HRSD scores among geriatric patients admitted to a psychiatric service. Mulsant et al. noted that the endorsement by older patients of somatic symptoms likely indicates greater depressive severity rather than physical illness and medical burden, a conclusion consistent with that reported earlier by Lyness, Caine, and Conwell (1993). Clinicians utilizing the HRSD may also wish to consider the "Extracted" version of this instrument, developed by Rapp, Smith, and Britt (1990), in which the rating scale's items have been modified to match the DSM-based Schedule for Affective Disorders, thus enhancing its capacity to serve case identification and diagnostic purposes within the standard psychiatric nomenclature.

A key premise underlying the formulation of a diagnosis using any of the previously described instruments is that a diagnosis indicates the need for treatment. Gurland, Cross, and Katz (1996) suggest, however, that while a psychiatric diagnosis conveys significant information about the disorder's clinical course when treated or untreated, this bundled information is too imprecise for purposes of developing a patient-specific treatment plan. They, therefore, have constructed a multidimensional Index of Affective Suffering, with its three most severe levels arguing strongly for the need to carefully evaluate and likely treat patients meeting their criteria (see Gurland et al., Chapter 15, this volume).

TREATMENT STRATEGIES

Our review of procedures available for the diagnosis of physical illness and/or depression suggested that symptoms of a mood disorder warrant treatment regardless of whether they are associated with, or independent of, physical illness. That physical illness per se is not a deterrent to effective treatment of late-life mood disorder is supported by the report of Miller, Paradis et al. (1996), who found no relationship between baseline degree of chronic physical illness and acute treatment outcome, including time to recovery among older patients treated for recurrent major depression. However, nonresponders were more likely to perceive their baseline physical health as only fair to poor. Therefore, Miller, Schulz, et al. (1996) caution physicians about the need to overcome therapeutic nihilism when treating this subgroup of their patients.

Our previously described model of the relationship between physical illness, disability, and depression (see Figure 11.1) suggests that different classes of treatments vary in their mechanisms for achieving specific outcomes. Thus, medications alleviate depressive symptoms more directly and rapidly by impacting on physiological pathways. Psychotherapies help older persons reduce depression by identifying interpersonal or practical problems, and exploring and implementing solutions, a process that leads to the alteration of selection and compensation processes. One could expect that older depressed persons also experiencing physical illnesses would improve more rapidly with medication than psychotherapy, but we know of no pertinent evidence. The remainder of this chapter, therefore, reviews: (1) the ever-increasing evidence regarding the safety and efficacy of diverse treatments for late-life major depression, and (2) issues associated with the delivery of high-quality interventions in the primary care practice setting.

Antidepressant Medication

The National Institutes of Health (NIH) Consensus Panel Development Conference (1992) and the Depression Guideline Panel of the Agency for Health Care Policy and Research (AHCPR) (1993) concluded that pharmacological interventions effectively reduce symptoms and signs of depression in older adults. However, the NIH Consensus Panel noted that its conclusion was based on studies conducted with patients younger than 70 years of age, and the AHCPR Panel cautioned that nearly all drug treatment trials to that time with depressed older patients had recruited medically healthy subjects in psychiatric facilities. Salzman (1994) also emphasized the ambiguous efficacy of continuation phase pharmacotherapy in late-life, since it remained unclear whether depression with extensive physical comorbidity required medications of different types or durations.

Further research during the mid- and late 1990s has reduced these concerns. A meta-analysis of 21 drug–placebo comparisons resulting from studies conducted with diverse populations found a mean posttreatment difference of –5.78 points on the HRSD favoring the active medication (McCusker, Cole, Keller, Bellavance, & Berard, 1998). Most investigators and clinicians presently would agree that antidepressants, regardless of class, improve depressive symptomatology in approximately 60% of older patients, while the placebo response rate is only 30–40% (Alexopoulos, 1998; Mittman et al., 1997). Small (1998) suggests, therefore, that choice of a specific medication should depend in part on its side-effects profile. For example, sedating drugs such as nefazodone might best be prescribed for older patients experiencing restlessness and insomnia, while less sedating compounds such as desipramine should be considered for those patients displaying psychomotor retardation. Zisook and Downs (1998) caution that because of pharmacokinetic and pharmacodynamic changes associated with aging, lower doses of medication and more gradual dose increases are required in treating older depressed patients. Additionally, medications should be selected that have minimal antihistamine, anticholinergic, and antiadrenergic effects; minimal cardiovascular risk; and minimal drug–drug interactions.

Tricyclic and heterocyclic antidepressants remain useful in treating depressed older primary care patients, but presently there is general agreement that the medications of choice are selective serotonin reuptake inhibitors (SSRIs; Kamath, Finkel, & Moran, 1996; Newhouse, 1996). Although their overall efficacy is similar to that of the earlier drug classes, the SSRIs have the major advantage of a markedly better side-effects profile. Therefore, they can generally be used safely in patients with benign prostatic hypertrophy, constipation, glaucoma, and cardiovascular disease. Even when poorly tolerated, the SSRIs typically produce subjective discomfort rather than medical risks. It should be noted, however, that the SSRIs interact with other medications frequently prescribed the elderly for physical illnesses. These interactions mainly result from inhibition of P450 cytochrome isoenzymes and protein displacement of other drugs such as warfarin or digitoxin.

Psychotherapies

The several available psychotherapies constitute a second possible treatment for older primary care patients experiencing a major depression, since many such patients have medical contraindications to an antidepressant drug and/or are experiencing life stresses such as loss of social support requiring a social treatment. The body of research supporting psychotherapy's efficacy with older patients is small but growing. Scogin and McElreath (1994) meta-analyzed several adequately designed studies conducted with such persons experiencing major depression and found psychotherapy's mean effect size to be .76, a clinical outcome similar to that obtained with midlife depressed patients. These findings and others led Niederehe (1996) to conclude that old age alone does not contraindicate the use of psychotherapy, nor does aging predict decreased responsiveness to psychosocial treatments.

Which psychotherapies are applicable to the treatment of older depressed primary care patients who also are possibly troubled by a co-occurring physical illness? The best validated appear to be cognitive-behavioral psychotherapy, interpersonal psychotherapy, and problem solving therapy. Cognitive-behavioral therapy emphasizes the importance of distorted thoughts and the lack of pleasurable activities in the development of affective disorders. Older adults respond well to the highly structured manner in which this therapy seeks to eliminate distorted thought systems or to increase the number of enjoyable daily activities (Thompson, 1996). Many older persons also like the treatment's focus on "here and now" issues rather than problems from the past. Treatment modifications such as presenting the material more slowly and maintaining an active learning process are indicated with older patients, particularly when they are experiencing difficulties with information processing.

Interpersonal psychotherapy (IPT) acknowledges the many losses and role transitions, and isolation experienced by older persons and focuses, therefore, on the manner in which these social developments contribute to the onset and prolongation of a depressive episode. As with younger patients, IPT concentrates on one or more problem areas: grief, interpersonal disputes, role transitions, and interpersonal deficits. Frank et al. (1993) suggest several modifications of IPT needed to engage the older depressed patient. They include limiting the length of the standard psychotherapy session when the patient is physically frail; helping the patient tolerate rather than resolve difficult relationships with friends and family; and maintaining an active rather than neutral therapeutic stance in working with the patient.

Problem-solving therapy considers depressive symptoms in late-life to be associated with functional limitations and psychological problems. Thus, depressive symptoms will resolve if these limitations and problems are approached and resolved in a structured and planned way. Even when a problem is only partially resolved, Mynors-Wallis (1996) speculates that the very act of tackling a problem helps the patient gain a sense of reasserting control over his or her life, which in turn likely improves mood.

In addition to the psychosocial interventions directed at depressed older patients, it also is well recognized that their families play critical roles in the treatment process. Thus, psychoeducational workshops involving family members are increasingly utilized as part of a comprehensive effort to resolve the older patient's depression. Sherrill, Frank, Geary, Stack, and Reynolds (1997) have found that high workshop attendance by family members was associated with a lower rate of patient dropout during continuation treatment and better compliance with prescribed medication.

Combined Treatments

There has been much conjecture during the past decade about the relative efficacy of antidepressant medications and psychotherapies. Niederehe (1996) considers the question unresolved and emphasizes that in routine clinical practice, combined therapy rather than monotherapy is quite common and perhaps should constitute the accepted standard of care. This principle appears particularly applicable to patients with histories of recurrent depressive episodes, and those who are 70+ years old and experience excellent short-term but brittle long-term response (Reynolds, Frank, Dew, et al., 1999). In such populations, the combination of the antidepressant nortriptyline and IPT led to recurrence rates of only 20% over 3 years, an outcome significantly better than that achieved with IPT alone and placebo alone, and trending toward superior efficacy over nortriptyline alone (Reynolds, Frank, Perel, et al., 1999; Reynolds, Miller, Mulsant, Dew, & Pollack, Chapter 13, this volume). Thus, when feasible, combination therapy should be the intervention of choice for depressed older adults.

Implementing Efficacious Treatments

The availability of efficacious treatments for the resolution of late-life major depression makes rather distressing the repeated finding that such pharmacological and psychosocial interventions are not routinely delivered in primary care practice. For example, Glasser and Gravdal (1997) determined that primary care physicians consider treatment of depression in older persons important. However, these physicians often find such patients frustrating and lack adequate time to work with them on psychiatric illnesses. In a related study, Callahan, Dittus, and Tierney (1996) found significant inconsistencies between physicians' frequent formulations of treatment plans for late-life depression and their infrequent implementation of these plans. Indeed, treatment was received by fewer than half of the older depressed patients for whom an intervention was intended.

The gap between the availability of efficacious treatments and their poor implementation in routine practice has been attributed by Unutzer, Katon, Sullivan, and Miranda (1999) to barriers associated with the patient, the provider, and the policies governing health care delivery. Patient factors obstructing the treatment process include negative attitudes by older persons toward psychiatric morbidity and their concerns about the stigma associated with acknowledging a depression; misattribution of depressive symptoms to the normal aging process; physical illnesses that can intensify the side effects of antidepressant medication; and sensory and cognitive impairments that contribute to decreased adherence to recommended treatments. Provider barriers to initiating efficacious interventions include the previously reviewed complexities in diagnosing depression when it co-occurs with physical illness; the primary care physician's reluctance to pursue the management of a mood disturbance given the extensive time which this will require; and a sense of inadequate consultative support from mental health specialists. Health policy barriers to the primary care physician's pursuing needed care of his or her depressed older patients include insurance policies that "carve out" psychiatric services to the mental health specialist sector; poor reimbursement to primary care physicians for the time spent counseling patients; and higher copayments by the patient for mental health compared to medical services. Resolution of these problems will not be easily achieved. Unutzer et al. (1999), therefore, recommend a broad array of strategies that over time will produce a cumulative impact and narrow the gap between our state of knowledge regarding efficacious management of late-life depression and the inadequate manner in which this disorder presently is treated.

SUMMARY

Depression in older primary care patients often co-occurs with physical illness and thereby creates diagnostic and treatment complexities for the clinician managing such morbidity. Thus, discerning the etiology of co-occurring mood disturbance and physical illness for purposes of differential diagnosis has been the subject of much attention. Various strategies can be employed in this assessment process and include ever more sophisticated laboratory procedures and psychiatric rating scales of the self-administered and objective-rater types. Simultaneous with the refinement of assessment tools for differential diagnostic purposes has been a growing sense that affective distress requires treatment regardless of whether its etiology is psychiatric or organic in nature. This perspective has merit given the present availability of efficacious antidepressant medications and psychotherapeutic interventions that can be feasibly delivered in the primary care setting. Monotherapies or combined therapy have been shown to produce better clinical and functional outcomes than placebo; and it is lamentable, therefore, that these interventions are utilized infrequently rather than routinely. The challenge of the new millennium, therefore, is to continue perfecting treatments for the depressed older primary care patient with co-occurring physical illnesses but also to overcome the patient, provider, and system barriers constraining their delivery in routine ambulatory medical practice.

REFERENCES

Alexopoulos, G. (1998, July). *Epidemiology, nosology, and treatment of geriatric depression.* Paper presented at Hartford Foundation Conference on Exploring Opportunities To Advance Mental Health Care For An Aging Population, Rockville, MD.

Alexopoulos, G., Abrams, R., Young, R., & Shamoian, C. (1988a). Cornell Scale for depression in dementia. *Biological Psychiatry, 23,* 271–284.

Alexopoulos, G., Abrams, R., Young, R., & Shamoian, C. (1988b). Use of the Cornell Scale on non-demented patients. *Journal of American Geriatric Society, 36,* 230–236.

Alexopoulos, G., Meyers, B., Young, R., Campbell, S., Silbersweig, D., & Charlson, M. (1997). "Vascular depression" hypothesis. *Archives of General Psychiatry, 54,* 915–922.

Alexopoulos, G., Vrontou, C., Kakuma, T., Meyers, B., Young, R., Klausner, E., & Clarkin, J. (1996). Disability in geriatric depression. *American Journal of Psychiatry, 153,* 877–885.

American Psychiatric Association. (1995). *Diagnostic and statistical manual of mental disorders: (4th ed.), Primary Care Version.* Washington, DC: Author.

Armenian, H., Pratt, L., Gallo, J., & Eaton, W. (1998). Psychopathology as a predictor of disability: A population-based follow-up study in Baltimore, Maryland. *American Journal of Epidemiology, 148,* 269–275.

Banazak, D. (1996). Late-life depression in primary care. How well are we doing? *Journal of General Internal Medicine, 11,* 163–167.

Blazer, D., Busse, E., Craighead, E., & Evans, D. (1996). Use of the laboratory in the diagnostic workup of older adults. In E. Busse & D. Blazer (Eds.), *Textbook of Geriatric Psychiatry* (2nd ed., pp. 191–209). Washington, DC: American Psychiatric Press.

Blazer, D., Hughes, D., & George, L. (1987). The epidemiology of depression in an elderly community population. *Gerontologist, 27,* 281–287.

Caine, E., Lyness, J., & Conwell, Y. (1996). Diagnosis of late-life depression: Preliminary studies in primary care settings. *American Journal of Geriatric Psychiatry, 4*(Suppl. 1), S45–S50.

Callahan, C., Dittus, R., & Tierney, W. (1996). Primary care physicians' medical decision making for late-life depression. *Journal General of Internal Medicine, 11,* 218–225.

Callahan, C., Hendrie, H, Dittus, R., Brater, D., Hui, S., & Tierney, W. (1994). Improving treatment of late-life depression in primary care: A randomized clinical trial. *Journal of American Geriatric Society, 42,* 839–846.

Callahan, C., Kesterson, J., & Tierney, W. (1997). Association of symptoms of depression with diagnostic test charges among older adults. *Annals of Internal Medicine, 126,* 426–432.

Callahan, C., Nienaber, N., Hendrie, H., & Tierney, W. (1992). Depression of elderly outpatients: Primary care physician's attitudes and practice patterns. *Journal of General Internal Medicine, 7,* 26–31.

Callahan, C., Wolinsky, F., Stump, T., Nienaber, N., Hui, S., & Tierney, W. (1998). Mortality, symptoms, and functional impairment in late-life depression. *Journal of General Internal Medicine, 13,* 746–752.

Chochinov, H., Wilson, K., Enns, M., & Lander, S. (1997). "Are you depressed?" Screening for depression in the terminally ill. *American Journal of Psychiatry, 154,* 674–676.

Cohen-Cole, S., & Stoudemire, A. (1987). Major depression and physical illness: Special considerations in diagnosis and biologic treatment. *Psychiatric Clinics of North America, 10,* 1–17.

Covinsky, K., Fortinsky, R., Palmer, R., Kresevic, D., & Landefeld, S. (1997). Relation between symptoms of depression and health status outcomes in acutely ill hospitalized older persons. *Annals of Internal Medicine, 126,* 417–425.

Creed, F., & Ash, G. (1992). Depression in rheumatoid arthritis: Aetiology and treatment. *International Review of Psychiatry, 4,* 23–34.

Cummings, J. (1992). Depression and Parkinson's disease: A review. *American Journal of Psychiatry, 149,* 443–454.

Deitch, J., & Zetin, M. (1983). Diagnosis of organic depressive disorders. *Psychosomatics, 24,* 971–979.

Depression Guideline Panel. (1993). *Depression in primary care: Vol. 2. Treatment of major depression. Clinical practice guideline, No. 5.* Rockville, MD: U.S. Department of Health and Human Services, Public Health Service, Agency For Health Care Policy and Research. Publication No. 93-0551.

Dew, M. (1998). Psychiatric disorder in the context of physical illness. In B. Dohrenwend (Ed.), *Adversity, stress, and psychopathology* (pp. 177–186). New York: Oxford University Press.

Dexter, P., & Brandt, K. (1994). Distribution and predictors of depressive symptoms in osteoarthritis. *Journal of Rheumatology, 21,* 279–286.

Endicott, J. (1984). Measurement of depression in patients with cancer. *Cancer, 53,* 2243–2247.

Frank, E., Frank, N., Cornes, C., Imber, S., Miller, M., Morris, S., & Reynolds, C. (1993). Interpersonal psychotherapy in the treatment of late-life depression. In G. Klerman & M. Weissman (Eds.), *New applications of interpersonal psychotherapy* (pp. 167–198). Washington, DC: American Psychiatric Press.

Garrard, J., Rolnick, S., Nitz, N., Luepke, L., Jackson, J., Fischer, L., Leibson, C., Bland, P., Heinrich, R., & Waller, L. (1998). Clinical detection of depression among community-based elderly people with self-reported symptoms of depression. *Journals of Gerontology: Series A, Biological Sciences and Medical Sciences, 53,* M92–M101.

Glasser, M., & Gravdal, J. (1997). Assessment and treatment of geriatric depression in primary care settings. *Archives of Family Medicine, 6,* 433–438.

Gurland, B., Cross, P., & Katz, S. (1996). Epidemiological perspectives on opportunities for treatment of depression. *American Journal of Geriatric Psychiatry, 4*(Suppl. 1), S7–S13.

Heckhausen, J., & Schulz, R. (1995). A life-course theory of control. *Psychological Review, 102,* 284–304.

Hendrie, H., Callahan, C., Levitt, E., Jui, S., Musick, B., Austrom, M., Nurnberger, J., & Tierney, W. (1995). Prevalence rates of major depressive disorders: The effects of varying the diagnostic criteria in an older primary care population. *American Journal of Geriatric Psychiatry, 3,* 119–131.

Kamath, M., Finkel, S., & Moran, M. (1996). A retrospective chart review of antidepressant use, effectiveness, and adverse effects in adults age 70 and older. *American Journal of Geriatric Psychiatry, 4,* 167–172.

Katon, W. (1982). Depression: Somatic symptoms and medical disorders in primary care. *Comprehensive Psychiatry, 23,* 274–287.

Katon, W. (1996). The impact of major depression on chronic medical illness. *General Hospital Psychiatry, 18,* 215–219.

Katon, W., Von Korff, M., Lin, E., Unutzer, J., Simon, G., Walker, E., Ludman, E., & Bush, T. (1997). Population-based care of depression: Effective disease management strategies to decrease prevalence. *General Hospital Psychiatry, 19,* 169–178.

Katz, I. (1996). On the inseparability of mental and physical health in aged persons: Lessons from depression and medical comorbidity. *American Journal of Geriatric Psychiatry, 4,* 1–16.

Kennedy, G., Kelman, J., & Thomas, C. (1990). The emergence of depressive symptoms in late-life: The importance of declining health and increasing disability. *Journal of Community Health, 15,* 93–104.

Koenig, H., George, L., Peterson, B., & Pieper, C. (1997). Depression in medically ill hospitalized older adults: Prevalence, characteristics, and course of symptoms according to six diagnostic schemes. *American Journal of Psychiatry, 154,* 1376–1383.

Koenig, H., & Kuchibhatla, M. (1999). Use of health services by medically ill depressed elderly patients after hospital discharge. *American Journal of Geriatric Psychiatry, 7,* 48–56.

Koenig, H., Pappas, P., Holsinger, T., & Bachar, J. (1995). Assessing diagnostic approaches to depression in medically ill older adults: How reliably can mental health professionals make judgments about the course of symptoms? *Journal of American Geriatric Society, 43,* 472–478.

Krisham, K. (1993). Neurotomic substrates of depression in the elderly. *Journal of Geriatric Psychiatry and Neurology, 1,* 39–58,

Krishnan, H., & Gadde, K. (1996). The pathophysiological basis for late-life depression: Imaging studies of the aging brain. *American Journal of Geriatric Psychiatry, 4*(Suppl. 1), S22–S33.

Krishnan, K., Hays, J., & Blazer, D. (1997). MRI-defined vascular depression. *American Journal of Psychiatry, 154,* 497–501.

Lavretsky, J. (1998, March). Late-life depression: Risk factors, treatment, and sex differences. *Clinical Geriatrics, 6,* 13–24.

Lyness, J., Bruce, M., Koenig, H, Parmelee, P., Schulz R., Lawton, P., & Reynolds, C. (1996). Depression and medical illness in late-life: Report of a symposium. *Journal of American Geriatric Society, 44,* 198–203.

Lyness, J., Caine, E., & Conwell, Y. (1993). Depressive symptoms, medical illness, and functional status in depressed psychiatric inpatients. *American Journal of Psychiatry, 150,* 910–915.

Lyness, J., Noel, T., Cox, C., King, D., Conwell, Y., & Caine, E. (1997). Screening for depression in elderly primary care patients: A comparison of the CES-D and the GDS. *Archives of Internal Medicine, 157,* 449–454.

Magni, F., Schifano, F., deLeo, D. (1986). Assessment of depression in an elderly medical population. *Journal of Affective Disorders, 11,* 121–124.

Mahoney, J., Drinka, T., Able, R., Gunter-Hunt, G., Matthews, C., Gravenstein, S., & Carnes, M. (1994). Screening for depression: Single question versus GDS. *Journal of American Geriatrics Society, 42,* 1006–1008.

McCusker, J., Cole, M., Keller, E., Bellavance, F., & Berard, A. (1998). Effectiveness of treatments of depression in older ambulatory patients. *Archives of Internal Medicine, 158,* 705–712.

Medicare Payment Advisory Commission. (1998, June). *Report to Congress: Context for change in medical programs.* Washington, DC: Author.

Miller, M., Paradis, C., Houck, P., Rifai, H., Mazumdar, S., Pollock, B., Perel, J., Frank, E., & Reynolds, C. (1996). Chronic medical illness in patients will recurrent major depression. *American Journal of Geriatric Psychiatry, 4,* 281–290.

Miller, M., Schulz, R., Paradis, C., Houck, P., Mazumdar, S., Frank, E., Dew, M., & Reynolds, C. (1996). Changes in perceived health status of depressed elderly patients treated until remission. *American Journal of Psychiatry, 53,* 1350–1352.

Mittmann, N., Herrmann, N., Einarson, T., Busto, U., Lanctot, K., Liu, B., Shulman, K., Silver, I., Naranjo, C., & Shear, N. (1997). The efficacy, safety and tolerability of antidepressants in late-life depression: A meta-analysis. *Journal of Affective Disorders, 46,* 191–217.

Mookhoeg, E., Sterrenburg, V., & Nieuwegiessen, I. (1998). Screening for somatic disease in elderly psychiatric patients. *General Hospital Psychiatry, 20,* 102–107.

Morrison, J. (1997). *When psychological problems mask medical disorders: A guide for psychotherapists.* New York: Guilford Press.

Mulsant, B., Sweet, R., Rifai, A., Pasternack, R., McEachran, A., & Zubenko, G. (1994). The use of the Hamilton Rating Scale for Depression in elderly patients with cognitive impairment and physical illness. *American Journal of Geriatric Psychiatry, 2,* 220–229.

Mynors-Wallis, L. (1996). Problem-solving treatment: Evidence for effectiveness and feasibility in primary care. *International Journal of Psychiatry in Medicine, 26,* 249–262.

National Institutes of Health Consensus Panel Development Conference. (1992). Diagnosis and treatment of depression in late-life. *Journal of American Medical Association, 268,* 1018–1024.

Newhouse, P. (1996). Use of serotonin selective reuptake inhibitors in geriatric depression. *Journal of Clinical Psychiatry, 57*(Suppl. 5), 12–22.

Niederehe, G. (1996). Psychosocial treatments with depressed older adults. *American Journal of Geriatric Psychiatry, 4*(Suppl. 1), S66–S78.

Ormel, J., Kempen, G., Penninx, B., Brilman, E., Beekman, A., & Van Sonderen, E. (1997). Chronic medical conditions and mental health in older people: Disability, and psychosocial resources mediate specific mental health effects. *Psychological Medicine, 27,* 1065–1977.

Parmelee, P., Katz, I., & Lawton, M. (1992a). Incidence of depression in long-term care settings. *Journal of Gerontology: Medical Sciences, 46,* M189–M196.

Parmelee, P., Katz, I., & Lawton, M. (1992b). Depression and mortality among institutionalized aged. *Journal of Gerontology: Psychological Sciences, 47,* P3–P10.

Radloff, L. (1977). The CES-D scale: A self-report depression scale for research in the general population. *Applied Psychological Measurement, 1,* 385–401.

Rapp, S., Smith, S., & Britt, M. (1990). Identifying comorbid depression in elderly medical patients: Use of the Extracted Hamilton Depression Rating Scale. *Psychological Assessment, 2,* 243–247.

Reynolds, C., Frank, E., Dew, M., Houck, P., Miller, M., Mazumdar, S., Perel, J., & Kupfer, D. (1999). Treatment of 70+-year-olds with recurrent major depression. *American Journal of Geriatric Psychiatry, 7,* 64–69.

Reynolds, C., Frank, E., Perel, J., Imber, S., Cornes, C., Miller, M., Mazumdar, S., Houck, P., Dew, M., Stack, J., Pollock, B., & Kupfer, D. (1999). Nortriptyline and interpersonal psychotherapy as maintenance therapies for recurrent major depression. *Journal of American Medical Association, 281,* 39–45.

Salzman, C. (1994). Pharmacological treatment of depression in elderly patients. In L. Schneider, C. Reynolds, & B. Lebowitz (Eds.), *Diagnosis and treatment of depression in late-life* (pp. 181–244). Washington, DC: American Psychiatric Press.

Schrader, G. (1997). Subjective and objective assessments of medical comorbidity in chronic depression. *Psychosomatics, 66,* 258–260.

Schulberg, H., McClelland, M., & Burns, B. (1987). Depression and physical illness: The prevalence, causation, and diagnosis of comorbidity. *Clinical Psychology Review, 7,* 145–167.

Schulberg, H., Mulsant, B., Schulz, R., Rollman, B., Houck, P., & Reynolds, C. (1998). Characteristics and course of major depression in older primary care patients. *International Journal of Psychiatry in Medicine, 28,* 461–476.

Schulz, R., & Heckhausen, J. (1996). A life span model of successful aging. *American Psychologist, 51,* 702–714.

Schulz, R., Heckhausen, J., & O'Brien, A. T. (1994). Control and the disablement process in the elderly. In D. S. Dunn (Ed.), Psychosocial perspectives on disability [Special issue]. *Journal of Social Behavior and Personality, 9*(5), 139–152.

Schwenk, S., Coyne, J., & Fechner-Bates, S. (1996). Differences between detected and undetected patients in primary care and depressed psychiatric patients. *General Hospital Psychiatry, 18,* 407–415.

Scogin, F., & McElreath, L. (1994). Efficacy of psychosocial treatments for geriatric depression: A quantitative review. *Journal of Consulting and Clinical Psychology, 62,* 69–74.

Seidman, S., & Walsh, B. (1999). Testosterone and depression in aging men. *American Journal of Geriatric Psychiatry, 7,* 18–33.

Sherrill, J., Frank, E., Geary, M., Stack, J., & Reynolds, C. (1997). Psychoeducational workshops for elderly patients with recurrent major depression and their families. *Psychiatric Services, 48,* 76–81.

Shua-Haim, J., Shua-Haim, V., Comsti, E., & Ross, J. (1998, April). Depression in the hospitalized elderly. *Clinical Geriatrics, 6,* 43–56.

Small, G. (1998). Treatment of geriatric depression. *Depression and Anxiety, 8*(Suppl. 1), 32–42.

Spencer, K., Tompkins, C., & Schulz, R. (1997). Assessing depression in patients with brain pathology: The case of stroke. *Psychological Bulletin, 122,* 132–152.

Stoudemire, A. (Ed.). (1995). *Psychological factors affecting medical conditions.* Washington, DC: American Psychiatric Press.

Thompson, L. (1996). Cognitive-behavioral therapy and treatment for late-life depression. *Journal of Clinical Psychiatry, 57*(Suppl. 5), 29–37.

Timberlake, N., Klinger, L., Smith, P., Venn, G., Treasure, T., Harrison, M., & Newman, S. (1997). Incidence and patterns of depression following coronary artery bypass graft surgery. *Journal of Psychosomatic Research, 43,* 197–207.

Turner, R., & Noh, S. (1988). Physical disability and depression: A longitudinal analysis. *Journal of Health and Social Behavior, 29,* 23–37.

Turrima, C., Caruso, R., Esta, R., Lucchi, F., Fazzari, G., Dewey, M., & Ermentini, A. (1994). Affective disorders among elderly general practice patients. *British Journal of Psychiatry, 165,* 533–537.

Unutzer, J., Katon, W., Sullivan, M., & Miranda, J. (1999). Treating depressed older adults in primary care: Narrowing the gap between efficacy and effectiveness. *Milbank Quarterly, 77,* 225–256.

Unutzer, J., Patrick, D., Simon, G., Grembowski, D., Walker, E., Rutter C., & Katon, W. (1997). Depressive symptoms and the cost of health services in HMO patients aged 65 years and older. *Journal of American Medical Association, 277,* 1618–1623.

Valenstein, M., Kales, H., Mellow, A., Dalack, G., Figueroa, S., Barry, K., & Blow, F. (1998). Psychiatric diagnosis and intervention in older and younger patients in a primary care clinic: Effect of a screening and diagnostic instrument. *Journal of American Geriatrics Society, 46,* 1499–1505.

Van Marwijk, H., Wallace, P., de-Bock, G., Hermans, J., Kaptein, A., & Mulder, J. (1995). Evaluation of the feasibility, reliability, and diagnostic value of shortened versions of the Geriatric Depression Scale. *British Journal of General Practice, 45,* 195–199.

Wells, K., Stewart, A., Hays, R., Burnam, M., Rogers, W., Daniels, M., Berry, S., Greenfield, S., & Ware, J. (1989). The functioning and well-being of depressed patients: Results from the Medical Outcomes Study. *Journal of American Medical Association, 262,* 914–919.

Williams, J., Kerber, C., Mulrow, C., Medina, A., & Aguilar, C. (1995). Depressive disorders in primary care: Prevalence, functional disability, and identification. *Journal of General Internal Medicine, 10,* 7–12.

Williamson, G., & Schulz, R. (1992a). Physical illness and symptoms of depression among elderly outpatients. *Psychology and Aging, 7,* 343–351.

Williamson, G., & Schulz, R. (1992b). Pain, activity restriction, and symptoms of depression among community-residing elderly adults. *Journal of Gerontology, 47,* 367–372.

Williams-Russo, P. (1996). Barriers to diagnosis and treatment of depression in primary care settings. *American Journal of Geriatric Psychiatry, 4*(Suppl. 1), S84–S90.

Yesavage, J., Brink, T., Rose, T., & Adey, M. (1983). The Geriatric Rating Scale: Comparison with other self-reports and psychiatric rating scales. In T. Crook, S. Gerris, & R. Bartus (Eds.), *Assessment in geriatric psychopharmacology* (pp. 153–165). New Canaan, CT: Mark Powley Associates.

Zisook, S., Downs, N. (1998). Diagnosis and treatment of depression in late-life. *Journal of Clinical Psychiatry,* 59(Suppl. 4), 80–91.

12

The Relationship of Major Depressive Disorder to Alzheimer's Disease

MYRON F. WEINER and RAMESH SAIRAM

In this chapter, the authors explore controversies in the relationship between major depressive disorder (MDD) and Alzheimer's disease (AD). The initial sections deal with the physiology and pathology of MDD and AD, and the differentiation of MDD from AD. The discussion then includes the cognitive and functional impact of MDD on persons with intact brain function and those with AD, the possible predisposing role of MDD to AD, and AD to MDD; instruments for quantifying MDD in AD; and the types and effects of treatments for MDD in persons with AD. The concluding section deals at length with confounds in the diagnosis of MDD and of AD and proposes criteria specifically for the diagnosis of MDD in persons with dementing illness.

PHYSIOLOGY AND PATHOLOGY

Major Depressive Disorder

MDD is an episodic illness that affects individuals throughout the life span, although it is less common in old age than middle age (Myers et al., 1984). A profound disturbance of mood (see Table 12.1 for a list of symptoms), its symptoms usually abate with or without treatment, but a substantial number of late-life-onset cases do not improve significantly even when treated (Murphy, 1983). Little is known about the pathophysiology of MDD. However, it is likely that MDD is a psychosomatic–somatopsychic disorder in which mental processes impinge on physiological processes and vice versa. In

---

MYRON F. WEINER and RAMESH SAIRAM • Department of Psychiatry, University of Texas Southwestern Medical Center, Dallas, Texas 75235.

Physical Illness and Depression in Older Adults: A Handbook of Theory, Research, and Practice, edited by Gail M. Williamson, David R. Shaffer, and Patricia A. Parmelee. Kluwer Academic/Plenum Publishers, New York, 2000.

Table 12.1. Comparison of DSM-IV Criteria for Major Depressive Disorder (MDD) and Alzheimer's Disease (AD)

MDD	AD
Five (or more) of the following symptoms have been present during the same 2-week period and represent a change from previous function. At least one is depressed mood **or** loss of interest or pleasure.	The development of multiple cognitive deficits manifested by both:
1. Depressed mood most of most days.	1. Impaired learning/memory.
2. Loss of pleasure most of most days.	2. One or more of the following cognitive disturbances: (a) aphasia, (b) apraxia, (c) agnosia, d) impaired executive function (planning, organizing, sequencing, abstracting)
3. Loss of 5% of body weight in a month	3. The deficits in criteria 1 and 2 significantly impair social or occupational function and result in a significant decline in function.
4. Too much or too little sleep nearly every day.	4. Gradual onset and continual cognitive decline.
5. Psychomotor agitation or retardation nearly every day.	5. Not due to other demonstrable condition.
6. Fatigue or loss of energy nearly every day.	
7. Feelings of worthlessness or guilt nearly every day.	
8. Diminished ability to think or concentrate nearly every day	
9. Recurrent suicidal ideas, attempt, or plan.	

MDD, a genetically determined vulnerability or an innate susceptibility seems to be triggered and reinforced by the interaction of psychological and physiological factors. A contributing factor may be an imbalance or dysregulation of norepinephrine neurotransmitter systems (Siever & Davis, 1985). As with the early-onset cases, MDD in older adults is associated with a family history of mood disorder, but the association is weaker than in early-onset MDD. Many elders who become depressed have had prior episodes of depression, suggesting that one episode of depression increases susceptibility to further episodes. Many individuals appear to suffer MDD as a consequence of disorders affecting brain function, such as stroke or Parkinson's disease. However, these individuals often have a history of earlier depressive episodes. For example, 20% of persons with poststroke depression have a history of earlier depressive episodes (Schwartz et al., 1993). There are no known pathological hallmarks of MDD, but a single neuropathological study has associated depression in AD with loss of neurons in the portion of the brain stem nucleus (locus ceruleus) that supplies norepinephrine to the cerebral cortex (Zubenko & Moosy, 1988).

Alzheimer's Disease

AD, a frequent disease of late life, characterized by progressive, unremitting impairment of memory and other cortical functions, is frequently accompanied by emotional and behavioral disturbances. Prevalence estimates for AD range from 2% to 10% in persons age 65 years and older to 13% to 48% in persons age 85 years or older (Weiner & Gray, 1996). Although still the subject of controversy (Neve & Robakis, 1998), the development of AD appears related to abnormal degradation and processing of the normal cell-wall component *amyloid precursor protein* (APP). The normal degradation and processing of this molecule produces a group of soluble compounds that are easily transported from the brain. Abnormal degradation and processing of the molecule leads to deposition in brain tissue of the insoluble, presumably toxic protein *beta amyloid*. This process is most evident in persons with Down's syndrome (trisomy 21), who uniformly develop AD by the fifth decade of life because their APP gene, located on chromosome 21, is reduplicated (Wisniewski, Wisniewksi, & Wen, 1985). The resultant overproduction of APP presumably overwhelms the normal degradative process and results in the production of beta amyloid. In other cases, abnormalities of APP processing appear due to mutations in genes on chromosomes 1, 14, and 21 (Morrison-Bogorad, Weiner, Rosenberg, Bigio, & White, 1997). An additional risk factor for AD is inheritance of the ε4 allele of the cholesterol-transporting molecule apolipoprotein E (Corder, Saunders, & Strittmatter, 1993).

There are known physiological and anatomic changes in the brain of persons with AD. For example, death of specific types of cells (neurons) in specific brain regions significantly reduces availability of the neurotransmitter acetylcholine in many parts of the brain and may also reduce availability of the neurotransmitter norepinephrine (DeKosky, 1996). Acetylcholine is important in memory function (Bartus, Dean, Pontecorvo, & Flicker, 1985). Reduced availability of this neurotransmitter may explain the early and prominent memory loss in AD. On the other hand, the deficiency of norepinephrine might increase vulnerability to depression (Burns, 1991), as it is thought to be important in mood regulation (Schildkraut & Kety, 1967). Finally, the clinical diagnosis of AD can be confirmed by postmortem brain examination. The outer layer of brain tissue (the cortex) thins because neurons die. In addition, large numbers of characteristic microscopic lesions occur in the cerebral cortex and other brain regions (Khachaturian, 1985).

It is not surprising that persons with MDD develop AD. AD is a highly prevalent illness, and there is no reason to believe that MDD protects against AD. Persons with AD may also develop MDD, but interestingly, the prevalence of MDD in AD (as will be discussed later) is much lower than in brain disorders such as stroke, with a reported rate of 25% within 2 years (Robinson & Forrester, 1987), or Parkinson's disease, in which 47% of patients in a large survey rated themselves as significantly depressed on the Beck Rating Scale for Depression (Mayeux, Stern, Rosen, & Leventhal, 1981).

DIFFERENTIATION OF MAJOR DEPRESSIVE DISORDER FROM ALZHEIMER'S DISEASE

Often, persons with AD are diagnosed mistakenly as MDD or as AD with MDD, and rarely, persons with MDD are misdiagnosed as AD. Ordinarily, MDD is easily differentiated from AD (Weiner, 1996). The development of MDD is relatively acute, over days or weeks. The development of AD is usually insidious, over months or years, often becoming obvious only in retrospect. The cognitive complaints of MDD plateau; they reach a nadir and do not worsen. The cognitive difficulties in AD may plateau for months but eventually continue to progress. Persons with MDD frequently complain bitterly of difficulty with memory and concentration, while persons with AD are often unaware of their deficits. Persons with MDD usually experience sustained depressed mood that is manifest on direct examination. Persons with AD usually deny, and do not show, sustained depressed mood on direct examination. They often express sadness or cry when confronted with a task they can no longer perform well, but the sadness and crying disappear when they are distracted from the frustrating task. Persons with MDD express feelings of guilt and lack of worth in terms of past misdeeds. Persons with AD may express similar feelings, but they are concerned with their present inability to function at a previously attainable level.

Neuropsychological testing of persons with depressed mood shows memory deficits attributable to impaired effort or concentration, such as greater impairment in immediate than in delayed recall (Dannenbaum, Parkinson, & Inman, 1988; Hart, Kwentus, Wade, & Hamer, 1987), while persons with AD show greater impairment of delayed recall and multiple other neuropsychological deficits, including language and visuospatial functioning (Paulman, Koss, & MacInnes, 1996).

Persons with depression often have a history of similar episodes, with similar type and degree of cognitive impairment. Family history may not help in differentiating MDD from AD. In one series, half of AD patients with MDD had at least one first-degree relative with MDD as compared with only 11% of nondepressed AD patients (Pearlson et al., 1990). In most cases, neuroimaging procedures such as computerized axial tomography or magnetic resonance imaging (MRI) that reveal brain structure are not useful in distinguishing MDD from AD. Imaging studies related to brain function such as positron-emission tomography (PET) and single-photon emission computerized tomography (SPECT) are more useful in ruling out AD than in making a positive diagnosis of MDD. PET is a direct measure of brain metabolic activity based on its uptake of radioactive glucose. SPECT measures brain blood flow, an indirect measure of brain metabolic activity. PET studies of regional cerebral metabolism show possible decrease of prefrontal lobe metabolism in MDD (Baxter et al., 1989), while in AD, PET studies show a pattern of temporoparietal hypometabolism (Duara et al., 1986). In AD, SPECT studies show reduction of temporoparietal blood flow (Bonte, Weiner, Bigio, & White, 1997).

Electroencephalography (EEG) can also help to distinguish between MDD and AD. The EEG is usually normal in MDD. In AD, there is often diffuse slowing of cortical activity (Robinson et al., 1994).

COGNITIVE IMPAIRMENT IN MAJOR DEPRESSIVE DISORDER

Depressed persons complain frequently of impaired concentration, memory, and problem solving. Indeed, depressed mood impairs attention, perception, speed of cognitive response, problem solving, memory, and learning (Miller, 1975; Weingartner, Cohen, Murphy, Martello, & Gerdt, 1981). In severe cases, this impairment has been termed *pseudodementia* (Kiloh, 1961; Wells, 1979), or the dementia of depression (Folstein & McHugh, 1978). Although most depressed persons function as well as nondepressed persons when dealing with highly structured and organized material, they deal less well with unstructured material. These changes appear related to a general deficit in sustaining effort. Additionally, there appears to be a relationship between severity of depressive symptoms and associated motor and cognitive impairment (Cohen, Weingartner, Smallberg, Pickar, & Murphy, 1982). Clinical experience in the University of Texas Southwestern Medical School's Clinic for Alzheimer's and Related Disorders confirms the impact of depression on subjective cognitive function. Of 19 individuals evaluated for memory complaints, in whom there was not significant evidence of dementing illness, depression of varying degree was present in 42% (Weiner, Bruhn, Svetlik, Tintner, & Hom, 1991).

The cognitive impairments of MDD are largely reversible with treatment, including memory (Sternberg & Jarvik, 1976) and general neuropsychological function (Fromm & Schopflocher, 1984; Henry, Weingartner, & Murphy, 1973). Thus, in theory, treatment of depression in depressed AD patients should improve cognitive performance in addition to overall function.

EFFECT OF MAJOR DEPRESSIVE DISORDER ON COGNITION IN ALZHEIMER'S DISEASE

Depression has been proposed as an important cause of excess disability in AD. In a comparison of 30 community-dwelling persons with AD who were not depressed and 20 community dwelling persons with AD who also met DSM-III (American Psychiatric Association, 1980) criteria for MDD, Pearson, Teri, Reifler, and Raskind (1989) showed depression to have a significant effect on function of persons with AD, based on caregiver reports of patients' ability to perform activities of daily living (ADL). Depressed AD patients were significantly less functional in ADL than nondepressed AD patients with the same severity of cognitive impairment on the Mini-Mental State Exam (MMSE); Folstein, Folstein, & McHugh, 1978).

Pharmacological treatment of depression in AD has been shown to improve ADL performance (Reifler, Larson, Teri, & Poulsen, 1986; Taragano, Lyketsos, Mangone, Allegri, & Comesana-Diaz, 1997) and cognitive functioning (Alexopoulos, Meyers, Young, Mattis, & Kakuma, 1993; Greenwald et al., 1989). These findings, together with MDD-associated decrement in function of both nondemented persons and persons with AD, provide strong support for the notion that MDD worsens both cognitive and ADL functioning in persons with AD. Fortunately, there is no evidence that MDD produces greater decline in cognitive function over time (Lopez, Boller, Becker, Miller, & Reynolds, 1990).

MAJOR DEPRESSIVE DISORDER AS A RISK FACTOR OR PRODROME FOR ALZHEIMER'S DISEASE

A number of studies have shown a high prevalence of AD among persons diagnosed with late-life MDD and suggest that depressive episodes may increase vulnerability to AD. In one study, 57% of depressed nondemented elderly patients developed full-blown dementia symptoms within 3 years (Reding, Haycox, & Blass, 1985); in another study, 6 of 14 survivors (43%) became demented at 15 to 45-month follow-up (Bulbena & Berrios, 1986), and in still another study, 91% of 22 persons became demented over a follow-up period of 4–18 years (Kral & Emery, 1987). None of these studies included careful neuropsychological testing at the time of the depressive episode or diagnosis of the type of dementia that developed subsequently, but the majority were presumably AD. Jorm et al. (1991) performed a pooled analysis of case-control studies to test the association between history of depression and subsequent development of AD. After controlling for age, sex, and education, medically treated depression was associated with the onset of AD, with an overall odds ratio of about 1.8, the association being greater for AD symptoms beginning after 65 years of age. The association also held for episodes occurring more than 10 years before the onset of dementia. Another case-control study examined the relationship between any type of treatment for depression and the subsequent development of AD (Speck et al., 1995). This latter study showed for depressive episodes occurring more than 10 years before dementia onset an odds ratio of 2.0; 95% confidence interval (CI) = 0.9 – 3.5), and for depressive episodes within 10 years of AD onset, an odds ratio of 0.9 (95% CI = 0.2 – 3.0). The greater risk associated with depression earlier in life suggests the possibility of a critical period in physiological development during which depression can predispose to the later development of AD.

The possibilities that depression may predispose to the development of AD, or that depression may be an early concomitant of dementing illness are both supported by a review of the 9 AD patients seen in our dementia clinic who were also diagnosed at initial evaluation with depression. In 4 subjects, depression antedated AD by several years, and in the remaining 5, depression appeared to coincide with the development of AD (Weiner et al., 1991). In another study, which involved 57 elderly patients with MDD treated in hospital, 23 met criteria for "reversible dementia"; that is, they met DSM-III-R (American Psychiatric Association, 1987) criteria for both MDD and dementia (AD), and had MMSE scores < 24 on admission. After treatment, they no longer met criteria for dementia, had MMSE scores > 23, and had scores < 12 on the Hamilton Rating Scale for Depression (HRSD; Hamilton, 1967). These individuals were compared with 34 persons with successfully treated MDD who did not meet criteria for dementia on admission. Subjects were followed at yearly intervals for an average of 34 months. Dementia developed in the group with reversible cognitive impairment (AD) significantly more frequently than in the group with MDD alone (43% vs. 12%, $p < .01$). From this study, one could infer that among the reversibly demented depressed patients, there were a number with early AD whose cognitive symptomatology was worsened by depression and that among the depressed group that did not meet criteria for dementia, MDD heralded the onset of AD. The findings of Zubenko and Moosy (1988) may provide a neurobiological explanation of MDD heralding AD. They found that persons with AD and depression had significantly more changes in parts of the brain thought to be important in the etiology of depression (locus ceruleus, basal ganglia) than persons with AD who had no history of depression. In some individuals, changes in parts of the brain that relate to depression may precede the pathological changes of AD that affect cognitive function.

In our experience, persons later diagnosed with AD are commonly treated first for depression; 25% of AD patients seen at two large tertiary care centers had previously received treatment with antidepressant medication (Weiner, Doody, Risser, & Liao, unpublished data). This suggests that persons with AD are commonly treated for depression in the belief or hope that they are suffering from depression rather than AD, or with the idea that some of the presenting symptoms are those of depression in addition to AD.

PREVALENCE AND INCIDENCE OF MAJOR DEPRESSION DISORDER IN ALAZHEIMER'S DISEASE

Estimates of the prevalence of MDD in AD vary depending on the source of information and the criteria employed for the diagnosis of MDD. Based on family caregiver reports concerning 175 persons with AD, 86% met DSM-III (American Psychiatric Association, 1980) criteria for MDD (Merriam, Aronson, Gaston, Wey, & Katz, 1988). Rovner, Broadhead, Spencer, Carson, and Folstein (1989) interviewed caregivers and AD patients with a modified Present State Examination (Wing, Cooper, & Sartorius, 1974) and found a 17% prevalence of MDD. However, on direct examination of AD patients, Mackenzie, Robiner, and Knopman (1989) found that 4.3% of 46 AD patients met DSM-III-R (American Psychiatric Association, 1987) criteria for MDD. Cummings, Ross, Absher, Gornbein, and Hadjiaghai (1995) found, in a series of 33 AD patients, that 6% met DSM-III criteria for MDD. Our own prevalence figures for MDD in AD (based on direct patient examination) were only 1.5%, because we required the presence of depressed mood on direct examination for the diagnosis of MDD (Weiner, Edland, & Luszczynska, 1994). Prospective and naturalistic studies have found higher frequencies of symptoms associated with depression (such as sleep and appetite disturbance) than of diagnosable depressive disorders (Rubin & Kinscherf, 1989). Furthermore, the depressive symptoms were often transient but recurrent, and rarely persisted on follow-up (Devenand et al., 1997). It appears that caregivers overestimate depression compared to clinicians who directly examine patients. One might speculate that this is so because depression carries a more hopeful prognosis, or that caregivers project onto their care recipients their own depressed mood. Moye, Robiner, and Mackenzie (1993) found that caregiver ratings of AD patients as depressed increased with the number of hours that caregivers spent with AD patients and among nonspousal caregivers. Their data did not suggest that ratings of AD patients as depressed reflected how caregivers might feel were they in the patients' shoes. And neither Moye et al. (1993) nor Weiner, Svetlik, and Risser (1997) found that caregivers confused increasing cognitive impairment with depression. Thus, in evaluating studies of MDD in AD, the source of observations must be considered in addition to the life circumstances of the observers.

ALZHEIMER'S DISEASE AS A RISK FACTOR FOR MAJOR DEPRESSIVE DISORDER

Does AD increase vulnerability to MDD? The widely varying prevalence studies reviewed here. They suggest a modest increase in prevalence of MDD in AD when compared with figures from a large population study that showed a 6-month prevalence of MDD of 1.7%, including MDD associated with bereavement (Weissman et al., 1985). When MDD associated with bereavement was excluded, 6-month prevalence dropped to 1.1%. Incidence figures for major depression in community-dwelling elders range from 2.4% to 10% (Snowdon, 1994). Our review of 1,095 cases in the database of the Consortium to Establish a Registry for Alzheimer's Disease (which excluded persons with MDD at entry) found a 1.3% two-year incidence of MDD by DSM-III-R criteria when depressed mood was one of the criteria. With the same criteria for MDD, we found no new cases of MDD in 153 AD patients followed at our own center for an average of 3 years (Weiner et al., 1994). Using Feighner criteria (Feighner et al., 1972), Burke, Rubin, Morris, and Berg (1988) found no new cases of MDD developing in 44 subjects with mild AD who were followed over 34 months (final $N = 30$).

How do we reconcile the finding of higher prevalence and lower incidence of MDD in community-dwelling persons with AD than in nondemented community-dwelling elders? In persons with AD who are depressed, depression often antedates their dementia diagnosis. They never become incident cases. The greatest risk factor for depression in elders is perceived poor health (Murphy, 1982). Elderly persons with the multiple health problems that appear to predispose to depression are not seen frequently in our clinic. More often, our patients are healthy individuals brought in by their families with concern about cognitive impairment. It is possible, therefore, that families pursue more vigorously cognitive deficits that occur in otherwise healthy family members and that studies of depression in AD patients evaluated at tertiary care centers may suffer from selection bias. On the other hand, these patients may more accurately reflect the impact of AD on mood because they are not suffering from other illnesses.

QUANTIFYING MAJOR DEPRESSIVE DISORDER IN PERSONS WITH ALZHEIMER'S DISEASE

Objective assessment of depressive symptoms in AD patients using rating instruments poses a number of problems. Comparing studies that have used these instruments is difficult because of significant methodological differences. Studies vary in their diagnostic criteria, patient population (AD vs. mixed dementia, community vs. institutionalized), severity of cognitive impairment, exclusionary criteria, and source of information (patient self-report, caregiver assessment, or clinician interview). Many instruments have been developed; some are well established for use with nondemented persons, and others were designed specifically for use with demented patients.

The Geriatric Depression Scale (GDS; Yesavage et al., 1983) is a 30-item self-rated yes–no questionnaire that examines subjective feelings relevant to both mood and cognition, with little emphasis on somatic symptoms and behavior. It has been validated in a nondemented elderly population (Brink, Yesavage, & Lum, 1982), but Brodaty and Luscombe (1996) found poor overall concordance in a mixed dementia population between GDS and the Hamilton Rating Scale for Depression (HRSD; Hamilton, 1960). Others also have found problems with this instrument's validity (Burke, Houston, Boust, & Roccaforte, 1989) and sensitivity (Kafonek et al., 1989) in demented patients.

The HRSD is a 24-item scale with scores derived from a semistructured interview. Heavily weighted for vegetative symptoms, it is widely used in studies of elderly persons and has been shown to have high interrater reliability in AD patients (Gottleib, Gur, & Gur, 1988). However, in a study by Logsdon and Teri (1995), the HRSD had low sensitivity in detecting depression as diagnosed by the Schedule for Affective Disorders and Schizophrenia interview (Endicott & Spitzer, 1978) and based on Research Diagnostic Criteria (Spitzer, Endicott, & Robins, 1978) and DSM-III-R (American Psychiatric Association, 1987) criteria.

The Dementia Mood Assessment Scale (DMAS; Sunderland et al., 1988) and the Cornell Scale for Depression in Dementia (CSDD; Alexopoulos, Abrams, Young, & Shamoian, 1988) use information obtained from both the patient and a caregiver. By utilizing reports from different sources, these instruments attempt to improve the reliability of information obtained. The DMAS is a 24-item scale, with the first 17 items measuring mood via sadness and depression indicators. The developers of this instrument found high correlations with global measures of depression and sadness, but they also acknowledged that these highly significant correlations do not establish its validity in measuring mood in dementia patients (Alexopoulos et al., 1988). The CSDD is a 19-item scale that has good correlation with depressive subtypes of various intensity classified by Research Diagnostic Criteria.

Limitations to the use of these measuring instruments begin with the absence of universally accepted diagnostic criteria ("gold standard") to establish their validity. Little normative or psychometric information is available on diagnostic criteria in the various editions of the American Psychiatric Association's DSM. Given the significant overlap between somatic and behavioral symptoms used in the inventories and changes associated with AD, the current measures do not distinguish between symptoms attributable to MDD or AD. Relying on information obtained from AD patients and/or their caregivers is also fraught with problems. As mentioned earlier, Moye et al. (1993) found low correlation between patient and caregiver reports of depressive symptoms. Rather, caregivers' ratings of severity of depression correlated with hours spent with the patient and tended to be based on observable behavior rather than intrapsychic distress, and this may cause such ratings to be erroneously high. On the other hand, it is possible that AD patients underreport depressive symptoms. These individuals are often unaware of their cognitive deficits, and it is conceivable that their illness in some way impairs other areas of self-awareness. However, Moye et al. found no relationship between severity of cognitive impairment and report of depressive symptoms, and these findings were confirmed by Weiner et al. (1997). Teri and Wagner (1991) examined 75 persons who met DSM-III-R criteria for AD, of whom 22 (29%) also met DSM-III-R criteria for MDD. Scoring on the HRSD by AD patients, caregivers, and clinicians was reviewed. In the nondepressed group, clinicians and caregivers rated 50% of patients as having depressed mood, while depressed mood was reported by only 23% of patients. In the depressed group, clinicians and caregivers rated approximately 90% of patients as having depressed mood in contrast to 68% of the patients. Within the depressed group, patients reported significantly less often than caregivers the following four of the 17 HRSD items: middle insomnia, change in interests, psychic anxiety, and energy change. The difference between AD patients and clinicians was even greater. AD patients reported significantly less often than clinicians seven of the 17 HRSD items: depressed mood, suicidal feelings, middle insomnia, change in interests, psychic and somatic anxiety, and energy change. Thus, it appears that AD patients may underreport many depressive symptoms, and that a clinical evaluation for depression in persons with AD should include ancillary information from caregivers. One way to improve diagnostic accuracy in this situation is to use scales such as the CSDD and DMAS, which include information from both patients and caregivers, and also direct clinical observation.

TREATMENT OF MAJOR DEPRESSIVE DISORDER

Antidepressant medications are the standard treatment for MDD (see also Reynolds, Miller, Mulsant, Dew, & Pollack, Chapter 13, this volume). Individual drugs within this group vary greatly in chemical structure but seem to have as their basic mode of action prolongation of the action in brain of the monoamine neurotransmitters norepinephrine and serotonin. These drugs are not always effective. For severe MDD that is refractory to antidepressant medication, electroconvulsive therapy is the treatment of choice. It is also the treatment of choice for individuals who are rapidly losing weight or markedly suicidal. As with any treatment, antidepressant drugs and electroconvulsive therapy are not completely efficacious. In a few cases, there is no remission of symptoms; in other cases, there is only partial response to treatment. For milder cases of MDD, both interpersonal psychotherapy (Klerman, Rounsaville, Chevron, & Weismann, 1984) and cognitive-behavioral therapy (Beck, Rush, Shaw, & Emery, 1979) have proven effective. The former focuses on enhancing interpersonal relationships and the latter on eliminating self-defeating thoughts and behaviors (see also Klausner, Snyder, & Cheavens, Chapter 14, this volume).

TREATMENT OF MAJOR DEPRESSIVE DISORDER IN ALZHEIMER'S DISEASE

As noted earlier, pharmacological treatment of MDD in AD improves cognitive functioning (Greenwald et al., 1989) and ADL performance (Reifler et al., 1986; Taragano et al., 1997), in addition to improving symptoms of depression. In the Taragano et al. study, there was a mean 9.4 point reduction in HRSD scores over 45 days, but 58% of patients treated with one drug (amitriptyline) and 22% of those treated with fluoxetine dropped out before completion of the study. Amitriptyline is a drug with numerous unpleasant side effects including sedation, lowered blood pressure, dry mouth, and constipation. Fluoxetine has far fewer side effects but occasionally produces nausea or excessive stimulation. An 8-week study by Reifler and Larson (1989) showed no difference in response of AD patients with MDD between the antidepressant imipramine and a placebo condition in which patients were seen in weekly visits, but showed a sizeable improvement for both conditions (mean decline of 40% in the 17-item HRSD), suggesting that support given to patient and caregiver was as effective as medication in relieving depressive symptoms in persons with AD.

Teri and Gallagher-Thompson (1991) proposed the use of cognitive therapy for depressed persons with mild AD and behavioral management techniques for patients with greater cognitive deficit. The primary aim of the cognitive therapy was to challenge patients' negative cognitions, thereby reducing distortions and enabling patients to view specific situations and events more adaptively. AD patients were generally seen for 16–20 visits, with homework assigned and contact with family members maintained by telephone. Thus far, this proposed treatment technique has not been objectively validated in persons with dementia. As it relies heavily on memory, it seems inappropriate for all but the most mildly cognitively impaired individuals.

Helping caregivers to tailor environmental demands to patient's abilities and increasing the pleasant events available to AD patients appears to reduce depressive symptoms (Teri, 1994). In one study (Teri, Logson, Uomoto, & McCurry, 1997), caregivers and AD patients were enrolled in a 9-week intervention program. Inclusion criteria for AD patients included meeting DSM-III-R criteria for MDD or minor depression and a HRS-D score > 9. The distribution of diagnoses was approximately 75% MDD and 25% minor depression. One intervention group (Behavior Therapy—Pleasant Events; BT-PE) involved 9 one-hour, highly structured sessions teaching caregivers behavioral strategies for decreasing patient depression by increasing pleasant events and using behavioral problem-solving strategies to alter contingencies related to depression and associated behavioral problems. A second intervention group (Behavior Therapy—Problem Solving; BT-PS) focused on problem-solving patient depressive behaviors of specific concern to caregivers, was less highly structured, involved more give and take, and did not emphasize maximizing pleasant events. Two comparison groups were included. One received typical clinic patient follow-up and the other was a waiting-list group. The amount of therapist contact and the number and length of sessions were the same in all conditions except for the waiting-list group. There were three outcome measures, the HRSD, Cornell Scale for Depression in Dementia (CSDD; Alexopoulos et al., 1988) and the Beck Depression Inventory (BDI; Beck, Ward, Mendelson, Mock, & Erbaugh, 1961). The former are clinician-administered scales and the latter is a self-report scale. The dropout rate was 18%. There was no change in AD patient HRSD score in the typical follow-up or waiting-list group. The groups were equivalent in initial symptomatology. The mean baseline HRSD scores for the two intervention groups were approximately 16 ± 5, in the mild depression range, as were the CSDD and BDI scores. In the BT-PE and BT-PS groups, there were 33% and 24% reductions in patient HAM-D scores ($p < .0001$, compared with control groups), 28% and 25% reductions, respectively, in patient CSDD scores ($p < .0001$, compared with control groups), and 8% and 21% reductions, respectively, in patient BDI scores ($p < .01$). Improvement was considered to be clinically significant if patients who were diagnosed by DSM-III-R criteria for MDD or minor depression no longer met those criteria. By these criteria, 60% of patients in the BT-PE and BT-PS conditions improved compared to 20% in the control conditions. No patient in the BT-PE and BT-PS conditions got worse, but 10% of those in the control condition worsened. At 6-month follow-up, 69% of the individuals in the BT-PE and BT-PS treatment groups maintained their improvement, while 31% relapsed. There was no significant difference between these two treatments.

The results of this study are not as impressive as the studies of various antidepressant agents in cognitively intact persons with MDD, in whom drops of 50% or more occur on standardized scales for measuring depressive symptomatology. Rather, the relapse rates are comparable to relapse rates in persons who discontinue their antidepressant medication. There are insufficient studies of antidepressant drug effects in AD patients with MDD to make a comparison with the behavioral treatments described earlier, but the consensus in the field is that pharmacotherapy is warranted for depressive symptoms, even if patients fail to meet criteria for MDD, because there is little risk in the use of the newer antidepressants (Small et al., 1997).

DIAGNOSTIC CRITERIA

It is clear from the preceding discussion that identification of MDD in AD is plagued by considerable uncertainty. Major conceptual and methodological differences must be resolved in order to reconcile the significant discrepancies in prevalence reports and to enable comparison of treatments. Psychiatric diagnosis has been vastly improved over the years by the evolution of the American Psychiatric Association's DSM. Paradoxically, these same criteria, by virtue of the gradual shift in nomenclature, have tended to confound the distinction between AD and MDD. The American Psychiatric Association began publishing criteria for the diagnosis of mental disorders in 1952. The first and second editions of the *Diagnostic and Statistical Manual* (DSM-I, DSM-II; American Psychiatric Association, 1952, 1968) indicated that the primary disturbance in persons with depression was depressed mood. DSM-III (American Psychiatric Association, 1980) modified criteria for MDD by indicating that the mood disturbance must be prominent and relatively persistent but not necessarily the most dominant symptom. In fact, DSM-III, DSM-III-R (American Psychiatric Association, 1987), and DSM-IV (American Psychiatric Association, 1994) do not require depressed mood for the diagnosis of MDD. Instead, they require depressed mood *or* loss of interest or pleasure in all, or almost all, usual activities and pastimes. DSM-III-R broadens the criterion of loss of interest or pleasure to include observation by others of apathy most of the time; DSM-IV requires only observation of loss of interest or pleasure by family members or other nonclinical observers. Thus, a diagnosis of MDD requires neither depressed mood nor subjective complaint of loss of interest or pleasure.

An examination of DSM-IV criteria for MDD and AD (Table 12.1) reveals the large overlap in diagnostic criteria. Activities that were formerly pleasurable for persons with AD are no longer so, because they become sources of confusion and frustration. As they lose pleasure and experience increasing frustration, persons with AD often withdraw from work, hobbies, and chores. Because of their cognitive impairment, they are unable to learn substitute activities and become even more frustrated. Executive function deficits in planning and organization make it difficult for persons with AD to initiate activities. Hence, increased passivity is common. Because they can no longer follow shifting themes in conversations, they tend to withdraw from social activities and to become increasingly passive. Persons with AD frequently forget to eat or become distracted while eating. As a consequence, weight loss is very frequent in AD. AD patients may oversleep because it is a comfortable retreat from a confusing world. They frequently have interrupted nighttime sleep. Unlike the interrupted sleep of severe depression, with early morning awakening and dread of the day ahead, AD patients awaken in the middle of the night thinking it is time to arise and get about the business of the day. Psychomotor agitation is common in AD, especially when persons with AD are involved with activities they can no longer master, such as making bank deposits or writing checks. Psychomotor retardation occurs in AD patients who have extrapyramidal (Parkinsonian) symptoms. Complaints of fatigue and reduced energy are common in AD; it takes much more psychological energy to initiate, organize, and perform simple everyday acts such as grooming and dressing. Guilt is not a common accompaniment of AD, but patients often express feelings of worthlessness because of their inability to maintain themselves independently or to perform ordinary ADL such as dialing a telephone without help. Diminished ability to think and to concentrate are direct cognitive effects of AD. Thoughts of suicide are less frequent, arising at times of frustration, and tend to abate along with the source of environmental frustration. Depression is strongly linked to suicide. On rare occasions, persons with AD will plan and carry out a lethal suicide attempt (Rohde, Peskind, & Raskind, 1995). We have had one suicide in the more than 1,000 persons we have diagnosed with AD. Family members had reported no signs of depression in that individual but had heard the person with AD indicate that he did not wish to live in a state in which he was totally dependent on others.

It is clear, then, that AD and the emotional reactions to the disease process can mimic most of the symptomatology of MDD. Much of the difference between symptoms caused by MDD and those resulting from AD is in the stability and persistence of depressive symptoms (most of the time, for 2 weeks or more) and the simultaneous presence of multiple depressive symptoms in MDD.

From the standpoint of clinical differentiation of MDD from AD symptoms, several characteristics of MDD patients are far less common in AD. Persons with MDD actively resist pleasurable activities, feeling both unable to participate and unworthy of participation. Persons with AD offer little resistance to pleasurable activities that are within their ability to comprehend and perform. Persons with MDD have little interest in food, while persons with AD eat well if they receive adequate cuing. Sexual appetite diminishes in MDD but frequently remains in persons with AD, often dismaying partners who are uncomfortable having sexual relations with a person who no longer functions as a spouse in other ways (Weiner & Svetlik, 1996). MDD patients awaken early in the morning after a poor night of sleep, ruminating about their failures, unable to understand how they will get through the coming day. In MDD, there is often a pattern of diurnal mood variation, with improvement in the latter part of the day, in contrast to the increasing confusion (so-called "sundowning") that often occurs in persons with dementia as they fatigue toward the latter part of the day. Persons with MDD often present with physical symptoms, while AD patients tend to underreport physical symptoms. Persons with AD may become seriously ill without seeming to notice, the signs being increased confusion or agitation or sudden loss of one or more ADL.

Can there really be depression without depressed mood? This has been suggested on numerous occasions. Lesse (1968) coined the term *masked depression* to describe a variety of symptoms thought to disguise an underlying depressed mood. More recently, Fogel and Fretwell (1985) introduced the term *depletion syndrome* to describe a state of depression in which older adults did not endorse sadness. Gallo, Rabins, Lyketsos, Tien, and Anthony (1997) have also suggested that depression may occur without dysphoria. In this latter study, persons with depressive symptoms without dysphoria were at greater risk of death than those with a depressive syndrome accompanied by dysphoria. This is of interest, for if one reviews the criteria for depression in Table 12.1, it becomes apparent that many symptoms of depression are nonspecific symptoms associated with any type of illness, including loss of pleasure, loss of appetite, loss of weight, sleep disturbance, fatigue, and diminished ability to think or concentrate. Thus, so-called "nondysphoric depression" might actually be the nonspecific manifestation of a life-endangering physical disease.

How can we know if a person without depressed mood is "really" depressed? One might reason that response to treatment with antidepressant medication would tend to confirm a diagnosis of underlying depression. But antidepressants are not specific for depression. They are useful in other psychiatric disorders, such as obsessive–compulsive disorder, and for potentiation of analgesics in the relief of chronic pain. One class of antidepressants, the serotonin reuptake inhibitors, have been reported to alleviate anxiety symptoms, aggression, and irritability associated with dementia (Karlsson,

Table 12.2. Comparison of DSM-IV Criteria for Major Depressive Disorder (MDD) and Suggested Criteria for MDD in Alzheimer's Disease (AD)

DSM-IV	MDD in AD
Five (or more) of the following symptoms have been present during the same 2-week period and represent a change from previous function. At least one is depressed mood **or** loss of interest or pleasure.	Five (or more) of the following symptoms have been present during the same 1-week period and represent a change from previous function. At least one is sadness.
1. Depressed mood most of most days.	1. Sadness most of the day, nearly every day, that is frequently unresponsive to distraction.
2. Loss of pleasure most of most days.	2. No apparent enjoyment when engaged by others in ordinarily pleasurable activities; or active refusal, nearly every day.
3. Loss of 5% of body weight in a month.	3. Little interest in eating when presented with food and/or helped to engage in eating at most meals.
4. Too much or too little sleep nearly every day.	4. Increased irritability nearly every day.
5. Psychomotor agitation or retardation nearly every day	5. Psychomotor agitation or retardation nearly every day; not attributable to medication or neurological dysfunction
6. Fatigue or loss of energy nearly every day.	6. Low energy nearly every day.
7. Feelings of worthlessness or guilt nearly nearly every day.	7. Feelings of worthlessness or crying nearly every day.
8. Diminished ability to think or concentrate every day.	8. Function below expected for level of cognitive impairment nearly every day.
9. Recurrent suicidal ideas, attempt, or plan.	9. Expresses wish to die, plans, or attempts suicide.

1996). Is depressed mood central to depression or only one by-product of an underlying pathophysiological process that may manifest more prominently with other symptoms? Since we have no definitive knowledge about the pathophysiology of depression, it would seem more reasonable to insist on the full-blown syndrome, including mood disturbance.

We would argue that because there is so much overlap in symptomatology, the criteria for the diagnosis of MDD in AD should require depressed mood. The more we dilute the criteria for depression, the less able we will be to maintain it as a meaningful construct. Hence, the authors offer criteria for the diagnosis of MDD in AD as indicated in Table 12.2. Our proposed criteria for MDD in AD are largely observational because of dementia patients' frequent loss of language and inability to conceptualize mood. Additionally, persons with dementia may lack the biological substrate to maintain a constantly depressed mood over a period of weeks but may, instead, suffer a "fragmentary" depression, much as they experience fragmented sleep. On the other hand, there must be some evidence of sustained lowering of mood or emotional symptomatology to differentiate depression from the ordinary frustration experienced by persons with dementia as they attempt to function at their premorbid level. Thus, a person with dementia might cry on failing to be able to write a check and transiently feel worthless. For these reasons, a time window of 1 week has been chosen instead of 2 weeks. With regard to individual criteria, a point-by-point rationale follows:

1. Because persons with dementia frequently cannot express their mood state in words, there must be observable sadness, as manifested by dejected expression and hand wringing. In those with preserved language, verbalization of depression or sadness is required.
2. Because of the difficulty initiating activities that is associated with dementia, this criterion requires attempts by others to engage persons in activities that were formerly pleasurable.
3. Persons with dementia often lose weight because they forget to eat or become distracted while eating. They may have lost the ability to use eating utensils. In addition, loss of the sense of smell frequently contributes to loss of appetite. This criterion requires the person with dementia to be presented with food and helped to engage in eating in a manner appropriate to the stage of dementia.
4. Persons with AD often have fragmented sleep that appears related to fragmentation of sleep architecture. They often sleep during the day in apparent response to inability to organize their activities or as a retreat from complex environmental demands. For these reasons, it was decided to substitute irritability, a well-known correlate of depression and a symptom of AD that often responds to the use of antidepressant drugs that inhibit the reuptake of the neurotransmitter serotonin (Sultzer, Gray, Gunay, Berisford, & Mahler, 1997).
5. Psychomotor retardation due to extrapyramidal symptoms or other neurological abnormality should be noted but not counted.
6. Fatigue should not be confused with slowing due to extrapyramidal symptoms (AD-related muscular stiffness).
7. Crying is substituted for guilt because of the relative absence of guilt in late-life depression. Crying may be either spontaneous or associated with attempts to perform tasks.
8. This highly subjective evaluation is based on the well-documented observation that depression impairs cognitive function and function in ADL (Weingartner et al., 1981).
9. Persons with AD are able to contemplate suicide and, on rare occasions, make successful attempts (Rohde et al., 1995).

CONCLUSIONS

MDD and AD, while differing greatly in clinical course and pathophysiology, have important interactions. MDD in late-life appears to be a risk factor for the development of AD for reasons that are not apparent. The prevalence of AD in MDD is probably higher than in the general elderly population, but MDD does not appear to be a frequent complication of AD. However, MDD in AD patients worsens both cognition and function in ADL. MDD appears to be overdiagnosed in AD, especially when the diagnosis is based on caregiver reports. Much of what are thought to be symptoms of MDD are, in fact, direct consequences of AD, such as difficulty with initiating activities and sustaining attention. When MDD occurs in AD, it is treatable by conventional pharmacological means with concomitant simplification of environmental demands and development of a structured routine. The major confound in the scientific investigation of MDD in AD is differentiating between symptoms engendered directly by AD and symptoms of MDD. The omission in DSM-IV of depressed mood as a requisite criterion for the diagnosis of MDD will add still more difficulty. The authors propose an alternative to DSM-IV criteria for MDD for the diagnosis of MDD in persons with dementia. These alternative criteria shorten the period of time during which symptoms must manifest to 1 week, make depressed mood a requirement, allow for lack of initiative, and substitute irritability for sleep disturbance. Finally, given the relatively benign side-effects profile of modern antidepressant agents, it is reasonable that persons with AD whose behavior suggests that they may be depressed be given a trial of antidepressant medication.

REFERENCES

Alexopoulos, G. S., Abrams, R. C., Young, R. C., & Shamoian, C. A. (1988). Cornell Scale for Depression in Dementia. *Biological Psychiatry, 23,* 271–284.

Alexopoulos, G. S., Meyers, B. S., Young, R. C., Mattis, S., & Kakuma, T. (1993). The course of geriatric depression with "reversible dementia:" A controlled study. *American Journal of Psychiatry, 150,* 1693–1699.

American Psychiatric Association. (1952). *Diagnostic and statistical manual: Mental disorders.* Washington, DC: Author.

American Psychiatric Association. (1968). *Diagnostic and statistical manual of mental disorders* (2nd ed.). Washington, DC: Author.

American Psychiatric Association. (1980). *Diagnostic and statistical manual of mental disorders* (3rd ed.). Washington, DC: Author.

American Psychiatric Association. (1987). *Diagnostic and statistical manual of mental disorders* (3rd ed., rev.). Washington, DC: Author.

American Psychiatric Association. (1994). *Diagnostic and statistical manual of mental disorders* (4th ed.). Washington, DC: Author.

Bartus, R. T., Dean, R. L., Pontecorvo, M. J., & Flicker, C. (1985). The cholinergic hypothesis: A historical overview, current perspective and future directions. *Annals of the New York Academy of Science, 44,* 332–358.

Baxter, L. R., Schwartz, J. M., Phelps, M. E., Mazziotta, J. V, Guze, B. H., Selin, C. E., Gerner, R. H., & Sumida, R. M. (1989). Reduction of prefrontal cortex glucose metabolism common to three types of depression. *Archives of General Psychiatry, 46,* 243–250.

Beck, A. T., Rush, A. J., Shaw, B. F., & Emery, G. (1979). *Cognitive therapy of depression.* New York: Guilford Press.

Beck, A. T., Ward, C. H., Mendelson, M., Mock, J., & Erbaugh, J. (1961). An inventory for measuring depression. *Archives of General Psychiatry, 4,* 561–571.

Bonte, F. J., Weiner, M. F., Bigio, E. H., & White, C. L., III. (1997). Brain blood flow in the dementias: SPECT with histopathological correlation in 54 patients. *Radiology, 202,* 793–797.

Brink, T. L., Yesavage, J. A., & Lum, O. (1982). Screening tests for geriatric depression. *Clinical Gerontology, 1,* 37–43.

Brodaty, H., & Luscombe, G. (1996). Depression in persons with dementia. *International Psychogeriatrics, 8,* 609–622.

Bulbena, A., & Berrios, G. (1986). Pseudodementia: Facts and figures. *British Journal of Psychiatry, 148,* 87–94.

Burke, W. J., Houston, M. J., Boust, S. J., & Roccaforte, W. H. (1989). Use of the Geriatric Depression Scale in dementia of Alzheimer type. *Journal of the American Geriatrics Society, 37,* 856–860.

Burke, W. J., Rubin, E. H., Morris, J. C., & Berg, L. (1988). Symptoms of "depression" in dementia of the Alzheimer type. *Alzheimer Disease and Associated Disorders, 2,* 356–362.

Burns, A. (1991). Affective symptoms in Alzheimer's disease. *International Journal of Geriatric Psychiatry 6,* 371–376.

Cohen, R. M., Weingartner, H., Smallberg, S. A., Pickar, D., & Murphy, D. L. (1982). Effort and cognition in depression. *Archives of General Psychiatry, 39,* 593–597.

Corder, E. H., Saunders, A. M., & Strittmatter, W. J. (1993). Gene dose of apoliprotein E type 4 allele and the risk of Alzheimer's disease in late onset families. *Science, 261,* 921–923.

Cummings, J. L., Ross, W., Absher, J., Gornbein, J., & Hadjiaghai, L. (1995). Depressive symptoms in Alzheimer disease: Assessment and determinants. *Alzheimer Disease and Associated Disorders, 9*, 87–93.

Dannenbaum, S., Parkinson, S., & Inman, V. (1988). Short-term forgetting comparison between patients with dementia of the Alzheimer type, depressed and normal elderly. *Cognitive Neuropsychology, 5*, 213–233.

DeKosky, S. T. (1996). Advances in the biology of Alzheimer's disease. In M. F. Weiner (Ed.), *The dementias: Diagnosis, management, and research* (2nd ed., pp. 313–330). Washington, DC: American Psychiatric Press.

Devenand, D. P., Jacobs, D. M., Ming-Xin, T., Del Castillo-Castaneda, C., Marder, K., Bell, K., Bylsma, F. W., Brandt, J., Albert, M., & Stern, Y. (1997). The course of psychopathological features in mild to moderate Alzheimer's disease. *Archives of General Psychiatry, 54*, 267–263.

Duara, R., Grady, C. L., Haxby, J. V., Sundaram, M., Cutler, N. R., Heston, L., Moore, A., Schlageter, N., Larson, S., & Rapaport, S. I. (1986). Positron emission tomography in Alzheimer's disease. *Neurology, 36*, 879–887.

Endicott, J., & Spitzer, R. L. (1978). A diagnostic interview: The schedule for affective disorders and schizophrenia. *Archives of General Psychiatry, 35*, 837–844.

Feighner, J. P., Robins, E., Guze, S. B., Woodruff, R. A., Jr., Winokur, G., & Munoz, R. (1972). Diagnostic criteria for use in psychiatric research. *Archives of General Psychiatry, 26*, 57–63.

Fogel, B. S., & Fretwell, M. (1985). Reclassification of depression in the medically ill elderly. *Journal of the American Geriatrics Society, 33*, 446–448.

Folstein, M. F., Folstein, S. E., & McHugh, P. R. (1978). Mini-Mental State: A practical method for grading the cognitive state of patients for the clinician. *Journal of Psychiatric Research, 12*, 189–198.

Folstein, M. F., & McHugh, P. R. (1978). Dementia syndrome of depression. In R. Katzman, R. D. Terry, & K. L. Bick (Eds.), *Alzheimer's disease: Senile dementia and related disorders* (pp. 87–93). New York: Raven Press.

Fromm, D., & Schopflocher, D. (1984). Neuropsychological test performance in depressed patients before and after drug therapy. *Biological Psychiatry, 19*, 55–72.

Gallo, J. J., Rabins, P. V., Lyketsos, C. G., Tien, A. Y., & Anthony, J. C. (1997). Depression without sadness: Functional outcomes of nondysphoric depression in late life. *Journal of the American Geriatrics Society, 45*, 570–578.

Gottlieb, G. L., Gur, R. E., & Gur, R. C. (1988). Reliability of psychiatric scales in patients with dementia of the Alzheimer's type. *American Journal of Psychiatry, 145*, 857–860.

Greenwald, B. S., Kramer-Ginsberg, E., Marin, D. B., Laitman, L. B., Hermann, C. K., Mohs, R. C., & Davis, K. L. (1989). Dementia with coexistent major depression. *American Journal of Psychiatry, 146*, 1472–1478.

Hamilton, M. (1960). A rating scale for depression. *Journal of Neurology, Neurosurgery, and Psychiatry, 23*, 56–62.

Hamilton, M. (1967). Development of a rating scale for primary depressive illness. *British Journal of Social and Clinical Psychology, 6*, 278–296.

Hart, R. P., Kwentus, J. A., Wade, J. B., & Hamer, R. M. (1987). Digit symbol performance in mild dementia and depression. *Journal of Consulting and Clinical Psychology, 55*, 236–238.

Henry, G. M., Weingartner, H., & Murphy, D. L. (1973). Influence of affective states and psychoactive drugs on verbal learning and memory. *American Journal of Psychiatry, 130*, 966–971.

Jorm, A. F., Van Duijn, C. M., Chandra, V., Fratiglioni, L., Graves, A. B., Heyman, A., Kokmen, E., Kono, K., Mortimer, J. A., Rocca, W. A., Shalat, S. L., Soinnen, H., & Hofman, A. (1991). Psychiatric history and related exposures as risk factors for Alzheimer's disease: A collaborative re-analysis of case-control studies. *International Journal of Epidemiology, 20*, 43–47.

Kafonek, S., Ettinger, W. H., Roca, R., Kittner, S., Taylor, N., & German, P. S. (1989). Instruments for screening for depression and dementia in a long-term care facility. *Journal of the American Geriatrics Society, 37*, 29–34.

Karlsson, I. (1996). Treatment of non-cognitive symptoms in dementia. *Acta Neurologica Scandinavica, 168*, 93–95.

Kiloh, L. G. (1961). Pseudo-dementia. *Acta Psychiatrica Scandinavica, 37*, 336–351.

Khachaturian, Z. S. (1985). Diagnosis of Alzheimer's disease. *Archives of Neurology, 42*, 1097–1105.

Klerman, G. L., Rounsaville, B., Chevron, E., & Weissman, M. (1984). *Interpersonal psychotherapy of depression.* New York: Basic Books.

Kral, V. A., & Emery, O. (1987, August). *Long-term followup of depressive pseudodementia in the aged.* Paper presented at the Third Congress of the International Geropsychiatric Association, Chicago, IL.

Lesse, S. (1968). Masked depression—a diagnostic and therapeutic problem. *Diseases of the Nervous System, 29*, 169–173.

Logsdon, R. G., & Teri, L. (1995). Depression in Alzheimer's disease patients: Caregivers as surrogate reporters. *Journal of the American Geriatrics Society, 43*, 150–155.

Lopez, O. L., Boller, F., Becker, J. T., Miller, M., & Reynolds, C. F., III. (1990). Alzheimer's disease and depression: Neuropsychological impairment and progression of the illness. *American Journal of Psychiatry, 147*, 855–860.

Mackenzie, T. B., Robiner, W. N., & Knopman, D. S. (1989). Differences between patient and family assessments of depression in Alzheimer's disease. *American Journal of Psychiatry, 146*, 1174–1178.

Mayeux, R., Stern, Y., Rosen, J., & Leventhal, J. (1981). Depression, intellectual impairment and Parkinson's disease. *Neurology, 31*, 645–650.

Merriam, A. E., Aronson, M. K., Gaston, P., Wey, S. L., & Katz, I. (1988). The psychiatric symptoms of Alzheimer's disease. *Journal of the American Geriatrics Society, 36*, 7–12.

Miller, W. (1975). Psychological deficit in depression. *Psychological Bulletin, 82*, 238–260.

Morrison-Bogorad, M., Weiner, M.F., Rosenberg, R. N., Bigio, E., & White, C. L., III. (1997). Alzheimer's disease. In R. N. Rosenberg, S. B. Prusiner, S. DiMauro, & R. L. Barchi (Eds.), *The molecular and genetic basis of neurological disease* (2nd ed., pp. 581–600). Boston: Butterworth-Heinemann.

Moye, J., Robiner, W. N., & Mackenzie, T. B. (1993). Depression in Alzheimer patients: Discrepancies between patient and caregiver reports. *Alzheimer Disease and Associated Disorders, 7*, 187–201.

Murphy, E. (1982). Social origins of depression in old age. *British Journal of Psychiatry, 141*, 135–142.

Murphy, E. (1983). The prognosis of depression in old age. *British Journal of Psychiatry, 142*, 111–119.

Myers, J. K., Weissman, M. M., Tischler, G. L., Holzer, C. E., Leaf, P. J., Orvaschel, H., Anthony, J. C., Boyd, J. H., Burke, T. D., & Kramer, M. (1984). Six-month prevalence of psychiatric disorders in three communities. *Archives of General Psychiatry, 41*, 959–957.

Neve, R. L., & Robakis, N. K. (1998). Alzheimer's disease: A re-examination of the amyloid hypothesis. *Trends in Neurosciences, 21*, 15–19.

Paulman, R. G., Koss, E., & MacInnes, W. D. (1996). Neuropsychological evaluation of dementia. In M. F. Weiner (Ed.), *The dementias: Diagnosis, management, and research* (2nd ed., pp. 211–232). Washington, DC: American Psychiatric Press.

Pearlson, G. D., Ross, C. A., Lohr, W. D., Rovner, B. W., Chase, G. A., & Folstein, M. F. (1990). Association between family history of affective disorder and the depressive syndrome of Alzheimer's disease. *American Journal of Psychiatry, 147*, 452–456.

Pearson, J. L., Teri, L., Reifler, B. V., & Raskind, M. A. (1989). Functional status and cognitive impairment in Alzheimer's patients with and without depression. *Journal of the American Geriatrics Society, 37*, 1117–1121.

Reding, M., Haycox, J., & Blass, J. (1985). Depression in patients referred to a dementia clinic. *Archives of Neurology, 42*, 894–896.

Reifler, B. V., & Larson, E. (1989). Excess disability in dementia of the Alzheimer's type. In E. D. Light & B. D. Lebowitz (Eds.), *Alzheimer's disease treatment and family stress: Directions for research.* (pp. 363–397). Rockville, MD: National Institute of Mental Health.

Reifler, B. V., Larson, E., Teri, L., & Poulsen, M. (1986). Dementia of the Alzheimer's type and depression. *Journal of the American Geriatrics Society, 34*, 855–859.

Robinson, R. G., & Forrester, A. W. (1987). Neuropsychiatric aspects of cerebrovascular disease. In R. E. Hales & S. C. Yudofsky (Eds.), *The American Psychiatric Press textbook of neuropsychiatry* (pp. 191–208). Washington, DC: American Psychiatric Press.

Rohde, K., Peskind, E. R., & Raskind, M. A. (1995). Suicide in two patients with Alzheimer's disease. *Journal of the American Geriatrics Society, 43*, 187–189.

Rovner, B. W., Broadhead, J., Spencer, M., Carson, K., & Folstein, M. F. (1989). Depression and Alzheimer's disease. *American Journal of Psychiatry, 146*, 350–353.

Rubin, E. H., & Kinscherf, D. A. (1989). Psychopathology of very mild dementia of Alzheimer's disease. *American Journal of Psychiatry, 146*, 1017–1021.

Schildkraut, J., & Kety, S. (1967). Biogenic amines and emotion. *Science, 156*, 21–30.

Schwartz, J. A., Speed, N. M., Brunberg, J. A., Brewer, T. L., Brown, M., & Greden, J. F. (1993). Depression in stroke rehabilitation. *Biological Psychiatry, 33*, 694–699.

Siever, L. J., & Davis, K. L. (1985). Overview: toward a dysregulation hypothesis of depression. *American Journal of Psychiatry, 142*, 1017–1031.

Small, G. W., Rabins, P. V., Barry, P. P., Buckholtz, N. S., DeKosky, S. T., Ferris, S. H., Finkel, S. I., Gwyther, L. P., Khachaturian, Z. S., & Lebowitz, B. D. (1997). Diagnosis and treatment of Alzheimer's disease and related disorders: Consensus statement. *Journal of the American Medical Association, 278*, 1363–1371.

Snowdon, J. (1994). The epidemiology of affective disorders in old age. In E. Chiu & D. Ames (Eds.), *Functional psychiatric disorders of the elderly* (pp. 95–110). Cambridge, UK: Cambridge University Press.

Speck, C. E., Kukull, W. A., Brenner, D. E., Bowen, J. D., McCormick, W. C., Teri, L., Pfanschmidt, M. L., Thompson, J. D., & Larson, E. B. (1995). History of depression as a risk factor for Alzheimer's disease. *Epidemiology, 6*, 366–369.

Spitzer, R. L., Endicott, J., & Robins, E. (1978). Research Diagnostic Criteria: Rationale and reliability. *Archives of General Psychiatry, 35*, 773–782.

Sternberg, D. E., & Jarvik, M. E. (1976). Memory functions in depression: Improvement with antidepressant medication. *Archives of General Psychiatry, 33*, 219–224.

Sultzer, D., Gray, K. F., Gunay, I., Berisford, M. A., & Mahler, M. E. (1997). A double-blind comparison of trazodone and haloperidol for treatment of agitation in patients with dementia. *American Journal of Psychiatry, 5*, 60–69.

Sunderland, T., Alterman, I. S., Yount, D., Hill, J. L., Tariot, P. N., Newhouse, P .A., Mueller, E. A., Mellow, A. M., & Cohen, R. M. (1988). A new scale for the assessment of depressed mood in demented patients. *American Journal of Psychiatry, 145*, 955–959.

Taragano, F. E., Lyketsos, C. G., Mangone, C. A., Allegri, R. F., & Comesana-Diaz, E. (1997). A double-blind, randomized, fixed-dose trial of fluoxetine vs. amitriptyline in the treatment of major depression complicating Alzheimer's disease. *Psychosomatics, 38*, 246–252.

Teri, L. (1994). Behavioral treatment of depression in patients with dementia. *Alzheimer Disease and Associated Disorders, 8*, 66–74.

Teri, L., & Gallagher-Thompson, D. (1991). Cognitive-behavioral interventions for treatment of depression in Alzheimer's disease. *Gerontologist, 3*, 413–416.

Teri, L., Logsdon, R. G., Uomoto, J., & McCurry, S. M. (1997). Behavioral treatment of depression in dementia patients: A controlled clinical trial. *Journal of Gerontology, 52*, 159–166.

Teri, L., & Wagner, A. L. (1991). Assessment of depression in patients with Alzheimer's disease: Concordance among informants. *Psychology of Aging, 6*, 280–285.

Weiner, M. F. (1996). Diagnosis of dementia. In M. F. Weiner (Ed.), *The dementias: Diagnosis, management, and research* (2nd ed., pp. 1–42). Washington, DC: American Psychiatric Press.

Weiner, M. F., Bruhn, M., Svetlik, D. S., Tintner, R., & Hom, J. (1991). Experiences with depression in a dementia clinic. *Journal of Clinical Psychiatry, 52*, 234–238.

Weiner, M. F., Doody, R. S., Risser, R. C., & Liao, T. (2000). *Prevalence and incidence of major depressive disorder in Alzheimer's disease*. Unpublished manuscript.

Weiner, M. F., Edland, S. D., & Luszczynska, H. (1994). Prevalence and incidence of major depression in Alzheimer's disease. *American Journal of Psychiatry, 151*, 1006–1009.

Weiner, M. F., & Gray, K. F. (1996). Differential diagnosis. In M.F. Weiner (Ed.), *The dementias: Diagnosis, management, and research* (2nd ed., pp. 101–138). Washington, DC: American Psychiatric Press.

Weiner, M. F., & Svetlik, D. S. (1996). Dealing with family caregivers. In M. F. Weiner (Ed.), *The dementias: Diagnosis, management, and research* (2nd ed., pp. 233–249). Washington, DC: American Psychiatric Press.

Weiner, M. F., Svetlik, D. S., & Risser, R. C. (1997). What depressive symptoms are reported in Alzheimer's patients? *International Journal of Geriatric Psychiatry, 12*, 648–652.

Weingartner, H., Cohen, R. M., Murphy, D. L., Martello, J., & Gerdt, C. (1981). Cognitive processes in depression. *Archives of General Psychiatry, 38*, 42–47.

Weissman, M. M., Myers, J. K., Tischler, G. L., Holzer, C. E., Leaf, P. J., Orvaschel, H., & Brody, J. A. (1985). Psychiatric disorders (DSM-III) and cognitive impairment among the elderly in a U.S. urban community. *Acta Psychiatrica Scandinavica, 71*, 366–379.

Wells, C. E. (1979). Pseudodementia. *American Journal of Psychiatry, 136*, 895–900.

Wisniewski, K. E., Wisniewski, H. M., & Wen, G. Y. (1985). Occurrence of neuropathological changes and dementia of Alzheimer's disease in Down's syndrome. *Annals of Neurology, 17*, 278–282.

Yesavage, J. A., Brink, T. L., Rose, T. L., Lum, O., Huang, V., Adey, M., & Leirer, V. (1983). Development and validation of a depression screening scale: A preliminary report. *Journal of Psychiatric Research, 17*, 37–49.

Zubenko, G. S., & Moosy, J. (1988). Major depression in primary dementia: Clinical and neuropathologic correlates. *Archives of Neurology, 45*, 1182–1186.

13

Pharmacotherapy of Geriatric Depression

Taking the Long View

CHARLES F. REYNOLDS III, MARK D. MILLER, BENOIT H. MULSANT, MARY AMANDA DEW, and BRUCE G. POLLOCK

A PUBLIC HEALTH PERSPECTIVE ON DEPRESSION IN OLD AGE

A recent World Health Organization (WHO, 1996) study concluded that unipolar major depression and suicide accounted for 5.1% of the total global burden of disease in 1990 (with respect to a quality-of-life-based metric, disability-adjusted life years), making depression the fourth most important cause of global burden. The significance of illness burden attributable to depression increases with age weighting and is projected to grow further by the year 2020 based upon demographic shifts toward a greater proportion of elderly in the general population. Hence, finding ways of preventing the return of depression in elderly patients and of maintaining the gains of acute and continuation treatment would represent a significant treatment advance and contribution to public health. Finding better treatment strategies for depression in old age may also be, as suggested later, a matter of life and death. A further, major question is whether there are ways of predicting which older patients require combined treatment or medication alone, and which may be able to remain well on maintenance psychotherapy alone. Identifying such predictors of response will allow the cost-effectiveness of treatment choices to be maximized. There have been no studies of the cost-effectiveness (ratio of dollar costs of a treatment to quality-adjusted life years (QALYs) gained by the treatment) for depression interventions in the elderly. The issue is not whether a given treatment is more cost-effective than no treatment, but whether a

---

CHARLES F. REYNOLDS III, MARK D. MILLER, BENOIT H. MULSANT, MARY AMANDA DEW, and BRUCE G. POLLOCK • Department of Psychiatry, Western Psychiatric Institute and Clinic, University of Pittsburgh, Pittsburgh, Pennsylvania 15213.

Physical Illness and Depression in Older Adults: A Handbook of Theory, Research, and Practice, edited by Gail M. Williamson, David R. Shaffer, and Patricia A. Parmelee. Kluwer Academic/Plenum Publishers, New York, 2000.

given treatment (e.g., combined medication and psychotherapy) is more cost-effective than another treatment option (e.g., monotherapy). This has public health policy implications as well as implications for clinicians as they decide which treatment is most appropriate for their elderly patients (Katona, 1995; Livingston, Manela, & Katona., 1997).

Prevalence estimates for clinically significant depression range from about 10% for independently living elderly to approximately 25% of those with chronic illness, especially in persons with ischemic heart disease, stroke, cancer, chronic lung disease, arthritis, Alzheimer's disease, and Parkinson's disease (Beekman et al., 1995; Borson, 1995; Borson et al., 1986). These data underscore the inseparability of mental and physical health in aged persons and also highlight the need for clinical trials using agents likely to be safe and well-tolerated by elderly depressed patients burdened with chronic medical illnesses and depletion of psychosocial resources.

Certainly, the most compelling consequence of depression in later life is increased mortality from both suicide and medical illness. Elderly persons have the highest suicide rate of any age group (largely accounted for by older white males), with rates rising to 67.6 suicides per 100,000 in those age 85 and over, more than 5.5 times the overall national rate of 12 per 100,000 (National Center for Health Statistics, 1992). Suicide in the elderly is most likely to be a result of depression: In patients 75 years of age and older, 60–75% of suicide victims have diagnosable depression (Conwell, 1996). In addition, the connection between depression and nonsuicidal mortality is now well supported for myocardial infarction, where depression elevates mortality risk by a factor of five (Frasure-Smith, Lesperance, & Tajalic, 1993, 1995), and in nursing home patients, where major depression was found to increase the likelihood of mortality by 59%, independent of physical health measures (Rovner, 1993). Hence, the selection of treatment modalities that are both safe and effective for the long-term management of geriatric depression is, literally, a matter of life and death.

In this chapter, we describe the pharmacotherapy of old-age depression from two perspectives: (1) a supportive medication–clinic approach that does not utilize a specific psychotherapy; and (2) an approach that combines pharmacotherapy with a specific type of psychotherapy, interpersonal psychotherapy (IPT), as originally developed by Klerman, Weissman, Rounsaville, and Chevron (1984). We have operationalized both the medication–clinic approach and IPT for use with the elderly as part of our ongoing NIMH-sponsored research into maintenance therapy for late-life depression. For the clinically oriented reader, we describe in considerable detail how we actually carry out medication–clinic treatment using one of our study drugs, the selective serotonin reuptake inhibitor (SSRI), paroxetine. Our approach draws extensively from medication–clinic procedures originally developed by Fawcett, Epstein, Fiester, Ellein, and Anthony (1981). For the research-oriented reader, we summarize published data from our 10-year study of the tricyclic ant-depressant nortriptyline combined with IPT (Reynolds et al., 1999). Our data from the ongoing Pittsburgh studies suggest that an approach combining pharmacotherapy with IPT may be optimal for achieving and maintaining wellness in the elderly, particularly those above the age of 70. We also discuss the scientific and ethical justification for further placebo-controlled clinical trials in the very old and suggest a strategy for pursuing much-needed cost-effectiveness studies of combined treatment as compared with monotherapy in old-age depression. Thus, we have approached the writing of this chapter from the dual perspective of clinicians and researchers mindful of the large public health challenges posed by depressive illnesses in later life and the need to develop cost-effective treatments for this illness.

CONCEPTUAL FRAMEWORK FOR THE PHARMACOTHERAPY OF GERIATRIC DEPRESSION: THE MEDICAL MODEL APPROACH

Depression in late-life is often a chronic, relapsing illness that in many ways is similar to more traditional chronic illnesses treated by internists. The acute episodes are treated with active antidepressant medication and, following relief of symptoms, a period of continuation therapy is usually instituted. This therapy is frequently terminated after the patient is considered to have fully remitted from the illness. As with other medical patients, the prescription of medication, diet, exercise, and altered lifestyle does not guarantee successful outcome of the illness. However, it has been recognized that patients with chronic illnesses do better when there is an adequate social support system to help them deal with intercurrent stresses. This support system does not have to be provided solely by a physician and, in many instances, a clinic concept has arisen whereby nurse clinicians, social workers, or clinical psychologists provide the necessary counseling and support that enables a patient to deal with his or her illness and life issues.

In a similar conceptual model, psychiatric illnesses such as depression are regarded as diseases. The medical model operates under the premise that a specific etiology related to the functional anatomy of the brain will eventually be determined for psychiatric illnesses. Consistent with this theme, emphasis is placed upon the pathogenesis, symptomatology, differential diagnosis, and prognosis of specific psychiatric disorders.

The medical model traditionally makes use of supportive patient care in the clinical setting. The patient usually structures the content of each clinical contact and presents the clinician with a series of symptoms that the clinician must focus upon, expand, analyze, and ultimately, use to make a clinical diagnosis, to select a treatment, and, subsequently, to determine treatment response.

Long-term maintenance treatment is essential for chronic illnesses such as hypertension, diabetes mellitus, and old-age depression. Such treatment requires monitoring the clinical condition over the course of the illness. Chronic illness may not always require medication. For example, conservative treatment of diabetes mellitus consists of dietary restraints and regular clinic visits. The quality of the rapport established between the treatment team and the patient often influences treatment response. By design, a support system is incorporated into a medical model approach.

When medication is required for a chronic relapsing illness, additional supportive therapeutic maneuvers are also necessary. Continuing education of the patient and family is an important component of long-term maintenance treatment with medication. In addition, skillful, supportive reassurance promotes compliance and a smooth course.

The medication clinic model consists of several different components. First, the patient is engaged by the clinic in a treatment plan focusing on the relief of symptoms associated with the present illness. Second, the notion of a treatment team is also developed within the overall clinic milieu. As in most medical clinics, patients in our setting are cared for by a treatment team that includes two key figures (a primary clinician and a supervising psychiatrist) and several support personnel. Decisions concerning treatment and management are the responsibility of the primary treatment team that sees the patient on a regular basis. Specifically, in our Late Life Depression Prevention Clinic (LLDPC), patients are first greeted by the receptionist, who welcomes them and registers their visit. While waiting in the reception area, they are usually engaged in discussion of a general nature with the receptionist, and tea and coffee are freely available. Before being seen by their clinician, the patients are greeted by another member of the treatment team, who escorts them to an office where they are weighed and their blood pressure is measured. During this brief contact, some discussion about their illness or symptoms usually occurs. Patients then return to the reception area, where they may socialize with other patients before being greeted by their primary clinician.

CHRONOLOGICAL PHASES IN THE MEDICATION CLINIC APPROACH FOR THE TREATMENT OF DEPRESSION IN OLD AGE

Prior to initiation of treatment, patients may be withdrawn from any antidepressant medication that they have been taking at the time of initial evaluation in order to obtain a pretreatment baseline assessment, particularly if treatment has been unsuccessful. A drug-free period will also allow patients whose clinical depression is already subsiding to recover spontaneously. If the clinical condition is worsening and active treatment is needed, a baseline observation period is waived in the interest of the patient. During this time patients have a general physical examination and laboratory tests, including an EEG, urinalysis, complete blood count, serum electrolytes, blood urea nitrogen, serum creatinine, folate, and B^{12}. Thyroid status is monitored with determination of thyroid-stimulating hormone (TSH) levels. If the patient is at risk for thyroid or cardiovascular disease, or if special circumstances arise, additional clinical and laboratory evaluations are performed as indicated.

Pharmacotherapy is divided into two treatment phases: acute and continuation treatment (Phase I), followed by maintenance therapy (Phase II). The goal of the acute treatment phase is to treat the index episode to remission. The treatment of the index episode is used to stabilize the patient's clinical condition and assure remission of clinical symptomatology. Such remission is vital to reduce the incidence of reemergence of symptoms after maintenance assignment, and we have mandated a symptom-free period of 22 weeks (16 weeks of continuation phase and 6 weeks of transition to less frequent clinical contact) before the start of maintenance treatment. The maintenance phase is the longest phase of treatment. Its goal is to prevent or delay recurrence of major depression.

Phase I: Acute and Continuation Therapy

As noted in the introduction, we have chosen paroxetine to illustrate a manual-based, or operationalized practice, of pharmacotherapy for late-life depression. Acute-phase pharmacotherapy consists of daily administration of paroxetine, aiming for a modal dose of 20 mg. During the first week, patients are prescribed 10 mg daily; the daily dose is increased to 20 mg starting on Day 8 and remains at 20 mg through Day 21 (3 weeks). If the Hamilton depression rating (Hamilton, 1960) has not dropped 25% by the end of 3 weeks of treatment with paroxetine, the dose is increased to 30 mg/day on Day 22. If the Hamilton depression rating has not dropped by 50% by the end of 5 weeks of treatment (Day 35), dosage is increased to 40 mg (the maximal dose). Doses of paroxetine are also guided by a consideration of side-effect ratings and the judgment of the treating psychiatrist. Patients are allowed to remain in the acute phase of treatment up to 26 weeks. After a minimum of 8 weeks of therapy, patients whose Hamilton depression ratings have decreased to 10 or lower for 3 consecutive weeks move on to the continuation phase of therapy. The dosage of paroxetine may be adjusted downward if side-effect burden warrants. A daily dosage of 10 mg is considered the minimal acceptable dose.

The Initial Clinic Visit. At the patient's first medication clinic visit, the clinician reviews the independent clinical evaluation performed as part of the baseline assessment. A thorough review and reassessment of the present depressive symptoms are obtained, and attention is given to the establishment of target symptoms as a basis for ongoing clinical assessment and management. In addition, the clinician obtains a history of the patient's previous depressive episodes and psychiatric treatment (if any), family history, and medical history. An assessment of the patient's suicidality is of utmost importance during this initial clinic visit. The patient is also questioned about his or her use of alcohol and other drugs, including alternative medicine treatments (e.g., herbal therapies).

Following this review, the clinician explains to the patient his or her diagnosis and the rationale for treatment in a medication clinic. The patient is instructed as to the importance of taking the prescribed dosage of paroxetine, reassured that the medication is not addictive or dangerous, and informed about adjustment of the dosage that will be necessary to achieve a positive therapeutic response. It is important not to start the patient on a large dose of medication because of potential side effects. The patient is told to expect paroxetine to take 4–6 weeks for a therapeutic response and further instructed concerning those symptoms (e.g., sleep disturbance, agitation) that should first respond to treatment. The concept of gradual response is introduced so that the patient will not have an unrealistic expectation of a rapid and magical cure. This explanation may also decrease the probability of noncompliance and/or early dropout from treatment.

Patients are warned that potential side effects with paroxetine can include nausea, diarrhea, constipation, headache, tiredness, sedation, dry mouth, and sexual dysfunction. Patients are instructed that side effects are usually mild in intensity and diminish over time with continued administration. A comprehensive list of all possible side effects is elicited at each visit using the UKU side-effects scale as a guide (Lingjaerde, Ahlfors, Bech, Deneker, & Elgen, 1987). A careful determination is made to differentiate each somatic complaint that may be drug-related from a new-onset medical problem or exacerbation of an existing medical problem. Countermeasures to ameliorate drug-related, somatic complaints are encouraged, such as stool softeners and/or bulk laxatives for constipation, nighttime dosing instead of morning dosing for sedation, and acetaminophen or other appropriate analgesics for occasional headaches. The patient is instructed that although paroxetine has proven to be successful in the treatment of a high percentage of elderly depressed patients, the treatment approach will be changed if the patient shows a significant worsening of depressive symptoms or experiences severe side effects or adverse reactions that preclude continued treatment.

During this initial session, the patient is encouraged to ask questions or seek clarification about his or her illness and the use of paroxetine as the treatment of choice. Reassurance and optimism concerning the outcome of this treatment (e.g., "Depression is a very treatable illness") are conveyed to the patient by the clinician. The patient is instructed as to the frequency of future visits and the ongoing importance of his or her role in reporting significant changes in symptoms or problems with medication.

If family members have accompanied the patient at this initial session, the clinician summarizes the aforementioned discussion and explanations, and encourages the family's support and active involvement in the treatment approach. An ongoing task facing the clinician is that of educating both patient *and* family about strict adherence to the treatment schedule. Patients who miss appointments are contacted to facilitate cooperation and if this fails, the clinician will contact the family. A significant relative can provide assistance not only by encouraging the patient to take his or her prescribed medication but also by promptly reporting early signs of recurrence. The accompaniment of a family member during medication clinic visits is encouraged, particularly if the clinician suspects noncompliance, as part of a patient-focused, family-centered approach to the older patient. Such an approach allows family members and caregivers to be educated about the nature of late-life depression, ways of being helpful to the patient, and the rationale for treatment (to get well and to stay well).

Subsequent Clinic Visits. During the acute-phase treatment of symptomatology, we see patients once per week. The emphasis of clinic visits is the medication and its effect in modifying the core symptomatology of the illness. A typical visit in the acute phase commences with an open discussion of the patient's present clinical status, focusing on symptoms and medication side effects and their severity. Once the patient has remitted from acute clinical symptomatology (17-item Hamilton rating scale score of 10 or less), medication is continued for 16 weeks before entry into maintenance treatment. Toward the end of the initial treatment phase (continuation therapy), the necessary adjustments are made in anticipation of less frequent maintenance treatment.

Augmentation Pharmacotherapy: Protocol and Rationale. To provide an optimized best-practice treatment under each condition, the use of pharmacotherapy augmentation strategies is standardized and is *not* time-limited. Augmentation strategies are implemented in patients who have failed to show a 50% reduction in Hamilton-17 scores by week eight and in those not meeting criteria for remission (Hamilton-17 item score $\leq$ 10) at Week 12. We augment paroxetine initially with nortriptyline, starting at 10 mg daily for 7 days. If patients have not achieved the target Hamilton rating of 10 after 1 week of nortriptyline, the dose is raised to 25 mg qd (every day) for 14 days. Blood levels of nortriptyline are monitored weekly to target a steady level of 80–120 ng/ml. Patients failing to achieve response criteria after a total of 21 days of nortriptyline (at 80-120 ng/ml) are adjudged to be nonresponsive to nortriptyline augmentation and are offered augmentation treatment with bupropion (discussion to follow). Patients who achieve remission while on nortriptyline augmentation enter the continuation phase of treatment on the combination of paroxetine and nortriptyline used to bring about remission. *This decision to carry forward the augmentation regimen is based on our observation that discontinuation of adjunctive medication leads to relapse in 50% of patients* (Reynolds, Frank, Perel, Mazumdar, Dew, 1996). Continuing with augmentation pharmacotherapy for the 16-week period of stabilization therapy may thus permit a lower relapse rate and a higher recovery rate. Patients for whom nortriptyline is contraindicated, unsuccessful, or intolerable are offered augmentation with bupropion SR 150 mg/day that can be augmented to 150 mg/BID (twice daily) after 7 days. Bupropion augmentation is continued in patients who remit.

The rationale for the use of nortriptyline as an augmentation strategy is based several considerations: (1) The performance characteristics of nortriptyline as a maintenance strategy in geriatric depression have been established in controlled studies (Reynolds, Frank, Perel, Mazumdar, & Kupfer, 1995); (2) nortriptyline plasma levels can be measured, providing objective evidence of treatment adherence and adequacy; and (3) geriatric psychiatrists frequently prescribe a serotonin reuptake inhibitor and tricyclic antidepressant medication combination for treatment-resistant or refractory patients. For those patients in whom the use of nortriptyline may be contraindicated or not tolerated, the use of bupropion is a reasonable fallback strategy: it is well tolerated in elderly persons, effective, and linked to a different neurotransmitter system (i.e., dopamine) than paroxetine.

Continuation Treatment. Continuation-phase pharmacotherapy extends for 4 months beyond the end of acute-phase pharmacotherapy, using the final dose of paroxetine employed during acute-phase treatment. The same monitoring procedures are employed as during acute-phase treatment. Clinic visits are twice monthly rather than weekly, reflecting the patient's improved clinical status. Patients whose Hamilton depression scores remain at or below 10 for 4 months of continuation treatment are judged to have recovered from their index episodes and are eligible for maintenance therapy.

Phase II: Maintenance Therapy

In order for patients to enter maintenance treatment, several conditions must be satisfied: (1) The patient must have remained on a continuing steady dose of paroxetine for at least 16 consecutive weeks, and (2) the patient must have a continuing Hamilton Rating Scale score of 10 or less for 22 weeks (16 weeks continuation and 6 weeks of transition). Since 10–15% of patients may continue to demonstrate chronicity with respect to clinical symptoms, it is important to establish a relatively "symptom-free" interval of sufficient duration to enter patients into the maintenance phase. This will reduce the incidence of reemergence of symptoms and will also provide a clearer separation between persistence of clinical symptoms and development of a recurrent episode of depression.

Maintenance treatment is administered for at least 2 years, during which time patients are seen once every 4 weeks, with more frequent visits increased symptomatology or adverse reactions to medication occur. The dose of paroxetine used during maintenance treatment is the same as that used during acute-phase therapy to bring about remission. Patients and their families are instructed that "the dose that gets you well is the dose that keeps you well." For patients who recover with augmentation pharmacotherapy nortriptyline or bupropion, the augmentation strategy is continued during maintenance treatment as an added precaution against recurrence.

THE MEDICATION CLINIC SESSION DURING MAINTENANCE THERAPY: DIFFERENTIATION FROM INTERPERSONAL PSYCHOTHERAPY

The principal concern of the maintenance medication clinic is that each session is patient-centered and self-contained. While clinicians initiate the session by focusing the illness, they remain relatively passive, only encouraging the patient to expand upon issues as they arise. Thus, the issues introduced by the patient are of immediate concern, and while some of these may extend from visit to visit, they are explored only in the context within which they are presented, with an emphasis on the personal implications of these problem areas; that is, the visit is directed toward understanding how these problems affect the patient personally and relate to his or her symptoms. Most important, *no specific interpersonal problem areas or treatment goals are defined* by the clinician.

In the tradition of a medication clinic model, the patient is encouraged to present symptoms so that the session can be structured from this initial starting point. In the acute treatment phase, the patient becomes used to such questions as "How are you feeling now?", "What sort of a week have you had?", and "Have you noticed any changes in your mood?" The patient is therefore comfortable with such opening statements, and these can be easily used to open the maintenance medical clinic treatment session. One goal is clearly to aid the patient in recognizing symptoms of depression. From the focus on the illness, the clinic session can proceed in various directions. Clearly, the presence of symptoms or a response such as "Not as well as last month" requires further exploration, and again, this can be opened with a simple rejoinder (e.g., "in what way?"). It is important to avoid opening statements such as "How are you getting on with your husband or the children?" which immediately lead into interpersonal areas, since the emphasis in this model is strictly upon the patient and his or her symptoms. While it is acknowledged that interpersonal conflicts reflect the presence of a depressive illness and are a part of the depressive symptomatology, this is not the area of focus in this model.

At this point, a discussion of side effects of the medication is appropriate. Prior to the session, the patient is asked to complete a self-rating somatic symptom checklist. Again, this can be opened in a nondirective way (e.g., "Have you had any difficulties with the medication?") and if this does not bring about the required response, side effects could be specifically elicited using the standard side-effects checklist (UKU) completed during each treatment session. Again, this interaction should be used positively by the clinician to convey a general sense of concern for the patient's well-being and response to medication. Often, it is necessary to encourage the patient to continue with the medication, since compliance has proved to be a major area of difficulty in treating the elderly. Some side effects can be ascertained during the session by simple observation.

In the absence of any symptoms or side effects, the treatment session can progress in various ways. The simplest is to use a "diary" approach, in which the patient's activities since the last visit can be recounted. The patient is offered the opportunity to ventilate regarding areas that are creating personal difficulty. The role of the clinician is to be supportive and reassuring. The therapist must avoid the temptation to revert to material presented in previous sessions. Patients may tend to return repeatedly to familiar ground even though the significance of a particular issue may have substantially diminished since it was first presented. In those cases, patients may need encouragement and reassurance to move on to new topics.

In summary, this type of maintenance treatment emphasizes the following: (1) review of presence and absence of clinical symptoms of depression, (2) review of somatic complaints usually considered to be side effects of medication, (3) reassurance, and (4) avoidance of clarification with regard to psychological issues. Thus, when the patient is assessed monthly for approximately 15–20 minutes, attention is devoted to a review of clinical symptoms, possible adverse effects, physical complaints, and other issues that would normally be addressed in the context of reviewing medication management in a follow-up clinic. In the following section, we discuss combined treatment using both pharmacotherapy and psychotherapy, a clinical strategy tested in our recently completed maintenance studies.

THE LIMITATIONS OF MONOTHERAPY AND THE ADVANTAGES OF COMBINED TREATMENT APPROACHES USING BOTH PHARMACOTHERAPY AND PSYCHOTHERAPY

Our investigation of maintenance therapies in late-life depression has tested the hypothesis that long-term treatment with nortriptyline and interpersonal psychotherapy, either singly or in combination, is superior to placebo in preventing the recurrence of major depressive episodes. We recruited 187 elderly patients with recurrent, nonpsychotic, nondysthymic, unipolar major depression, of whom 180 actually began treatment (mean age = 67 years).

After signing an informed consent, patients were treated in a noblind fashion using a combination of nortriptyline (NT), with doses titrated to achieve plasma steady-state levels of 80-120 ng/ml, and weekly interpersonal psychotherapy (IPT), in order to achieve a remission of depressive symptoms as defined by a score of 10 or less for 3 consecutive weeks on the 17-item Hamilton Rating Scale for Depression. Following successful acute-phase treatment, patients entered a 16-week period of continuation therapy to ensure stability of remission and full recovery. Continuation treatment consisted of combined NT and IPT, using the same dose of NT as during acute therapy of the index episode but with frequency of IPT reduced to twice monthly. Patients whose remission was stable over 16 weeks were then randomly assigned to one of four maintenance therapy cells: (1) medication clinic with nortriptyline, (2) medication clinic with placebo, (3) monthly maintenance interpersonal psychotherapy (IPT-M) and full-dose nortriptyline (80–120 ng/ml), or (4) IPT-M and placebo. For patients randomly assigned to a placebo condition, nortriptyline was slowly discontinued over 6 weeks and placebo was gradually substituted under double-blind conditions. Patients remained in maintenance therapy for 3 years or until recurrence of a major depressive episode, whichever occurred first.

Major findings from the acute and continuation phases of treatment included the following observations:

- Older patients appear to benefit as much as midlife patients from treatment of recurrent unipolar major depression, but relapse rate during continuation treatment is higher among elderly patients as compared with midlife patients (15% vs. 7%) (Reynolds, Frank, Kupfer, et al., 1996a).
- Severe life events and comorbid anxiety disorders slow response times in depressed elderly patients (Karp et al., 1993), but chronic medical illness does not so long as it is optimally managed (Miller et al., 1996) .
- The median time to remission in elderly patients is 12 weeks (Reynolds et al., 1994), and the earliest point of statistically reliable discrimination of recovering and nonrecovering patients is 4 weeks (Reynolds, Frank, Dew, Perel, 1995). Almost one-third (30.5%) of patients showed rapid sustained response to combined treatment with NT and IPT (i.e., well by 4 weeks), 22.1% showed gradual sustained response (well by 8–10 weeks), 23.2% showed partial or mixed response, and 24.2% showed little or no evidence of response (Dew et al., 1997). Slower and more brittle treatment response is associated with higher pretreatment levels of anxiety, lower levels of social support, greater current age at index episode, and higher percentage of REM sleep before the initiation of treatment.
- Under conditions of open (i.e., nonblinded) combined treatment, we detected no difference in observed recovery rates between patients with different psychotherapeutic foci: interpersonal dispute, 68%; role transitions, 75%; or grief, 83%. Likewise, recovery rates did not vary between patients with multiple versus single IPT foci (Wolfson et al., 1997).
- Patients needing augmented pharmacotherapy to get well probably need continuation of adjunctive medication to remain well. Also, factors that lead to the use of augmentation in the first place (e.g., heightened anxiety) may increase the risk for relapse (Reynolds, Frank, Perel, Mazumdar, Dew, 1996).
- Although our study was not originally designed to include a full quality of life assessment, post hoc analyses showed that quality of life (perceived ability to cope and well being), as measured by the General Life Functioning scale (Elkin, Parloff, Hadley, & Anthony, 1985) improved significantly over the course of treatment in both recovered and nonrecovered patients, even after controlling for initial level of depression (Mazumdar et al., 1996). Quality of life was better in recovered than in nonrecovered patients. (Since this observation was made from a noncontrolled part of our protocol, we cannot exclude the role of a Hawthorne effect.)
- Clinicians should not be lured into a lesser degree of clinical vigilance if elderly depressed patients' suicidal ideation is "only" passive. Both groups of depressed elderly, those with active and those with passive ideation, appear to be more alike than different with respect to most indicators of affective morbidity, and both groups reports a greater sense of hopelessness than those without suicidal ideation (Szanto et al., 1996). Hence, it is vital that clinical and psychotherapeutic efforts in elderly depressed patients be directed at detecting and treating hopelessness. Higher levels of hopelessness in patients with a history of suicide

attempts persist into remission than is the case with ideators or nonsuicidal patients (Szanto et al., 1997).

- In order to examine the effect of *earlier versus later lifetime* age at onset of depressive illness (as distinct from *current* age at time of study entry) on rates of remission, relapse, recovery, and recurrence, we examined two groups of patients: those whose first lifetime episodes had occurred before age 60, and those with initial episodes at age 60 or later. The two groups did not differ in absolute rates of remission and recovery (during nonblinded combined treatment with NT and IPT), nor in rates of relapse or recurrence (whether on maintenance NT or IPT). However, subjects with earlier-onset depressive illness took on average 5–6 weeks longer to achieve remission of their index episodes, possibly a reflection of the greater number of prior lifetime episodes (chronicity) (Reynolds, Frank, Perel, Mazzumdar, & Kupfer, 1998). Because early-onset subjects also had a higher rate of past suicide attempts, such subjects needed especially careful surveillance during acute treatment.
- The rate of treatment resistance to combined treatment (NT pharmacotherapy with augmentation and IPT), as determined by failure to remit or relapse during continuation treatment and failure to recover, was 18% (Little et al., 1998).

We think that these findings, together with those presented in the following section, support combined treatment as a powerful clinical management strategy in late-life depression.

RECENT ADVANCES IN PHARMACOTHERAPY OF GERIATRIC DEPRESSION USING NORTRIPTYLINE COMBINED WITH INTERPERSONAL PSYCHOTHERAPY: SETTING THE STAGE FOR MAINTENANCE STUDIES IN THE VERY OLD

In the recently completed Pittsburgh study of maintenance therapies in late-life depression involving 180 patients, 67 participants were aged 70 or older at the time of study entry. Forty-nine of these 67 subjects (73.1%) remitted during open acute-phase combined treatment with NT, with dosage titration to 80–120 ng/ml and weekly interpersonal psychotherapy. The intent-to-treat remission rate in the 60- 69-year-olds was 91/113 (80.5%). These data suggest that the use of a combined acute-phase treatment approach (medication plus psychotherapy) is useful in ensuring a satisfactory rate of remission in patients age 70 and older, one that is comparable to that observed in 60-69-year-old patients (Reynolds, Dew, et al., 1998).

A similar percentage of subjects age 70 and older (40/67, or 59.7%) entered maintenance treatment as was observed in subjects age 60–69 (70/113, or 61.9%). Attrition in the age 70+ subgroup was related to treatment refusal, side effects, or the development of intercurrent medical illness. Hence, use of a selective serotonin reuptake inhibitor (SSRI) such as paroxetine, with its milder side-effect profile and greater safety in medically ill patients, may permit a lower attrition rate and hence ,a greater intent-to-treat recovery rate in the elderly. This is clearly an important issue in both the short- and long-term management of depressive illness those age 70 and older, who are more vulnerable both to side effects and to the development of incident medical illnesses that may complicate or even contraindicate the further use of a tricyclic antidepressant.

Despite almost identical recovery rates and percentages of subjects in both age groups entering maintenance treatment, the overall recurrence rate during the first year of maintenance treatment was 60.5% (23/40) in the subjects age 70 and older versus 30.4% (21/70) in those age 60–69 (log rank $X^2_1 = 8.07$; $p < .005$; Reynolds, Dew, et al., 1998; Reynolds, Frank, Perel, Mazumdar, & Kupfer, 1998). This observation suggests a more brittle, or variable, long-term response in the very old patient.

RECENT ADVANCES IN PHARMACOTHERAPY WITH THE SELECTIVE SEROTONIN REUPTAKE INHIBITORS: WHERE DO WE NEED TO GO FROM HERE?

In a recent extensive review of 20 randomized trials comparing the acute efficacy of tricyclic antidepressants (TCAs) and SSRIs, Schneider (1996), concluded that TCAs and SSRIs have similar acute-phase efficacy but that SSRIs are better tolerated, with dropout rates two-thirds to one-half less. In general, paroxetine is well tolerated by elderly patients (Dunner & Dunbar, 1992; Hutchinson, Tong, Moon, Vince, & Clarke, 1992; Boyer & Blumhardt, 1992), and our ongoing double-blind, randomized study comparing paroxetine with nortriptyline shows similar effectiveness in the *acute* pharmacotherapy of severely depressed, medically ill, elderly patients with a range of cognitive impairment (Mulsant et al., 1998). Moreover, a beneficial effect of paroxetine on cognitive function has also been demonstrated in depressed elderly patients (Schone & Ludwig, 1993). In contrast to in vitro data, our current work also shows that at therapeutic plasma concentrations, paroxetine has approximately one-fifth the anticholinergic potential of NT, making it a logical candidate for studies in the age 70+ subjects (Pollock et al., 1998). Thus, based on all these considerations, paroxetine appears to be an excellent candidate for short- and long-term therapy in the elderly. However, we know of no randomized, placebo-controlled, maintenance clinical trials of SSRIs in old-age depression.

As reported elsewhere (Walters, Reynolds, Mulsart, & Pollack, 1999), we have recently performed an open trial comparing paroxetine and NT in continuation treatment for 18 months following remission and recovery. Twenty-five of 27 subjects whose depression remitted during medication clinic treatment with paroxetine (including 5 subjects who crossed over to paroxetine after failing NT) elected to enter continuation treatment, while 15 of 16 NT responders agreed to continuation therapy with this agent. Paroxetine was held at the same dose used during acute-phase pharmacotherapy. To date, during an average follow-up interval of 11.4 months, 4 of 25 subjects have experienced a depressive relapse during continuation therapy with paroxetine (mean dose 20 mg/day). Three other subjects have left treatment for other reasons: 1 withdrew consent, 1 left due to sexual dysfunction, and 1 died.

Examination of paroxetine blood levels in patients who relapsed disclosed no difference in levels or in variability over time as compared with levels in patients who remained well. This observation suggests that relapse was not due to treatment noncompliance. Thus, paroxetine holds promise for longer-term maintenance treatment in severely depressed patients in their 70s as long as paroxetine dosage is maintained at the same dose used during acute-phase pharmacotherapy of the index episode. We observed that paroxetine and NT may have similar efficacy for preventing relapse during continuation and early maintenance treatment. If we consider a survival plot displaying time to all-cause termination, where the latter includes relapse/recurrence, side effects, medical illness, and death, then paroxetine appears to offer greater advantage than NT as a maintenance strategy.

Monitoring and Maximizing Compliance with Pharmacotherapy

Maximizing compliance with antidepressant treatment is critical to good patient care and to the success of the intervention studies. We have recently reported low rates of missed medication doses in the Pittsburgh study of maintenance therapies. Using data based on a semistructured interview, zero noncompliance occurred in 43% of subjects during acute and continuation therapy, one missed doses/month in 50.3%, 2 missed doses/month in 5.6%, and three or more missed doses/month in 1.1% (Miller et al., 1997). In clinical practice and research, we employ the following strategies to monitor and encourage compliance: (1) serum drug levels; (2) use of level-to-dose (L/D) ratios; (3) use of pill boxes and pill counts; (4) review of medication lists; (5) patient self-report checklist; (6) weekly compliance questionnaire administered by clinicians; (7) side effect surveillance and education; (8) close liaison with other physicians; (9) team coordination; (10) family psychoeducational workshops; and (11) courier service for medication delivery during bad weather.

Previously, it has been found with TCAs, that the longitudinal stability in quotients of drug plasma levels divided by dose is more applicable to monitoring adherence with pharmacotherapy than plasma levels alone, since dosage changes may limit the usefulness of steady-state plasma levels (Foglia & Pollock, 1997; Miller et al., 1997). Our preliminary data suggest paroxetine steady-state levels have comparable stability, permitting this same approach. A subject is determined to be nonadherent if the percentage coefficient of variation (CV) in his or her plasma level/dose values exceed his or her prior mean quotient by ±2 standard deviations in the absence of an interacting medication. In the paroxetine pilot study, the mean %CV was 19.2 (SD = 6.7, n =19) which is very comparable to that of NT-treated patients (M 17.6, SD =9.7, n =16). Subsequently, during maintenance treatment, 4 paroxetine-treated subjects were identified as being a least partially nonadherent. Thus, we believe this method complements pill counts and clinician interview in the assessment of pharmacotherapy treatment adherence.

Are Placebo-Controlled Studies Still Needed? Scientific and Ethical Considerations

We think that the use of a double-blind, placebo–control design is still necessary in studies of long-term efficacy (1) to enhance the interpretability of maintenance data in older study samples more heterogeneous than those investigated to date; (2) to meet the highest standards of scientific evidence in demonstrating the long-term efficacy and cost-effectiveness of medication and psychotherapy in those age 70 and above; and (3) to permit a rigorous demonstration of the long-term side-effect profile of paroxetine or other SSRIs in those age 70 and older. Absent a placebo–control condition, clinical practice in the most vulnerable and rapidly growing segment of elderly patients could not be as wellgrounded scientifically *or* ethically as the long-term management of depressive illness in young-old subjects or midlife subjects. Moreover, we think that the risk of placebo therapy can be substantially mitigated, because early detection of placebo-related recurrence and procedures to restabilize patients rapidly (including the use of a nonblind monitoring committee) has demonstrated feasibility. Patients in a placebo–control condition are getting more than if they were assigned to no treatment at all: they are still closely monitored on a frequent basis with clear backup plans for treatment (Reynolds et al., 1994). Finally, no patients are randomly assigned to a placebo maintenance condition until they have recovered from their index episode.

What Types of Clinical Trials Are Needed Now?

We believe that additional long-term clinical trials testing the maintenance efficacy of both antidepressant medication (particularly SSRIs) and interpersonal psychotherapy in patients age 70+ are strongly needed. A 2 × 2 factorial design is useful to test both main effects of drug and psychotherapy and also their hypothesized interaction. Ideally, future studies would also include a focus exclusively on subjects age 70+, who demonstrate the most brittle long-term response and for whom combined treatment may be the best clinical strategy. We therefore suggest studies of subjects age 70+ and above (those at highest risk for relapse/recurrence, development of cognitive impairment, intercurrent medical illness complicating the use of antidepressants, and suicide). We also advocate the use of an SSRI antidepressant (such as paroxetine) widely used in clinical practice and safer in medically burdened elderly patients and in overdose than traditional TCAs; the standardized use and continuation of pharmacotherapy augmentation in patients responding only partly to paroxetine by 8 weeks; evaluation not only of direct improvement in depressive illness but also treatment impact on quality of life, quality-adjusted life years, and associated cost-effectiveness; and the examination of moderators potentially related to maintenance treatment success and failure in age 70+ patients, to identify which patients need which treatment.

Cost-Effectiveness Issues Related to the Pharmacotherapy of Geriatric Major Depression

We know of no cost-effectiveness analysis (CEA) of combined therapy, monotherapy with medication, and monotherapy with IPT in treatment studies of depression in old age. We recommend that such analyses be linked to clinical trials and follow the guidelines for conducting a CEA, as recently outlined by U.S. Public Health Service Panel on Cost-Effectiveness in Health and Medicine (Russell, Gold, Siegel, Damiels, & Weinstein, 1996). The recommended procedures yield results as incremental cost-effectiveness ratios, with quality-of-life and health effects expressed in units of quality-adjusted life years (QALYs) and costs expressed in U.S. dollars. The measures of QALYs gained across treatments capture both morbidity and mortality during the period of study and can be directly calculated from such scales as the Quality of Well-Being Scale (QOWBS) and the utilities associated with status on this scales. The analysis is intended to inform policy decisions at the societal level. Thus, it utilizes a reference case format, and a societal perspective is adopted such that all costs and other consequences of treatment are taken into account (Siegel, Weinstein, Russell, & Gold, 1996). The standard discounting rate of 3% is applied to costs (Siegel et al., 1996).

Data included in the analysis can come from study respondents' repeated QALY assessments during maintenance treatment and from the specific costs estimated to be associated with providing a given treatment to the respondents for a given period. Estimated costs include dollar costs of a given treatment assignment if that treatment assignment were to be applied under "usual care" circumstances (e.g., prescribed medication costs, clinic space required, staff time, patient lab costs, and any costs incurred for symptom monitoring), patient dollar costs for non-health-care resources related to receiving the treatment (e.g., transportation to the treatment setting, parking charges, child- or adult-care arrangements), and time costs associated with the treatment (reflecting time spent in the treatment setting or in any activity related to treatment, and time spent by a family caregiver in assisting the patient to participate in treatment).

Standard results can be presented from the reference case analysis, including total cost, total effectiveness (QALYs), incremental cost (of one treatment assignment vs. another), incremental effectiveness (QALYs) and cost-effectiveness (incremental cost/incremental QALY ratio). An example of earlier CEA work from the Pittsburgh Study of maintenance treatment in midlife depression has been reported (Kamlet et al., 1995).

Challenges and Opportunities for the Next 15 Years

Data from naturalistic studies have identified several predictors of relapse and recurrence in geriatric depression, including a history of frequent episodes, first-episode onset after age 60, supervening medical illness, a history of myocardial infarction or vascular disease, high pretreatment severity of depression and anxiety, and cognitive impairment—especially frontal lobe dysfunction as signaled by difficulties with initiation and perseveration (Alexopoulos et al., 1996, 1997; Cole & Bellavance, 1997; Baldwin & Jolley, 1986; Hinrichsen, 1992; Burvill, Hall, & Stampfer, 1991; Keller, Lavori, Collins, & Klebman, 1983; Georgotas & McCue, 1989; Zis, Graf, & Webster, 1980). Moreover, such factors as these appear to interact with low treatment intensity in determining more severe courses of illness (Alexopoulos et al., 1996). Despite the evidence that high treatment intensity is effective in preventing relapse and recurrence, naturalistic studies have shown that the intensity of antidepressant treatment prescribed by psychiatrists begins to decline within 16 weeks of entry and approximately 10 weeks prior to recovery (Alexopoulos et al., 1996). In this context, residual symptoms of anxiety and excessive worry were found to predict early recurrence after tapering continuation treatment in elderly depressed patients (Meyers, Gabriele, Kakuma, Ippolito, & Alexopoulus, 1996). Because these observations about the moderators of a relapsing/recurrent course in late-life depression derive almost exclusively from naturalistic studies, they leave unanswered questions of great importance to patients and their families, clinicians, and health policy analysts. Can long-term antidepressant treatment affect the course of depression in the age 70+ group, reducing long-term treatment–response variability? If so, what types of treatment are effective in which patients? These two questions need to drive future studies, and they provide the essential rationale for a controlled, long-term efficacy study in subjects age 70+ and over, using contemporary antidepressant pharmacotherapy and interpersonal psychotherapy.

ACKNOWLEDGMENT: This research was supported by Grant Nos. P30 MH52247, R37 MH43832, K05 MH00295, K02 MH01509, and K01 MH01613.

REFERENCES

Alexopoulos, G. S., Meyers, B. S., Young, R. C., Kakuma, T., Feder, M., Einhorn, A., & Rosendahl, E. (1996). Recovery in geriatric depression. *Archives of General Psychiatry, 53*, 305–312.

Alexopoulos, G. S., Meyers, B. S., Young, R. C., Kakuma, T., Silbersweig, D., & Charlson, M. (1997). Clinically defined vascular depression. *American Journal of Psychiatry, 154*(4), 562–565.

Baldwin, R. C., & Jolley, D. J. (1986). The prognosis of depression in old age. *British Journal of Psychiatry, 149*, 574–583.

Beekman, A. T., Deeg, D. J., van Tilburg, T., Smit, J. H., Hooijer, C., & van Tilburg, W. (1995). Major and minor depression in later life: A study of prevalence and risk factors. *Journal of Affective Disorders, 36*(1–2), 65–75.

Borson, S. (1995). Psychiatric problems in the medically ill elderly. In H. I. Kaplan & B. J. Sadock (Eds.), *Comprehensive textbook of psychiatry* (pp. 2586–2593). Baltimore: Williams & Wilkins.

Borson, S., Barnes, R. A., Kukull, W. A., Okimoto, J. T., Veith, R. C., Inui, T. S., Carter, W., & Raskind, M. A. (1986). Symptomatic depression in elderly medical outpatients: I. Prevalence, demography, and health service utilization. *Journal of the American Geriatrics Society, 34*(5), 341–347.

Boyer, W. F., & Blumhardt, C. L. (1992). The safety profile of paroxetine. *Journal of Clinical Psychiatry, 53*(Suppl.), 61–66.

Burvill, P. W., Hall, W. D., & Stampfer, H. G. (1991). The prognosis of depression in old age. *British Journal of Psychiatry, 158*, 64–71.

Cole, M. G., & Bellavance, F. (1997). The prognosis of depression in old age. *American Journal of Geriatric Psychiatry, 5*(1), 4–14.

Conwell, Y. (1996). Suicide in elderly patients. In L. S. Schneider, C. F. Reynolds, & B. D. Lebowitz (Eds.), *Diagnosis and treatment of depression in late life* (pp. 397–418). Washington, DC: American Psychiatric Press.

Dew, M. A., Reynolds, C. F., Houck, P. R., Hall, M., Buysse, D. J., Frank, E., & Kupfer, D. J. (1997). Temporal profiles of the course of depression during treatment: Predictors of pathways toward recovery in the elderly. *Archives of General Psychiatry, 54*, 1016–1024.

Dunner, D. L., & Dunbar, G. C. (1992). Optimal dose regimen for paroxetine. *Journal of Clinical Psychiatry, 53*(Suppl.), 21–26.

Elkin, I., Parloff, M. B., Hadley, S. W., & Autry, J. H. (1985). NIMH Treatment of Depression Collaborative Research Program: Background and research plan. *Archives of General Psychiatry, 42*(3), 305–316.

Fawcett, J., Epstein, P., Fiester, S. J., Elkin, I., & Autry, J. H. (1981). Clinical management: Imipramine-placebo administration manual. NIMH treatment of depression collaborative Research Program. *Pharmacology Bulletin, 23*(2), 309–324.

Foglia, J. P., & Pollock, B. G. (1997). Medication compliance in the elderly. *Essential Psychopharmacology, 1*, 243–253.

Frasure-Smith, N., Lesperance, F., & Talajic, M. (1993). Depression following myocardial infarction: Impact on 6-month survival. *Journal of the American Medical Association, 270*, 1819–1825.

Frasure-Smith, N., Lesperance, F., & Talajic, M. (1995). Depression and 18-month prognosis after myocardial infarction. *Circulation, 91*, 999–1005.

Georgotas, A., & McCue, R. E. (1989). Relapse of depressed patients after effective continuation therapy. *Journal of Affective Disorders, 17*, 159–164.

Hamilton, M. (1960). A rating scale for depression. *Journal of Neurology, Neurosurgery and Psychiatry, 23*, 56–62.

Hinrichsen, G. A. (1992). Recovery and relapse from major depressive disorder in the elderly. *American Journal of Psychiatry, 149*(11), 1575–1579.

Hutchinson, D. R., Tong, S., Moon, C. A., Vince, M., & Clarke, A. (1992). Paroxetine in the treatment of elderly depressed patients in general practice: a double-blind comparison with amitriptyline. *International Clinical Psychopharmacology, 6*(Suppl 4), 43–51.

Kamlet, M. S., Paul, N., Greenhouse, J., Kupfer, D., Frank, E., & Wade, M. (1995). Cost utility analysis of maintenance treatment for recurrent depression. *Controlled Clinical Trials, 16*(1), 17-40.

Karp, J. F., Frank, E., Anderson, B., George, C. J., Reynolds, C. F., Mazumdar, S., & Kupfer, D. J. (1993). Time to remission in late-life depression: Analysis of effects of demographic, treatment, and life-events measures. *Depression, 1*, 250–256.

Katona, C. (1995). Rationalizing antidepressants for elderly people. *International Clinical Psychopharmacology, 10* (Suppl 1), 37–40.

Keller, M. B., Lavori, P. W., Collins, C. E., & Klerman, G. L. (1983). Predictors of relapse in major depressive disorder. *Journal of the American Medical Association, 250*, 3299–3304.

Klerman, G. L., Weissman, M. M., Rounsaville, B. J., & Chevron, E. (1984). *Interpersonal Psychotherapy of Depression*. New York: Basic Books Inc.

Lingjaerde, O., Ahlfors, U. G., Bech, P., Deneker, S. J., & Elgen, K. (1987). The UKU side effect rating scale: A new comprehensive rating scale for psychotropic drugs and a cross-sectional study of side effects in neuroleptic-treated patients. *Acta Psychiatrica Scandinavica Supplementum 334*, 1–100.

Little, J. T., Reynolds, C. F., Dew, M. A., Frank, E., Begley, A. E., Miller, M. D., Cornes, C. L., Mazumdar, S., Perel, J. M., & Kupfer, D. J. (1998). How common is resistance to treatment in recurrent, nonpsychotic geriatric depression? *American Journal of Psychiatry, 155*(8), 1035–1038.

Livingston, G., Manela, M., & Katona, C. (1997). Cost of community care for older people. *British Journal of Psychiatry, 171*, 56–59.

Mazumdar, S., Reynolds, C. F., Houck, P. R., Frank, E., Dew, M. A., & Kupfer, D. J. (1996). Quality of life in elderly patients with recurrent major depression: A factor analysis of the General Life Functioning (GLF) Scale. *Psychiatry Research, 63*, 183–190.

Meyers, B. S., Gabriele, M. S., Kakuma, T., Ippolito, L., & Alexopoulos, G. (1996). Anxiety and recurrence as predictors of recurrence in geriatric depression: A preliminary report. *American Journal of Geriatric Psychiatry, 4*, 252–257.

Miller, M. D., Foglia, J. P., Pollock, B. G., Begley, A., & Reynolds, C. F. (2000). Maximizing antidepressant compliance in depressed geriatric patients. *Essential Psychopharmacology, 20*, 93–99.

Miller, M. D., Paradis, C. F., Houck, P. R., Rifai, A. H., Mazumdar, S., Pollock, B., Perel, J. M., Frank, E., & Reynolds, C. F. (1996). Chronic medical illness in ambulatory elders with recurrent major depression: effects on acute treatment outcome. *American Journal of Geriatric Psychiatry, 4*(4), 281–290.

Miller, M. D., Pollock, B. G., Foglia, J. P., Begley, A., & Reynolds, C. F. (2000). Maximizing antidepressant compliance in depressed geriatric patients: Cumulative experience from the maintenance therapies in late-life depression study. *Directions in Psychiatry, 20*, 93–99.

Mulsant, B. H., Pollock, B. G., Nebes, R., Miller, M., Little, J. T., Stack, J. A., Houck, P. R., Bensasi, S., Mazumdar, S., & Reynolds, C. F. (1999). A double-blind randomized comparison of nortriptyline and paroxetine in the treatment of late-life depression: Six-week outcome. *Journal of Clinical Psychiatry, 60*(Suppl.), 16–20.

National Center for Health Statistics. (1992). Advance report of final mortality statistics, Report No. 40. In: *NCHS Monthly Vital Statistics Report*. Washington, DC: U.S. Department of Health and Human Services.

Pollock, B. G., Mulsant, B. H., Nebes, R., Kirshner, M. A., Begley, A. E., & Reynolds, C. F. (1998). Serum anticholinergicity in older patients treated with paroxetine or nortriptyline. *American Journal of Psychiatry, 155*, 1110–1112.

Reynolds, C. F., Dew, M. A., Frank, E., Begley, A. E., Miller, M. D., Cornes, C., Mazumdar, S., Perel, J. M., & Kupfer, D .J. (1998). Lifetime age at onset: Effects on treatment response and illness course in elderly patients with recurrent major depression. *American Journal of Psychiatry, 155*(6), 795–799.

Reynolds, C. F., Frank, E., Dew, M. A., Houck, P. R., Miller, M. D., Mazumdar, S., Perel, J. M., & Kupfer, D. J. (1999). The challenge of treatment in 70+ year olds with major depression: Excellent short-term but brittle long-term response. *American Journal of Geriatric Psychiatry, 7*(1), 64–69.

Reynolds, C. F., Frank, E., Dew, M. A., Perel, J. M., Mazumdar, S., Buysse, D. J., Begley, A., Houck, P. R., Miller, M. D., Cornes, C., & Kupfer, D. J. (1995). Discrimination of recovery in the treatment of elderly patients with recurrent major depression: Limits of prediction. *Depression, 2*, 218–222.

Reynolds, C. F., Frank, E., Kupfer, D. J., Thase, M. E., Perel, J. M., Mazumdar, S., & Houck, P. R. (1996). Treatment outcome in recurrent major depression: A post-hoc comparison of elderly ("young old") and mid-life patients. *American Journal of Psychiatry, 153*(10), 1288–1292.

Reynolds, C. F., Frank, E., Perel, J. M., Imber, S. D., Cornes, C., Miller, M. D., Mazumdar, S., Houck, P. R., Dew, M. A., Stack, J. A., Pollock, B. G., & Kupfer, D. J. (1999). Nortriptyline and interpersonal psychotherapy as maintenance therapies for recurrent major depression: A randomized controlled trial in patients older than 59 years. *Journal of the American Medical Association, 281*(1), 39–45.

Reynolds, C. F., Frank, E., Perel, J. M., Mazumdar, S., Dew, M. A., Begley, A., Houck, P. R., Hall, M., Mulsant, B., Shear, M. K., Miller, M. D., Cornes, C., & Kupfer, D. J. (1996). High relapse rates after discontinuation of adjunctive medication in elderly patients with recurrent major depression. *American Journal of Psychiatry, 153*(11), 1418–1422.

Reynolds, C. F., Frank, E., Perel, J., Mazumdar, S., & Kupfer, D. J. (1995b). Maintenance therapies for late-life recurrent major depression: Research and review circa 1995. *International Psychogeriatrics, 7*(Suppl), 27–40.

Reynolds, C. F., Frank, E., Perel, J., Mazumdar, S., & Kupfer, D. J. (1998). Maintenance therapies for late-life recurrent major depression: Research and review circa 1996. In J. C. Nelson (Ed.), *Geriatric psychopharmacology* (pp. 127–142). New York: Marcel.

Reynolds, C. F., Frank, E., Perel, J. M., Miller, M. D., Cornes, C., Rifai, A. H., Pollock, B. G., Mazumdar, S., George, C. J., Houck, P. R., & Kupfer, D. J. (1994). Treatment of consecutive episodes of major depression in the elderly. *American Journal of Psychiatry, 151*(12), 1740–1743.

Rovner, B. W. (1993). Depression and increased risk of mortality in the nursing home patient. *American Journal of Medicine, 94*(Suppl. 5A), 19S–22S.

Russell, L. B., Gold, M. R., Siegel, J. E., Daniels, N., & Weinstein, M. C. (1996). The role of cost-effectiveness analysis in health and medicine: Panel on Cost-Effectiveness in Health and Medicine. *Journal of the American Medical Association, 276*(14), 1172–1177.

Schneider, L. S. (1996). Pharmacologic considerations in the treatment of late-life depression. *American Journal of Geriatric Psychiatry, 4*(4, Suppl. 1), S51–S65.

Schone, W., & Ludwig, M. (1993). A double-blind study of paroxetine compared with fluoxetine in geriatric patients with major depression. *Journal of Clinical Psychopharmacology, 13*(6. Suppl. 2), 34S–39S.

Siegel, J. E., Weinstein, M. C., Russell, L. B., & Gold, M. R. (1996). Recommendations for reporting cost-effectiveness analyses: Panel on Cost-Effectiveness in Health and Medicine. *Journal of the American Medical Association, 276*(16), 1339–1341.

Szanto, K., Reynolds, C. F., Conwell, Y., Begley, A .E., & Houck, P. R. (1998). High levels of hopelessness persist in geriatric patients with remitted depression and a history of suicide attempt. *Journal of the American Geriatrics Society, 46*, 1401–1406.

Szanto, K., Reynolds, C. F., Frank, E., Stack, J., Fasiczka, A. L., Miller, M. D., Mulsant, B. H., & Mazumdar, S. (1996). Suicide in elderly depressed patients: Is "active" vs. "passive" suicidal ideation a clinically valid distinction? *American Journal of Geriatric Psychiatry, 4*(3), 197–207.

Walters, G., Reynolds, C. F., Mulsant, B. H., & Pollock, B. G. (1999). Continuation and maintenance pharmacotherapy in geriatric depression: An open-trial comparison of paroxetine and nortriptyline in patients over age 70. *Journal of Clinical Psychiatry, 60*(Suppl.), 38–44.

Wolfson, L. K., Miller, M., Houck, P. R., Ehrenpreis, L., Stack, J. A., Frank, E., Cornes, C., Mazumdar, S., Kupfer, D. J., & Reynolds, C. F. (1997). Foci of interpersonal psychotherapy (IPT) in depressed elders: Clinical and outcome correlates in a combined IPT/Nortriptyline protocol. *Psychotherapy Research, 7*(1), 45–55.

World Health Organization. (1996). Global health statistics: A compendium of incidence, prevalence and mortality estimates for over 200 conditions. In *The Global Burden of Disease* (pp. 000–000). Cambridge, MA: Harvard University Press.

Zis, A. P., Graf, P., & Webster, M. (1980). Predictors of relapse in recurrent affective disorders. *Psychopharmacology Bulletin, 16*, 47–49.

14

A Hope-Based Group Treatment for Depressed Older Adult Outpatients

ELLEN J. KLAUSNER, C. R. SNYDER,
and JEN CHEAVENS

INTRODUCTION

This chapter reviews a new model of hope that is posited to help in the understanding and amelioration of depression. The tenets of this new hope theory as well as its supposed role in the etiology of depression are examined. Furthermore, the epidemiology and treatments for depressions in later life are explored briefly, with the conclusion that hopelessness is implicated in such depressions. Because of the theoretical linkage of hope theory to depression, as well as data relating hopelessness to depression, it is concluded that a hope-based, goal-focused group psychotherapy (GFGP) for depression may be worth pursuing. Accordingly, a study in which depressed older adults are given either a hope-based (GFGP) or a reminiscence-based group psychotherapy experience is described, along with the results of this study and an in-depth description of one of the hope group clients. Finally, the implications of hope theory as an intervention for depressed older adults are discussed, along with ideas for future work in this area.

HOPE THEORY AND PSYCHOTHERAPY IMPLICATIONS

Dictionary definitions of hope describe it as the perception "that something desired may happen." Scholarly writings have taken this definition and emphasized the importance of goals. Thus, many have suggested that hope is a unidimensional construct

ELLEN J. KLAUSNER • Department of Psychiatry, Weill Medical College of Cornell University, White Plains, New York 10605. **C. R. SNYDER and JEN CHEAVENS** • Department of Psychology, University of Kansas, Lawrence, Kansas 66045.

Physical Illness and Depression in Older Adults: A Handbook of Theory, Research, and Practice, edited by Gail M. Williamson, David R. Shaffer, and Patricia A. Parmelee. Kluwer Academic/Plenum Publishers, New York, 2000.

involving an overall perception that goals can be met (e.g., Cantril, 1964; Erickson, Post, & Paige, 1975; Farber, 1968; Frank, 1968; Frankl, 1992; Melges & Bowlby, 1969; Menninger, 1959; Schactel, 1959; Stotland, 1969). According to these writers, expectancies for goal-attainment can be used to explain diverse behaviors, including those involving physical and mental health. Although these conceptualizations of hope have assumed that people are goal-directed and that such goal-directedness is adaptive, they have not examined the means by which goals are pursued. Expanding on these earlier views, Snyder and his colleagues have drawn upon goal concepts (see Lee, Locke, & Latham, 1989; Pervin, 1989) so as to elucidate the cognitive set of hope. In contrast to previous goal theorization, where the role of goals is portrayed as a means of enhancing motivation (Bandura, 1989; Locke, Shaw, Saari, & Latham, 1981; Mento, Steel, & Karren, 1987), the present hope theory not only suggests that goals are related to motivation (what is called *agency* in this new theory), but that these goals also have important cognitive ties to thoughts about how to reach both goals and subgoals (called *pathways thinking* in hope theory).

Initial Definition

According to this new theory, hope reflects a thinking process whereby a person has a sense of agency and pathways for goals (Snyder, 1994a, 1994b; Snyder, Harris, et al., 1991,1996; Snyder, Hoza, et al., 1997; Snyder, McDermott, Cook, & Rapoff, 1997). More specifically, hope has been defined as "a reciprocally derived sense of successful (a) agency (goal-directed determination) and (b) pathways (planning of ways to meet goals)" (Snyder et al., 1991, p. 571). The two components of hope—agency and pathways—are additive, reciprocal, and positively related, but they are not synonymous. To sustain movement toward life goals, both the sense of agency and pathways must be operative. Thus, agency and pathways are necessary, but neither is sufficient to define hope. Additionally, hope does not merely involve one iteration in which a person first assesses agency and then proceeds to an analysis of available pathways, thereafter eliciting goal-directed behaviors. Nor does one pathways analysis unleash the agency to eventuate in goal-directed behavior. Rather, agency/pathways and pathways/agency iterations continue throughout all stages of goal-directed behavior; as such, hope reflects the cumulative level of perceived agency and pathways thinking.

Goal Thoughts

The focus in this definition of hope is the goal, which is the desired mental target of human actions. We must think about goal objects in order to respond effectively to our environment. As noted psychotherapist Alfred Adler (1964, p. 68) put it, "We cannot think, feel, or act without the perception of a goal." High-hope people can conceptualize their goals clearly, while low-hope people are more ambiguous and uncertain about their goals (Snyder, 1994b). Whatever the individual differences may be in overall hope (Snyder et al., 1991, 1996; Snyder & Hoza, 1997; Sympson & Snyder, 1998), it should be emphasized that the pathway thinking of all individuals are aimed at such endpoints. These goals may represent a short-term target (e.g., getting a drink of water), or a long-term objective that subsumes several other subgoals (e.g., dieting so as to lose 30 pounds). To elucidate this further, goal-directed thinking is depicted visually in the mind's eye of the protagonist in Figure 14.1.

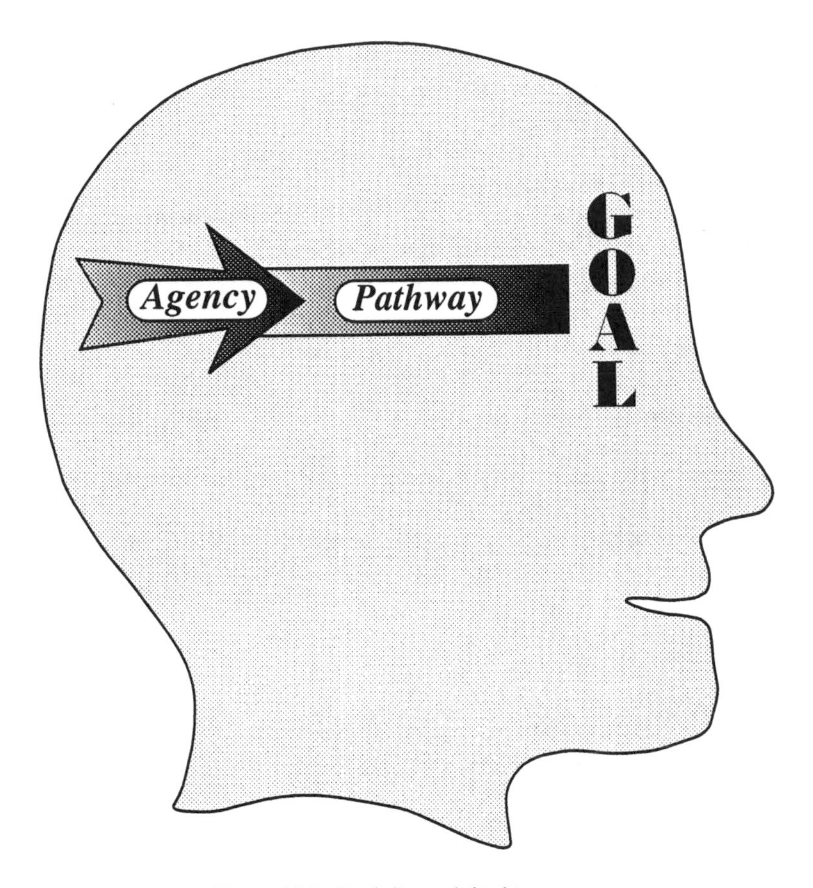

Figure 14.1. Goal-directed thinking.

PATHWAY AND AGENCY THOUGHTS

Goal thinking is accompanied by two additional crucial thought processes. First, there is a process known as pathway thinking, which taps people's perceptions of being able to produce one or more workable routes to goals. Pathway thinking thus reflects the capacity to produce mental maps to reach goal destinations; in most instances, there is one central route that is the focus for the mental journey to a particular goal.

The second process is known as agentic thought, which pertains to the person's perceived ability to begin, as well as to continue, movement along a selected pathway to a goal. Such thinking fills the person with motivational energy and determination to employ a pathway thought toward a goal. We see our protagonist in Figure 14.1 applying agentic thinking to the pathway thought about a particular goal.

Impediments

Given that hope theory is based on goal thoughts, it follows that anything blocking either pathway or agency thinking should play a major role in subsequent coping activities (for further discussions of the role of blockages in relation to hope theory, see Snyder, 1996, 1998). Hope theory posits that the unimpeded pursuit of goals should result in positive emotions, whereas blockages should produce negative feelings. Research in our laboratory utilizing both correlational and experimental designs shows that goal blockages yield negative emotional responses (Snyder et al., 1996). This premise of hope theory also is supported in the findings from other laboratories, namely, difficulties in the pursuit of important goals have been shown to undermine well-being (Diener, 1984; Emmons, 1986; Little, 1983; Omodei & Wearing, 1990; Palys & Little, 1983; Ruehlman & Wolchik, 1988). Furthermore, it appears that perceived blockages or lack of progress toward major goals may cause decreases in well-being rather than the inverse relationship (Brunstein, 1993; Little, 1989). To repeat this point: Goal-pursuit thoughts drive emotional experiences. This core notion is useful when applying hope therapy to actual psychotherapy interventions.

Expanded and Summarized Definition

Together, these three mental components–goals, pathways, agency–form the motivational concept of hope. It is important to emphasize the *cognitive* basis of this definition of hope because it stands in contrast to most older, emotion-based models (Averill, Catlin, & Chon, 1990; Farran, Herth, & Popovich, 1995; Snyder, 1994a, 1994b; Stotland, 1969). Hope theory owes its premises to the emerging field of cognitive psychology. (See Barone, Maddux, and Snyder's [1997] *Social Cognitive Psychology* for a historical explication of the cognitive roots of hope.) In this new model, hope is defined as a mental representational process whereby the protagonist has a goal, along with the perceived capability to produce the associated goal pathway(s), and the perceived agency to apply to at least one pathway. The goal, pathway, and agency thinking of hope theory can be readily translated for applications to the psychotherapy process; moreover, hope theory provides a framework for understanding those common processes that underlie successful processes across differing psychotherapy approaches (for reviews, see Snyder, Michael, & Cheavens, 1999; Snyder et al., 1998; Snyder et al., 2000).

HOPE AND DEPRESSION

Elsewhere (Snyder, 1994b; Snyder, LaPointe, Crowson, & Early, 1998), we have described how hope theory can be applicable to major depressive episodes. According to the *Diagnostic and Statistical Manual of Mental Disorders* (DSM-IV: American Psychiatric Association, 1994), there are nine markers of a major depressive episode. To meet the diagnostic criteria, the person needs to manifest at least five of the following: (1) diminished interest in activities most of the day, over several days; (2) psychomotor retardation; (3) loss of energy and fatigue; (4) inability to concentrate and focus; (5) depressed mood most of the day, over several days; (6) worthlessness feelings; (7) thoughts of, or attempts at, suicide; (8) weight loss or gain; and (9) insomnia or hypersomnia. As can be seen, this list contains markers that either directly tap a lack of agency (items 1, 2, and 3) and pathway thoughts (item 4), or feelings (items 5 and 6), thoughts (item 7), and behaviors (items 8 and 9) that are by-products of blocked or unsuccessful goal pursuits. Thus, the collapse of hopeful thinking undergirds depression.

We repeatedly have found strong inverse relationships between hope and depression in adults (Snyder et al., 1991); moreover, we have found that the thinking patterns of low-hope and depressed persons appear to be similar (Snyder, LaPointe et al., 1998). Based on this theoretical and empirical correspondence, therefore, depression appears to be a likely candidate for a hope-based, goal-focused group psychotherapy intervention. This leads to our targeting of depression for a specific sample of older adults, which we describe in the next section.

AN OVERVIEW OF LATE-LIFE DEPRESSION

Epidemiology

Although the conventional wisdom suggests that some depression is a natural course of aging, the prevalence estimates of these depression-related syndromes in older adults vary. The 1–4% prevalence estimate of major depressive episodes in older community residents has been found to be lower than the same prevalence estimate of major depressive episode in younger and middle-aged adults (Blazer, Hughes, & George, 1987). Estimates of elderly adults treated in primary care settings with diagnoses of major depression span from 6% to 9% (Oxman, Barrett, Barrett, & Gerber, 1990). A subset of aging individuals, those with chronic medical conditions and elders who reside in institutional settings, however, have evidenced a higher prevalence of major depression, ranging from 10% to 15% (Parmelee, Katz, & Lawton, 1992). The variability in prevalence estimates stems from reliance on assessment measures not specifically designed to evaluate elders, as well as the difficulties involved in separating the signs and symptoms of depression from somatic complaints related to chronic comorbid medical conditions (Weissman, Bruce, Leaf, Florio, & Holzer, 1991). Additionally, compared to younger subjects, older adults are less likely to endorse specific aspects of depressive syndromes, including dysphoria, changes in sleep patterns, guilt, and thoughts of death (Weissman, Bruce, Leaf, Florio, & Holzer, 1991).

Older adults who report symptoms that do not meet full criteria for the diagnosis of major depression subsequently may receive a diagnosis of minor depression. Prevalence estimates for such minor depressions range from 4% to 26% in older community residents (Blazer et al., 1987), 47% to 52% in elderly medical outpatients (Oxman et al., 1990), and 18% to 30.5% in nursing home patients (Parmelee et al., 1992). Thus, a multiplicity of factors appears to contribute to the disparity in prevalence estimates for depressive syndromes in older adults. Nevertheless, depression appears to be a problem for a sizable subset of older adults.

Consequences and Treatment

Late-life depression is a chronic and recurring illness. It has been demonstrated to have costly consequences for older adults, their families, and the community (National Institute of Health Consensus Conference, 1992). Depression in elders is linked to an increase in medical morbidity and mortality, impaired daily functioning, as well as increased utilization of health care services and institutionalization. In this regard, pharmacotherapy, psychosocial interventions, and the combination of both have demonstrated efficacy in the treatment and maintained remission of late-life depression (Niederehe, 1996). Psychosocial interventions have been demonstrated to reduce psychopathology, pain, and disability;[1] moreover, these interventions improve adherence to medication regimens. While psychotherapy research in older adults is limited, most studies suggest that it is effective. For example, in a meta-analysis examining the efficacy of 17 published studies of psychosocial interventions for late-life depression, which included cognitive, behavioral, supportive, interpersonal, reminiscence, and eclectic approaches, treatment was found to be more effective than no treatment or placebo with an overall mean effect size of .78, which is characterized as a large effect (Scogin & McElreath, 1994).

The most common, standardized psychosocial treatments used in the care of older adults include cognitive-behavioral, problem-solving, and interpersonal psychotherapies. Psychodynamic psychotherapy and reminiscence therapies also have been found to be effective in elders. (For a more detailed discussion of interventions for older adults, see Klausner et al., 1998; Klausner & Alexopoulos, 1999) Additionally, studies have shown that group therapy for depressed older adults is as effective as individual psychotherapy with this population (Dhopper, Green, Huff, & Austin-Murphy, 1993; Leung & Orrell, 1993). Wolfe, Morrow, and Fredrickson (1996) have suggested that a group format decreases isolation and hopelessness through social interactions of group members.

It should be noted, however, that psychotherapy, pharmacotherapy, and their combination for older adults have not always proven to be successful in decreasing all depressive symptoms or other outcomes, such as functional disability (Pearson, Reynolds, Kupfer, & Lebowitz, 1995; Schneider, 1996). Indeed, a recent review of the efficacy of antidepressant medication concluded by noting that a significant number of individuals with late-life depression remain unresponsive to treatment, while others suffer from residual depressive symptoms (Schneider, 1996). For example, depressed older adults with a history of suicide attempts were found to experience persistent feelings of hopelessness even after initial improvement with psychotherapy and medication (Rifai, George, Stack, Mann, & Reynolds, 1994). Furthermore, hopelessness predicted premature termination of these patients' treatment for depression, and such hopeless thinking also has been linked to eventual suicide in depressed individuals (Beck, Steer, Kovacs, & Garrison, 1985; Snyder, 1994b). In this latter regard, therefore, we see strong evidence linking the construct of hope to the crucial feelings and behaviors (e.g., suicide) of depressed persons.

HOPE-BASED, GOAL-FOCUSED GROUP PSYCHOTHERAPY WITH DEPRESSED OLDER ADULTS

Recent psychotherapy research has underscored the importance of identifying the specific symptoms and outcomes targeted by interventions (Koder, Brodaty, & Anstey,

[1] Residual depressive symptoms are only part of a larger array of adverse outcomes of late-life depression. Functional disability is another consequence of persistent depressive syndrome (Pearson et al., 1995). Increasing evidence suggests that a complex relationship exists between functional disability and depression in late-life, a relationship in which both conditions contribute to negative outcome. Disability can worsen quality of life and increase hospital admissions (Spector, Katz, Murphy, & Fulton, 1987) and morbidity of specific medical conditions (Pahor et al., 1994). Evidence linking depression and disability is further strengthened by studies demonstrating that changes in depression result in reduction of disability in both older and younger adults (Ormel et al., 1993; Von Korff, Ormel, Katon, & Lin, 1992).

1996). Within this context, hope theory was used to design an intervention for the specific residual symptoms of late-life depression. This study was conducted under the auspices of the Clinical Research Center in Geriatric Affective Disorders at the New York Hospital—Cornell Medical Center, Westchester Division. The National Institute of Mental Health (NIMH) funded Clinical Research Center was established in 1992 by George S. Alexopoulos and colleagues, and was designed to facilitate multidisciplinary research on the course, outcomes, and predictors of geriatric affective disorders. Developing and testing the hope model and the related manual within the Clinical Research Center permitted the systematic recruitment, clinical assessment, and follow-up of elderly research patients. A finalized version of the treatment manual may be utilized reliably to train mental health professionals to deliver the focused intervention in a variety of treatment settings. (Note: This manual may be obtained by contacting the senior author).

The main study was an outgrowth of an earlier pilot-study finding of an association between hope and suicidal ideation in depressed older adults enrolled in a longitudinal study of late-life depression (Klausner, Alexopoulos, Kakuma, Abrams, & Clarkin, 1996). In this pilot sample, individuals who expressed higher levels of hope about the future were less likely to report thoughts of death. This finding was based on a simple one-item and face-valid scale, but it nevertheless was consistent with earlier findings showing a relationship between higher hope and lower suicidal ideation (Beck et al., 1985; for review, see Snyder, 1994b). The study also was linked to the finding that specific depressive signs and symptoms were associated with functional disability in elders (Alexopoulos et al., 1996).

The hope-based, goal-focused group approach was tailored to reduce functional disabilities and those symptoms of depression that failed to achieve remission with medication or other types of psychotherapy. Special attention was paid to the persistent symptom of hopelessness which, as stated earlier, has been found to be linked to suicidal ideation and premature termination of treatment (Klausner et al., 1996; Rifai et al., 1994). Although the hope group syllabus was designed to include elements of psychological treatments such as cognitive-behavioral and problem-solving therapy, the overarching theme was the utilization of hope theory to increase participants' thoughts about the identification and achievement of individualized personal goals (Snyder, 1994b; Snyder, Cheavens, & Michael, 1999; Snyder, Michael, & Cheavens, 1999; Snyder et al., 1998). Wolfe and colleagues (1996) have addressed the importance of attention to the particular needs and skills of older adults with mood disorders in therapy.

The primary objective was to develop a manual-based group intervention for older adults diagnosed with major depression who, despite achieving remission with either medication or psychotherapy, continued to report symptoms of anxiety, hopelessness, suicidal ideation, psychomotor retardation, or some degree of decreased daily functioning. The treatment was designed to be brief, practical, and based on the identification and achievement of attainable, individual goals. The study also explored the feasibility of conducting a group intervention for those individuals diagnosed with late-life depression who resided in the community. Of specific interest was whether these elders, relative to persons in an appropriate comparison treatment group, could achieve their therapeutic goals.

Method

To be included in the pilot study, older adults were required to (1) meet the criteria for major depressive disorder; (2) have residual depressive symptoms despite pharmacotherapy for 11 weeks; and (3) be cognitively intact and free of psychosis. Most of the participants were enrolled in other Clinical Research Center protocols, and the majority were receiving antidepressant treatment. All participants were outpatients at the New York Hospital–Cornell Medical Center geriatric clinic and had mild functional disabilities. Most patients were able to drive to the clinic, but several needed to use a medical transport service.

First, a complete assessment of each participant who signed an informed consent was obtained. Using a variety of assessment measures, participants were interviewed with respect to level of daily functioning, affective state, and cognitive ability. A major screening device for depression was the Schedule for Affective Disorders and Schizophrenia (SADS; Spitzer & Endicott, 1978). In addition to assessment, each participant was asked to generate an attainable goal during this initial phase of the study. Goal formulation was a collaborative task arrived at by consensus between the investigator and each participant. When each potential research participant was assessed at baseline, he or she was asked to identify a daily activity performed prior to the onset of depression that was not being performed at present (despite any improvements). Examples of these day-to-day activities included household chores, financial management, self-care activities, meal preparation, grocery shopping, and transport outside one's neighborhood. Because level of functioning varied by individual, participants were encouraged to select modest and attainable goals that had a high likelihood of being accomplished at the end of the 11-week group therapy. Group leaders helped subjects prioritize and operationalize these goals when necessary.

Once the assessment phase was complete, participants were randomly assigned to one of two therapy groups: goal-focused group psychotherapy (GFGP) or reminiscence therapy (RT), which served as a comparison condition. For the control RT, Butler's (1974) model of life review was selected. This intervention centers on the recall of memories associated with different stages of life. Each week a different stage of life was discussed, and participants in this group were encouraged to bring memorabilia to share with others to facilitate discussion. This type of group was chosen for comparison with the hope treatment group because it has been found to be mildly effective in reducing the depressive symptoms of older adults (Arean et al., 1993). Despite the potential efficacy of reminiscence in treating depression, however, it does not utilize the specific strategies and techniques considered essential in hope theory, that is, clearly identifying goals, as well as increasing one's ability to find ways to those goals (pathways thinking), and energizing oneself in using those routes to desired goals (agency thinking). Analyses included clients from the initial treatment trial (GFGP n = 5; RT n = 5) as well as 3 new patients from a partial crossover group (N = 6). The partial crossover design treated patients who completed the RT control group with the experimental treatment and was dictated by moral and scientific reasons. For a more detailed account of the procedures and results of this study, see Klausner et al. (1998).

Both the hope-based GFGP and the RT groups met weekly for 1 hour over 11 weeks and were videotaped. Assessments of participants occurred each week prior to the group session, and a final assessment took place at the completion of the intervention. Instruments included both self-report and observer-rated measures. For a more detailed review of instruments and timetable, see Klausner et al. (1998). All of the group sessions were held mid-morning to facilitate transport and attendance. A light snack was provided to encourage social interaction and to prevent fatigue.

Components of the hope-based GFGP manual and syllabus included individualized goal-formulation, psychoeducation for late-life depression, and skills training. Specifically, the skills training segments integrated relaxation skills with cognitive restructuring and problem solving in order to reduce anxiety and hopelessness. The cognitive-restructuring and problem-solving components of the skills training were utilized to increase cognitive flexibility, thereby allowing the participants to incorporate later skills necessary for agency and pathways thoughts. Using the hope model, the principles of successful agency and pathways were gradually introduced to the participants as the group meetings unfolded.

The weekly GFGP group sessions were highly structured, beginning with the primary group leader taking attendance, making announcements, and reviewing the prior week's group topic and homework assignment. During group sessions, the primary group leader asked each member in turn to relate his or her experience in completing the homework assignment and to discuss any difficulties that may have arisen. These difficulties were then reviewed and analyzed in a supportive manner by the primary group leader, and group members were encouraged to respond, if possible, with suggestions to enhance daily functioning. At the close of each weekly group session, the primary group leader assigned individualized behavioral practice tasks for the following week, gave a "recap" of the major session themes, and answered questions.

To accommodate the changes in learning, memory, and sensory capacity that accompany aging, the group leaders organized each group session carefully in order to sit in close proximity to all participants, to speak slowly and audibly, and to repeat key concepts frequently. Large-type printed handouts describing the topic or skill discussed in the session were distributed each week, and homework assignments were outlined, printed, and distributed as well. Whenever possible, group leaders used mnemonics and personal, concrete examples to demonstrate the principles taught in a session.

Results

Both the hope-based GFGP and RT interventions were feasible in that patients mobilized themselves to attend weekly group sessions despite frequent impediments such as illness and inclement weather. One female subject from each of the two groups dropped out due to health reasons. Although a female participant in the RT group became psychotic and required hospitalization after the sixth group session, self-reports and clinician-rated assessments were obtained until Week 7. No significant differences were found between the two groups on demographic variables (see Table 14.1). There were two significant baseline differences between the groups, however. At the outset of treatment, individuals treated in the GFGP group were more active and less anxious than individuals in the RT group (see Table 14.2).

Overall, it appears that the hope-based GFGP treatment was associated with decreased disability scores and increased hope scores. Two of the three scales related to hope increased significantly after the hope treatment (i.e., the State Hope Total Scale and the Agency subscale) (Snyder, Harris, et al., 1991). The summed variable of hope

Table 14.1. Demographic Characteristics of the Hope-Based (GFGP) and Reminiscence Groups

Characteristic	Hope-based GFGP	Reminiscence
Race		
Caucasian	87.5%	80%
Hispanic	12.5%	20%
Gender		
Male	50%	40%
Female	50%	60%
Age	M = 66.5 (SD = 5.4)	M = 67.2 (SD = 8.6)
Vocabulary Score–WAIS	M = 13.9 (SD = 3.1)	M = 12.6 (SD = 4.9)

(created by adding measures of hope across scales) and family social functioning also increased. All measures of depression, as well as the summed variables of hopelessness, anxiety, and disability, decreased significantly from pre- to posttreatment. Specific within-group changes are shown in Table 14.3.

In contrast to these posttreatment changes in the hope-based GFGP group, the only variables demonstrating significant pre-post treatment decreases in the RT group were the Hamilton Rating Scale for Depression (HRSD; Hamilton, 1960) scores and the summed disability index (see Table 14.3). Therefore, while RT served as more than merely a control treatment, it did not appear to match the hope-based GFGP therapy in the amelioration of a wide range of outcomes.

To examine differences between treatments while adjusting for pretreatment symptoms, mean change scores between pre- and posttreatment were compared using t tests. Mean change scores were calculated by subtracting the posttreatment scores from the pretreatment scores. Two variables showed significant differences between the two treatment groups: the HRSD (Hamilton, 1960), and the summed variable of hopelessness. Both groups decreased in depression as measured by the HRSD, but the GFGP group dropped 15 points on average, moving the mean to levels usually associated with remission of depression. In contrast, the average change in the RT group was a decrease of only 4.6 points, with a high final HRSD score of 20.2, indicating the continued presence of major depression. No other change scores showed significant differences between the two treatment groups. For a more in-depth view of the results of this study, see Klausner et al. (1998).

Table 14.2. Significant Differences in Clinical Assessment Scores of Hope-Based (GFGP) and Reminiscence Groups Compared at Baseline (N = 13)

Measure	GFGP Mean ± SD	Reminiscence Mean ± SD	t
MAI Activities Scale	4.9 ± 12.0	33.6 ± 11.9	2.2*
State Anxiety	43.9 ± 12.4	57.0 ± 6.7	2.2*
Summed Variable: Anxiety	30.5 ± 6.8	44.2 ± 5.4	3.8*

Note. MAI = Philadelphia Multilevel Assessment Instrument (Lawton, Moss, Fulcomer, & Kleban, 1982); State anxiety = Spielberger State–Trait Anxiety Scale (Spielberger, Gorsuch, & Lushene, 1970); Summed variable: Anxiety = Items aggregated to tap anxiety for this study.

*$p < .05$

Table 14.3. Significant Differences in Clinical Assessment Scores of Older Adults in the Hope-Based (GFGP) and Reminiscence Groups before and after Treatment (N = 13)

	GFGP				Reminiscence			
Measures	Pre Mean ± SD	Post Mean ± SD	t	Effect size	Pre Mean ± SD	Post Mean ± SD	t	Effect size
HRSD	22.7 ± 5.7	7.6 ± 6.3	5.3*	2.51	24.8 ± 6.3	20.2 ± 9.4	2.6*	0.57
Montgomery–Asberg	19.8 ± 6.6	5.6 ± 6.9	3.5*	2.10	22.4 ± 7.4	20.6 ± 8.0	ns	n/a
BDI	15.5 ± 10.8	7.4 ± 6.5	2.9*	0.91	18.0 ± 3.9	11.5 ± 7.9	ns	n/a
State Hope—Total	22.8 ± 9.4	31.7 ± 11.8	2.5*	0.83	20.5 ± 6.2	27.5 ± 7.3	ns	n/a
State Hope—Agency	8.6 ± 5.9	15.25 ± 5.9	2.9*	1.13	7.75 ± 4.9	11.5 ±7 .9	ns	n/a
MAI	10.3 ± 4.4	15.3 ± 3.8	2.5*	1.22	10.8 ± 7.5	10.0 ± 5.6	ns	n/a
Summed variables:								
Depressed mood	8.25 ± 3.6	4.75 ± 2.4	2.8*	1.14	7.5 ± 1.7	7.0 ± 1.6	ns	n/a
Hope	27.75 ± 10.1	37.1 ± 11.6	2.4*	0.86	25.5 ± 7.1	30.8 ± 8.4	ns	n/a
Hopelessness	5.25 ± 4.16	2.25 ± 3.6	2.3*	0.77	3.75 ± 2.0	6.5 ± 3.1	ns	n/a
Anxiety	30.5 ± 6.8	24.1 ± 8.3	2.4*	0.84	43.5 ± 5.9	39.0 ± 13.0	ns	n/a
Disability	3.5 ± 1.6	1.4 ± 1.3	5.3*	1.44	3.6 ± 0.6	2.2 ± 0.8	5.7*	1.98

Note: HRSD = Hamilton Rating Scale for Depression (Hamilton, 1960)
Montgomery–Asberg = Montgomery–Asberg Scale for Depression (Montgomery & Asberg, 1979)
BDI = Beck Depression Inventory (Beck, Ward, Mendelson, Mock, & Erbaugh 1961)
State Hope—Total = State Hope Scale Summed Score (Snyder, Sympson, et al., 1996)
State Hope—Agency = State Hope Scale Agency Subscale Score (Snyder, Sympson, et al., 1996)
MAI = Philadelphia-Multilevel Assessment Instrument (Lawton et al., 1982)
Summed variables = Similar items across inventories that were aggregated for this study.
*$p < .05$

Case Study of Hope Group Participant: Planting the Roses

The following case study illustrates the successful treatment outcome of a depressed older man with the new GFGP intervention.

Mr. T is a 61-year-old man who works in sales. At entry to the pilot study, despite taking antidepressant medication for several weeks, he scored a 30 on the HRSD, which indicates the presence of major depression (Hamilton, 1960). He also reported passive suicidal ideation, hopelessness, worthlessness, and lack of energy. During the initial assessment meeting with the group leaders, he stated that his goals were to (1) increase both his energy and enthusiasm at work, and (2) have sufficient energy and motivation to return his garden to the source of pride it had been previously. Mr. T was randomized to the hope-based GFGP group.

Immediately after the group sessions began, Mr. T became an active, empathic, and creative group member. He frequently asked questions and volunteered his own experiences as examples for the group. He interacted well with other group members, validating their feelings whenever possible. In completing weekly assignments, he was thorough and meticulous, and managed to provide thoughtful and accurate depictions of weekly events for self-monitoring assignments.

When the group syllabus focused on skills training, he enthusiastically reported his success using the prescribed relaxation exercises and tape. He also reported experiencing less interpersonal friction at home as well as reduced "time urgency." As the concept of hope was introduced to the group, he operationalized his goal in terms of the agency and pathways concepts. His HRSD score continued to decline to a 9 in the early phase of group.

While the primary group leader was on a scheduled vacation in the midphase of group, Mr. T reported an increase in his negative sense of self relative to group members and to the other group leader. He openly discussed anger toward his boss, resentment for being treated like a "second class citizen" at work, and lamented, "Nothing good happens to me, everything bad does." At this point in time, his HRSD rose to a 15.

When the primary group leader returned from vacation, Mr. T again became an animated and active member of group. No interpretations were made of possible transference issues related to the primary group leader's absence. Instead, the focus remained on the manual-based syllabus and the scheduled skills lessons. During group therapy, when members commented to Mr. T that his mood and cognitions had improved over the prior week, he acknowledged feeling "much better." In subsequent weeks, he was able to conceptualize and perform the substeps necessary toward the completion of his two stated goals: (1) planting his rose bushes, which had been purchased for his garden many months earlier, and (2) organizing the list of clients he needed to contact at work. He expressed satisfaction with undertaking and completing these tasks.

Mr. T's HRSD score declined in the final group phase, and he continued to assist other group members in outlining the specific substeps necessary to achieve their respective goals. He participated with the other group members in planning the refreshments for the final group session. At the last session, Mr. T served group members and leaders slices from an elaborate cake he had purchased from a gourmet bakery. He recounted in detail the planning and purchase of this cake, and the pleasure he received in bringing it to the group. During the group session, he reported feeling more positive about himself, his work, and recovery from depression.

When assessed after completion of group therapy, Mr. T's total HRSD score was 3—a score associated with remission of depression. He reported no feelings of either hopelessness or helplessness, but he still felt residual feelings of worthlessness. He told group leaders that "approximately 85%" of his goals now had been accomplished. With pride, he explained that he was continuing to organize his desk and work papers to increase his interactions with clients, and he reported more motivation for contacting these clients. At home, he was far less irritable and reactive to the moods of others and again was able to enjoy his garden (with his planted and now blooming roses).

Although one participant in the hope-based GFGP group is described here, it should be noted that the other members of this intervention group also became quite proactive, whereas the members of the RT control group remained somewhat passive. For example, the GFGP intervention group members brought refreshments and snacks to the final session, whereas the RT control group's lack of enthusiasm is captured in one member's spontaneous announcement: "Well, I'm not baking anything!" At the end of the hope-based GFGP intervention, the members presented small gifts to each leader and thanked them for what had been accomplished. The hope-based group apparently had established a more supportive, active style of interacting with each other.

DISCUSSION AND FUTURE DIRECTIONS

This study suggests that hope-based, goal-focused group psychotherapy is a viable approach to treating specific symptoms and functional outcomes of late-life depression. While the RT group proved helpful in reducing some depressive symptoms and outcomes, its success was most likely due to a supportive and caring group format. A group format has the potential to reduce the isolation often present in older adults who are experiencing depression and disability, while also providing an opportunity for social interaction and emotional support. Likewise, group therapy can be implemented conveniently in settings where older adults gather.

One explanation for the hope-based GFGP patients faring better than the RT patients in a number of areas, including hope, anxiety, and social functioning, may be the skills training component of the intervention. GFGP, unlike RT, enabled elders to model their own adaptive coping behaviors for others while using the group setting to practice and refine newly learned skills. Encouraging older adults to focus on the flexibility, resourcefulness, and wisdom acquired during a lifetime when approaching the goals and challenges faced in the present, coupled with new coping techniques, may have led to the increased levels of hope and the improved social functioning. Indeed, hope theory suggests that such cognitive flexibility is one of the adaptive results of increased pathways thinking (Snyder, 1994b; Snyder et al., 1998).

While group psychotherapy in older adults appears promising, the literature contains few controlled treatment trials that test the efficacy of manual-based, focused treatments with this population. The manual approach used in this study, if replicated, could be utilized to target and successfully treat specific symptoms and outcomes of major depression in elders with different levels of functioning.

It should be noted that this was an small *n*, pilot study with pretreatment differences between the two groups in levels of anxiety and activity evident at baseline. There were no significant pretreatment differences in levels of depression, disability, hopelessness, family interaction, or hope. Higher anxiety and less activity may have contributed in some fashion to the lack of change in the RT group. Perhaps such individuals could benefit from additional treatment time or more intense skills training to achieve goals and improve daily functioning. For this reason, further study is needed.

SUMMARY

In conclusion, the results suggest the hope-based GFGP intervention, which targets specific residual symptoms and functional outcomes associated with late-life depression, is an effective basis for group treatment. Unlike the RT intervention, the hope-based intervention went beyond reducing depressed mood and demonstrated efficacy in reducing anxiety and hopelessness, as well as increasing hope and social functioning. These domains have the potential to improve the quality of life in this population. One limitation of this study is that it targeted depressed elders with mild functional disabilities. The selection of this population was the initial step in the development and controlled testing of the new treatment. Overall, the results demonstrated that older adults respond eagerly to a structured psychosocial intervention and can achieve reasonable goals. Further research is required with a broader range of patients, including those with personality disorders, chronic comorbid conditions, active suicidal ideation, and significant functional disabilities. Additional study with a more diverse sample of older adults can identify patient subgroups and characteristics associated with positive treatment outcome while highlighting the precise therapeutic "ingredients" beneficial to different populations of elders. For example, tailoring the GFGP intervention to treat fragile older adults with significant and chronic medical comorbidity may serve to improve pain management, strengthen coping mechanisms, and increase adherence to medical and rehabilitation regimens.

Future research also should address the types of psychosocial interventions that are effective with individuals in a variety of treatment settings, such as primary care and nursing homes. Many older adults are treated by primary care physicians and the integration of a standardized group intervention into primary care settings may prove useful in enhancing the functioning of large numbers of patients. In nursing home settings, focused group interventions such as GFGP enable depressed and disabled residents to model, practice, and refine newly learned skills in a supportive environment.

With the older adult population expected to rise substantially in the next century as the baby boomers move into this age cohort, focused psychosocial interventions using hope theory as a premise have the potential to play an expanded role in enhancing both the mental health and functioning of older adults as well as younger persons (Snyder et al., 1998; Snyder, Ilardi, et al., 2000; Snyder, Michael, & Cheavens, 1997). As such, interventions that increase hope should be equally available to persons of all ages as we move into the 21st century.

ACKNOWLEDGMENT: The goal-focused group psychotherapy intervention that served as the basis for this chapter was supported by the Dammann Fund, Inc.

REFERENCES

Adler, A. (1964). Individual psychology, its assumptions and its results. In H. M. Ruitenbeek (Ed.), *Varieties of personality theory* (pp. 57–73). New York: Dutton.

Alexopoulos, G. S., Vrontou, C., Kakuma, T., Meyers, B. S., Young, R. C., Klausner, E. J., & Clarkin, J. (1996). Disability in geriatric depression. *American Journal of Psychiatry, 153,* 877–885.

American Psychological Association. (1994). *Diagnostic and statistical manual of mental disorders* (4th ed.). Washington, DC: Author.

Arean, P. A., Perri, M. G., Nezu, A. M., Schein, R., Christopher, F., & Joseph, T. (1993). Comparative effectiveness of social problem-solving therapy and reminiscence therapy as treatments for depression in older adults. *Journal of Consulting and Clinical Psychology, 61,* 1003–1010.

Averill, J. R., Catlin, G., & Chon, K. K. (1990). *Rules of hope.* New York: Springer-Verlag.

Bandura, A. (1989). Human agency in social cognitive theory. *American Psychologist, 44,* 1175–1184.

Barone, D., Maddux, J., & Snyder, C. R. (1997). *Social cognitive psychology: History and current domains.* New York: Plenum Press.

Beck, A. T., Steer, R. A., Kovacs, M., & Garrison, B. (1985). Hopelessness and eventual suicide: A 10-year prospective study of patients hospitalized with suicidal ideation. *American Journal of Psychiatry, 142,* 559–563.

Beck, A. T., Ward, C. H., Mendelson, M., Mock, J., & Erbaugh, J. (1961). An inventory for measuring depression. *Archives of General Psychiatry, 4,* 561–571.

Blazer, D. G., Hughes, D. C., & George, L. K. (1987). The epidemiology of depression in an elderly community population. *Gerontologist, 27,* 281–287.

Brunstein, J. C. (1993). Personal goals and subjective well-being: A longitudinal study. *Journal of Personality and Social Psychology, 65,* 1061–1070.

Butler, R. (1974). Successful aging and the role of life review. *Journal of the American Geriatric Society, 22,* 529–535.

Cantril, H. (1964). The human design. *Journal of Individual Psychology, 20,* 129–136.

Dhopper, S. S., Green, S. M., Huff, M.,& Austin-Murphy, J. (1993). Efficacy of a group approach to reducing depression in nursing home elderly residents. *Journal of Gerontological Social Work, 20,* 87–100.

Diener, E. (1984). Subjective well-being. *Psychological Bulletin, 95,* 542–575.

Emmons, R. A. (1986). Personal strivings: An approach to personality and subjective well-being. *Journal of Personality and Social Psychology, 51,* 1058–1068.

Erickson, R. C., Post, R., & Paige, A. (1975). Hope as a psychiatric variable. *Journal of Clinical Psychology, 31,* 324–329.

Farber, M. L. (1968). *Theory of suicide.* New York: Funk & Wagnalls.

Farran, C. J., Herth, A. K., & Popovich, J. M. (1995). *Hope and hopelessness: Critical clinical constructs.* Thousand Oaks, CA: Sage.

Frank, J. D. (1968). The role of hope in psychotherapy. *International Journal of Psychiatry, 5,* 383–395.

Frankl, V. E. (1992). *Man's search for meaning: An introduction to logotherapy* (4th ed). Boston: Beacon.

Hamilton, M. (1960). A rating scale for depression. *Journal of Neurology and Neurosurgical Psychiatry, 23,* 56–62.

Klausner, E. J., & Alexopoulos, G. S. (1999). The future of psychosocial treatments for elderly patients. *Psychiatric Services, 50,* 1198–1204.

Klausner, E. J., Alexopoulos, G. S., Kakuma, T., Abrams, R., & Clarkin, J. F. (1996). Hope, environment, and suicidal ideation. *Abstracts of the Annual Meeting of the American Association for Geriatric Psychiatry,* p. 71.

Klausner, E. J., Clarkin, J. F., Spielman, L., Pupo, C., Abrams, R., & Alexopoulos, G. S. (1998). Late-life depression and functional disability: The role of goal-focused group psychotherapy. *International Journal of Geriatric Psychiatry,* 13, 707–716.

Koder, D., Brodaty, H., & Anstey, K. (1996). Cognitive therapy for depression in the elderly. *International Journal of Geriatric Psychiatry, 11,* 97–107.

Lawton, M. P., Moss, M., Fulcomer, M., & Kleban, M. (1982). A research and service oriented multilevel assessment instrument. *Journal of Gerontology, 37,* 91–99.

Lee, T. W., Locke, E. A., & Latham, G. P. (1989). Goal setting theory and job performance. In L. A. Pervin (Ed.), *Goal concepts in personality and social psychology* (pp. 291–326). Hillsdale, NJ.: Erlbaum.

Leung, S. N., & Orrell, M. W. (1993). A brief cognitive behavioural therapy group for the elderly: Who benefits? *International Journal of Geriatric Psychiatry, 8,* 593–598.

Little, B. R. (1983). Personal projects: A rationale and method for investigation. *Environment and Behavior, 15,* 273–309.

Little, B. R. (1989). Personal projects analysis: Trivial pursuits, magnificent obsessions, and the search for coherence. In D. M. Buss & N. Cantor (Eds.), *Personality psychology: Recent trends and emerging directions* (pp. 15–31). New York: Springer-Verlag.

Locke, E. A., Shaw, K. N., Saari, L. M., & Latham, G. P (1981). Goal setting and task performance: 1969–1980. *Psychological Bulletin, 90,* 125–152.

Melges, R., & Bowlby, J. (1969). Types of hopelessness on psychopathological processes. *Archives of General Psychiatry, 20,* 690–699.

Menninger, K. (1959). The academic lecture on hope. *American Journal of Psychiatry, 116,* 481–491.

Mento, A. J., Steel, R. P., & Karren, R. J. (1987). A meta-analytic study of the effects of goal setting on task performance. *Organizational Behavior and Human Decision Processes, 39,* 52–83.

Montgomery, S. A., & Asberg, M. (1979). A new depression scale designed to be sensitive to change. *British Journal of Psychiatry, 134,* 382–389.

National Institute of Health Consensus Conference. (1992). Diagnosis and treatment of depression in late life. *Journal of the American Medical Association, 268,* 1018–1024.

Niederehe, G. (1996). Psychosocial treatments with depressed older adults. *American Journal of Geriatric Psychiatry, 4,* S66–S78.

Omodei, M. M., & Wearing, A. J. (1990). Need satisfaction and involvement in personal projects: Toward an integrative model of subjective well-being. *Journal of Personality and Social Psychology, 59,* 762–769.

Ormel, J., Von Korff, M., Van den Brink, W., Katon, W., Brilman, E., & Oldehinkel, T. (1993). Depression, anxiety, and social disability show synchrony of change in primary care patients. *American Journal of Public Health, 83,* 385–390.

Oxman, T. E., Barrett, J. E., Barrett, J., & Gerber, P. (1990). Symptomatology of late-life minor depression among primary care patients. *Psychosomatics, 31,* 174–180.

Pahor, M., Guralnik, J., Slive, M., Chrischilles, E., Manot, A., & Wallace, R. (1994). Disability and severe gastrointestinal hemorrhage: A prospective study of community-dwelling older persons. *Journal of the American Geriatric Society, 42,* 816–825.

Palys, T. S., & Little, B. R. (1983). Perceived life satisfaction and organization of personal projects systems. *Journal of Personality and Social Psychology, 44,* 1221–1230.

Parmelee, P. A., Katz, I. R., & Lawton, M. P. (1992). Incidence of depression in long-term care settings. *Journal of Gerontology: Medical Sciences, 47,* M189–M196.
Pearson, J. L., Reynolds, C. F., Kupfer, D. J., & Lebowitz, B. (1995). Outcome measures in late-life depression. *American Journal of Geriatric Psychiatry, 3,* 191–197.
Pervin, L. A. (Ed.) (1989). *Goal concepts in personality and social psychology.* Hillsdale, NJ.: Erlbaum.
Rifai, A. H., George, C. J., Stack, J., Mann, J., & Reynolds, C. (1994). Hopelessness in suicide attempters after acute treatment of major depression in late life. *American Journal of Psychiatry, 151,* 1657–1690.
Ruehlman, L. S., & Wolchik, S. A. (1988). Personal goals and interpersonal support and hindrance as factors in psychological distress and well-being. *Journal of Personality and Social Psychology, 55,* 293–301.
Schactel, E. (1959). *Metamorphosis.* New York: Basic Books.
Schneider, L. S. (1996). Pharmacologic considerations in the treatment of late-life depression. *American Journal of Geriatric Psychiatry, 4*(4, Supp.), S51–S65.
Scogin, F., & McElreath, L. (1994). Efficacy of psychosocial treatments for geriatric depression: A quantitative review. *Journal of Consulting and Clinical Psychology, 62,* 69–74.
Snyder, C. R. (1994a). Hope and optimism. In V. S. Ramachandren (Ed.), *Encyclopedia of human behavior* (Vol. 2, pp. 535–542). San Diego: Academic Press.
Snyder, C. R. (1994b). *The psychology of hope: You can get there from here.* New York: Free Press.
Snyder, C. R. (1996). To hope, to lose, and hope again. *Journal of Personal and Interpersonal Loss, 1,* 1–16.
Snyder, C. R. (1998). A case for hope in pain, loss, and suffering. In J. H. Harvey, J. Owarzu, & E. Miller (Eds.), *Perspectives on loss: A sourcebook* (pp. 63–79). Washington, DC: Taylor & Francis.
Snyder, C. R., Cheavens, J., & Michael, S. T. (1999). Hoping. In C. R. Snyder (Ed.), *Coping: The psychology of what works* (205–231). New York: Oxford University Press.
Snyder, C. R., Harris, C., Anderson, J. R., Holleran, S. A., Irving, L. M., Sigmon, S. T., Yoshinobu, L., Gibb, J., Langelle, C., & Harney, P. (1991). The will and the ways: Development and validation of an individual differences measure of hope. *Journal of Personality and Social Psychology, 60,* 570–585.
Snyder, C. R., Hoza, B., Pelham, W. E., Rapoff, M., Ware, L., Danovsky, M., Highberger, L., Rubinstein, H., & Stahl, K. (1997). The development and validation of the Children's Hope Scale. *Journal of Pediatric Psychology, 22,* 399–421.
Snyder, C. R., Ilardi, S. S., Cheavens, J., Yamhure, L., Michael, S. T., & Sympson, S. (1998). *The role of hope in cognitive behavior therapies.* Unpublished manuscript, University of Kansas, Lawrence, Kansas.
Snyder, C. R., Ilardi, S., Michael, S., & Cheavens, J. (2000). Hope theory: Updating a common process for psychological change. In C. R. Snyder & R. E. Ingram (Eds.), *Handbook of psychological change: Psychotherapy processes and practices for the 21st century* (pp. 128–154). New York: Wiley.
Snyder, C. R., LaPointe, A. B., Crowson Jr., J. J., & Early, S. (2000). Preferences of high- and low-hope people for self-referential feedback. *Cognition and Emotion, 12,* 807–823.
Snyder, C. R., McDermott, D., Cook, W., & Rapoff, M. (1997). *Hope for the journey: Helping children through the good times and the bad.* San Francisco: Westview/Basic Books.
Snyder, C. R., Michael, S. T., & Cheavens, J. (1999). Hope as a psychotherapeutic foundation of common factors, placebos, and expectancies. In M. A. Huble, B. Duncan, & S. Miller (Eds.), *Heart and soul of change* (pp. 179–200). Washington, DC: American Psychological Association.
Snyder, C. R., Sympson, S. C., Ybasco, F. C., Borders, T. F., Babyak, M. A., & Higgins, R. L. (1996). Development and validation of the State Hope Scale. *Journal of Personality and Social Psychology, 70,* 321–335.
Spector, W., Katz, S., Murphy, J., & Fulton, J. (1987). The hierarchical relationship between activities of daily living and instrumental activities of daily living. *Journal of Chronic Diseases, 40,* 481–489.
Spielberger, C. D., Gorsuch, R. L., & Lushene, R. E. (1970). *Manual for the State-Trait Anxiety Inventory.* Palo Alto, CA: Consultants Psychologists Press.
Spitzer, R., & Endicott, J. (1978). *Schedule for affective disorders and schizophrenia.* Biometric Division, New York State Research Institute.
Stotland, E. (1969). *The psychology of hope.* San Francisco: Jossey-Bass.
Sympson, S. C., & Snyder, C. R. (1998). *Development and validation of the Domain Specific Hope Scale.* Unpublished manuscript, University of Kansas, Lawrence.
Von Korff, M., Ormel, J., Katon, W., & Lin, E. (1992). Disability and depression among high utilizers of health care. *Archives of General Psychiatry, 49,* 91–100.
Weissman, M. M., Bruce, M. L., Leaf, P. J., Florio, L. P., & Holzer, C. III. (1991). Affective disorders. In L. Robins & D. Regier (Eds.), *Psychiatric disorders in America* (pp. 53–81). New York: Free Press.
Wolfe, R., Morrow, J., & Fredrickson, B. L. (1996). Mood disorders in older adults. In L. L. Carstensen, B. A. Edelstein, & L. Dombrand (Eds.), *The practical handbook of clinical gerontology* (pp. 274–303). Thousand Oaks, CA: Sage.

15

Complex Unity and Tolerable Uncertainty

Relationships of Physical Disorder and Depression

BARRY GURLAND, SIDNY KATZ,
and ZACHARY M. PINE

In this chapter, we examine the interactive development of physical disorders and depression—a complex mix of subjective and objective causes and consequences that appears often to impair capacity for adapting to life's challenges, and hence may threaten the quality of living. Looking to clarify causal pathways that might explain the association of physical disorders and depression, we consider the potential contribution of longitudinal studies and critically reconsider the premise that temporal sequencing of interactive conditions is decisive in the assignment of cause-and-effect.

In the course of testing the limits of the information on cause-and-effect derived from longitudinal study, we resume our prior exploration of depressed mood, where we discussed ambiguities arising from varying views of depression as a diagnostic condition or a dimension of altered mood (Gurland & Katz, 1997). We there argued for a hierarchically organized *continuum* of "affective suffering" for adding to the understanding of the etiological and consequential associations of depression. In this chapter, we apply the continuum model to impaired functioning in the activities of daily living, so as to allow consideration of temporal ordering of the two continua: functioning and suffering. We discuss whether this patterning can throw light on cause–effect connections between depressed mood as a subjective condition, and physical functioning as an objective entity.

We are led by this exploration to conclude that conventional cause-and-effect thinking is *not* the most productive perspective on complex interactive conditions such as

BARRY GURLAND and SIDNEY KATZ • Columbia University, Stroud Center, New York, New York 10032. **ZACHARY M. PINE** • Geriatric Division, Department of Medicine, University of California at San Francisco, San Francisco, California 94143.

Physical Illness and Depression in Older Adults: A Handbook of Theory, Research, and Practice, edited by Gail M. Williamson, David R. Shaffer, and Patricia A. Parmelee. Kluwer Academic/Plenum Publishers, New York, 2000.

the concurrence of depression and physical disorder. Instead, we argue for the usefulness of accepting such relationships as whole-person phenomena, explained only as a complex web of interacting attributes: body, mind, and values; living and nonliving environment; and life experiences in space and time (Katz & Gurland, 1991).

As the chapter unfolds, its initial direction, that of solving cause-and-effect equations through longitudinal designs and hierarchically organized measurement continua, does not bring closure to conceptual and methodological dilemmas. We turn, therefore, to a renewed effort to approach the comorbidity of depression and physical disorder as a reciprocally interactive, subjective–objective process; a complex whole, *sui generis*. From that vantage point, we can gain a clearer sight of the uses of knowledge derived from temporal sequencing of patterns of depression and physical disorder against the background of their contextual associations.

At several points in this chapter, we use the terms *depression* and *physical disorder* in a deliberately loose manner to refer to certain heterogenous but related topic areas in a general field of study. At these points, we do not attempt to distinguish between disorder, syndrome, illness, and diagnostic category; we wish to keep all under consideration. As will be seen, those distinctions are not the basis for precision in our meanings and thinking.

CONTEMPORARY UNCERTAINTIES

For some lengthy period of time, the research literature has paid close attention to whether physical disorder and depression are reciprocal influences. A search of the MEDLINE bibliography between 1990 and late 1998, using the keywords *physical illness and depression*, turned up 441 citations in English; in 1997 alone, there were 71 references in English and 134 in all languages. Published reports under this rubric prior to 1990 include those of Aneshensel, Frerichs, and Huba (1984), Clark, von Ammon, Cavanaugh, and Gibbons (1983), Ouslander (1982) Salzman and Shader (1978), and Dorfman (1978). The duration of our personal efforts in this area goes back almost two decades (Gurland, Golden, Lantigua, & Dean, 1984) and is pointedly reflected in a paper entitled "Depression and Disability: Reciprocal Relations and Changes with Age" (Gurland, Wilder, & Berkman, 1988). Yet, for us, and probably for most others who have struggled with the underlying questions about relationships, the issue remains unresolved.

Given the prolonged and extensive study of this relationship, the lack of a decisive finding could be attributed to ambiguities surrounding the models of physical disorder and depression, and their interaction, that guide research design and measurement. These models typically incorporate a concept of cause linked explicitly with effect. Data for testing the models are, where feasible, generally derived from longitudinal observations or, more rarely, intervention studies. The reasons for the limitation on insights that this model has provided may reside in the nature of the available data or in fallacies hidden in its logic. Accordingly, we first reconstruct the logic behind the model and search for fallacies. We then present a method that offers a good chance of obtaining appropriate data. Finally, we evaluate whether even good data can overcome logical deficits.

CAUSE-AND-EFFECT MODELS

Time (chronological order) is the essence of unraveling the interactions of two conditions, exemplified by physical and emotional disorders in this instance, where a causal role is expected to be assigned to one condition (the actor) and an effect role is expected of the other (the reactor). Time enters into the question of whether the presence of one condition systematically precedes or follows the emergence of the other. Systematic precedence of one condition relative to another creates a case for the direction of causality. Sequences of presence and emergence, which oscillate between conditions over time, add to evidence in favor of reciprocal influence. A more dynamic model attends to temporally related *changes* in the acting and reacting conditions, not merely their presence or emergence.

Recognition of the presence or emergence of conditions at points in time is tied to criteria that essentially set diagnostic or other classification thresholds of intensity and extensity (i.e. severity) of the features of the conditions. Recognition of changes in two or more conditions also requires identification of points of time marking the onset and duration of changes, but the changes can occur at any level of severity. Both kinds of recognition (i.e., presence, emergence, or changes in severity) are faced with formidable methodological hurdles.

Accuracy in placing the onset of occurrence or change depends on the type of the event. An event may occur naturally or by accident in the course of living, or be contrived by experiment or therapeutic intervention. A contrived event in one condition is more readily placed within the timed context of an event in another condition than is a naturally occurring event. As a practical matter, contrived change is generally restricted to ethically valid therapeutic interventions; in the present context, that would mean inducing change in the physical disorder by treatment and noting resultant changes in the emotional state, or vice versa. While this strategy may produce useful findings on the effectiveness of treatment, it does not directly cast light upon the pathways that naturally lead to comorbid conditions and determine their course.

In the natural (uncontrolled) course of events, it is conventional to examine causality by measuring and sequencing the presence, emergence, or changes in the conditions of interest in a longitudinal cohort design. It may be that the sequences are so closely spaced that the interval between them, and thus their chronological relationship, is missed by the chosen methods of longitudinal observation. Nevertheless, it is premised that there is a lag between the actions and reactions in physical disorder and depression in the period of old age that ranges from days to months, even years—a sufficient average interval to allow detection by a longitudinal observational method. A greater problem is that the order of onsets may be distorted by difficulties in ascertaining thresholds for presence–emergence or points of onset of changes.

It is implicit in an interaction model that there is a specific time for determining presence–emergence or for placing the onsets of changes in the actor and reactor; this specific timing is the sine qua non of relative timing. Given the state of measurement, if the threshold of recognition for presence–emergence or the point of onset in one condition is harder to detect than in another condition, the former may be recognized later in its development than the latter. Stated in other terms, the presence or change of the true actor may only be picked up after the presence or change of the true reactor. Under those circumstances, timing will incorrectly assign the roles of actor and reactor (see Figure 15.1). The sequences can then be confounded further by a reciprocal system wherein a serried oscillation of acting and reacting roles is set in motion, but the relative points of onsets are blurred in time.

Vagaries surrounding identification of the point of onset may seemingly be circumvented by arbitrarily keying the deemed onsets to transitions across defined thresholds of severity. This is demonstrated in observations of the relative temporal occurrence of diagnostic entities within the sphere of depression (such as major depression) and physical disorder (such as ischemic heart disease). Points of onset are viewed against operational diagnostic criteria. Yet the acting and reacting elements in depression and physical disorder may be at work at levels of severity that do not cross the threshold demanded by diagnosis. Even more damaging to the logical integrity of the cause-and-effect model is that the levels of severity sufficient for acting or reacting with the other condition may vary between the conditions relative to the thresholds for diagnosis.

Misinterpretations of temporal relationships arising from discrepancies between conditions in recognition (or detection) thresholds might be avoided by referencing changes anywhere along a continuum of severity. In this approach, changes along a continuum of degree of influence by one condition, acting on another, might avoid the logical traps laid for recognition of points of onset and thresholds for diagnostic and other categorical entities. Correspondingly, a continuum of susceptibility of one con-

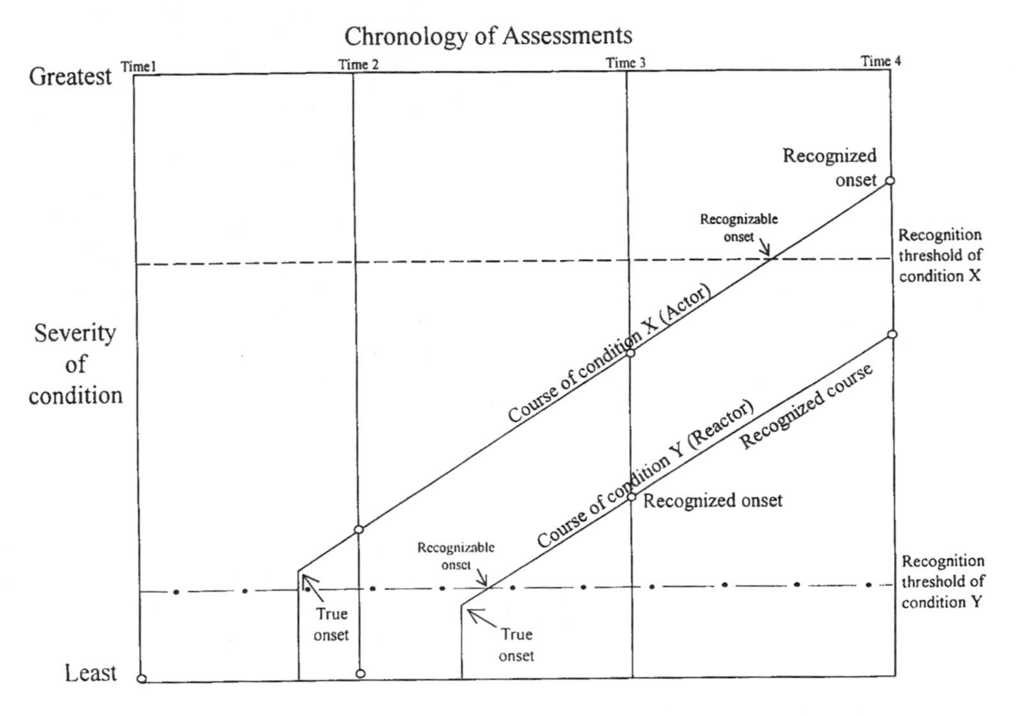

Figure 15.1. Differences between true and apparent sequences of onsets can confound assignment of cause-and-effect roles. (The onset and course of condition X actually precedes that of condition Y; that is, X is the true actor and Y the true reactor, but because of varied recognition thresholds, the order of the potentially recognizable onsets, or of the actually recognized onsets at the times of assessment, is the opposite; that is, Y appears to be the actor and X appears as the reactor.)

dition for reacting to another condition could be constructed. Thus, cause-and-effect interpretation of the time sequences of two conditions would be clarified by each being recognizable (detectable) at all stages of its development. Changes along any segments of the respective continua could be observed for action and reaction.

Such continuous measures as are available in this field of study are almost always constructed to be continua of severity, not of influence–susceptibility. Development of the former requires only information on the condition of primary interest; the latter must evolve from information on the interaction of the conditions that are related by influence–susceptibility. The two types of continua may diverge in form because of surges in influence or susceptibility along graded segments of the severity continua.

Aiming at the delineation of continua of influence and susceptibility leads to a sharper look at the relevant elements (i.e., active ingredients of influence or susceptibility) within depression and physical disorder. There is likely more than one acting or reacting element in each of those conditions, introducing uncertainty into observations of relationships. The most prominent and distinctive elements are those that are respectively subjectively and objectively loaded. Clinical and quantitative experience support the assumption that both of these elements are found in each of the spheres of depression and physical disorder (Paradiso, Ohkubo, & Robinson, 1997; Schrader, 1997). Relationships may build between physical and emotional conditions through either objective or subjective pathways, leading to the subjective–objective elements in one condition altering the subjective and/or objective elements of the other condition. The direction of influence along these pathways may be one way, either way, or both ways (Gurland et al., 1988). A theoretical inventory of subjective–objective actor–reactor patterns could take account of eight possible sets (see Figure 15.2). Typically, the objective physical → subjective emotional and subjective emotional → objective physical pathways receive the lion's share of attention. However, the other possibilities can and do complicate the interpretation of relationships.

Even a solid attempt to hew to either subjective or objective elements of a condition may be subverted by expediency. It may transpire that one aspect is better able to measure the milder levels of influence, while the other aspect best captures the more severe end. In that case, the continuum may be stitched together with a hybrid combination of subjective and objective measures. For example, objective performance measures may be more sensitive, specific, and reproducible than self-reports in detecting the earliest and the most advanced signs of physical impairment, while self-reports

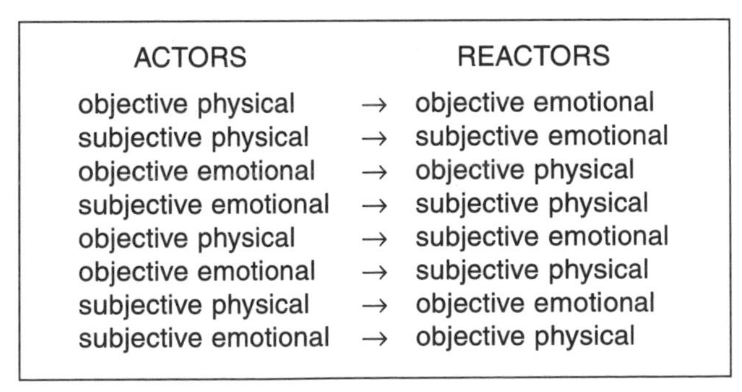

Figure 15.2. A spectrum of actor–reactor patterns.

with an unknown subjective loading may be the most practical way of recording progressive physical impairment in its middle stages. In relating such a hybrid (composed of a mix of subjective and objective elements) continuum of physical impairment to correspondingly mixed emotional measures, uncertainty may arise as to findings on associations, whether these spring from progression (increasing severity) of the respective conditions, or could be attributed to a shift from more objective to heavily subjective segments of the continuous measure.

For this reason, but also because of the differing implications of findings on subjective and objective elements, purity of measurement in this respect would be ideal. Yet information collected by an interview technique can only be as accurate as allowed by the clarity of the language exchanged between interviewer and respondent. In colloquial exchanges, clarification of meaning may be achieved through incrementally more pointed questions and answers. In contrast, interview question-and-answer combinations are customarily laconic, in order to save time. This constricted communication may result in uncertainty as to whether the information collected is objective or subjective, or a given mix of the two elements.

Take the question–answer set consisting of a question ("Do you have *difficulty* walking?") and answer ("Yes!"). Behind this answer could lie a range of clarifications. A relatively objective example is "I must be having difficulty if I cannot walk more than 4 blocks without resting." Or, relatively subjective is "I experience difficulty in that it requires more effort to walk than I feel is usual for me." Other explanations of what *difficulty* means might be composed of varying mixes of objectivity and subjectivity: "It takes more time to cover a certain distance than usual or normal. I have to use a device to walk. It causes me pain, even though I continue to walk normally. I have to take great care to avoid falling, even though I continue to walk normally."

Similarly, a variety of explanations along the objective–subjective spectrum could be implied by an affirmative answer to the question "Is walking a *problem* for you?" Walking could be deemed a problem because it is difficult, as just described, it takes up too much energy and time, it is uncomfortable, it is embarrassing, it prevents the respondent from doing the things he or she desires to do, it makes for dependency on others, it places a worrisome burden on others, it is costly to treat, or its treatment is painful or has unpleasant side effects.

In pointing to the need for clarification of the degree of subjectivity and objectivity in the bits of information entering into a continuum of severity of the acting and reacting conditions, and in raising an issue about the detail required for understanding meaning, we question the ability of existing measurement to separate adequately subjective and objective elements. Some otherwise tried and tested standard instruments are not designed to dissect the subjective and objective meanings of information about the conditions embodied in the interaction between physical and emotional states. Study of the connections between the two conditions would be simplified if there were available distinct measures of the objective and subjective elements of the physical and emotional domains.

ILLUSTRATIVE PROTOTYPES OF CONTINUUM MEASURES

As a next step in advancing a model of reciprocal influence between depression and physical disorder, we demonstrate two continuum measures, the one primarily subjective, and the other principally objective. As already stated, measures in this field of study need (1) to cover the full spread of the strength (i.e., range of severity) of the elements in physical and emotional conditions that enable each to influence the other (i.e., that property of a condition that can act upon, or react to, the other condition); and (2) to distinguish between the subjective and objective elements of each condition.

Earlier, in Figure 15.2, we listed eight potential sets of relationships between depression and physical disorder. It will suffice for the argument we are presenting to limit the discussion of potentially useful insights to that which arises in investigating just two of the possible combinations: objective physical → subjective emotional, subjective emotional → objective physical. We review the principles and provisional application involved in a hierarchically organized index of the subjective severity of depression, the Index of Affective Suffering (IAS), and thereafter apply those principles to measuring objective elements in physical disorder. With two continua representing respectively the conditions of interest, we model their interaction and bring to the surface our remaining reservations.

THE INDEX OF AFFECTIVE SUFFERING

Our intention in constructing this index was to draw attention to the essential therapeutic challenge of depressive disorders, namely, the relief of suffering (Gurland & Katz, 1997). By introducing the IAS, we targeted the need for a new measure of severity of disorder burden salient to affective quality of life. This abstraction of the heavily subjective aspects or elements of depression strikes a contrast with a diagnostic category, which is a composite of both objective and subjective elements.

The IAS satisfies one side of the "subjective emotional → objective physical" equation that we have proposed to examine. It (the IAS) is arranged on a continuum of increasing severity, based on its intensity (depth of distress) and extensity (variety and duration of affected events) (see Figure 15.3). At one extreme, the suffering is so intense that it is intolerable, and it is incessant in time and omnipresent in situations. At the other extreme, suffering is never present; all situations are free of it, and positive affect is predominant.

So that changes along all segments of the continuum can be identified, the continuum is divided into levels of severity, operationally defined. Each level can be derived from the kind of information collected at interview from the subject. The items defining a level are descriptions that transparently convey the degree of severity of the suffering; these terms that can be related to likely service demands, including urgency of the need for relief.

A concentration on subjectivity is preserved by the use of commonly understood terms. Also, since meaning resides partly in context, the number of hierarchically organized levels is restricted to a number (seven) that can be held in visual memory, so that any level can be viewed against the background of alternatives. As detailed elsewhere (Gurland, Katz, & Chen, 1997), the continuum of severity is supported not only by face validity but also by linear associations with observed activity in searching for relief. The latter is reflected in use of sedatives, psychotropics, and other medications for emotional problems.

There are strong associations with the criteria for the DSM-IV diagnosis of major

GENERAL LEVEL	INTENSITY CRITERIA	EXTENSITY CRITERIA
Extreme	**Intolerable, agonizing**	**Incessant**

Language: Thoughts of suicide *or* wishes to be dead/life not worth living *and* depressed all/ most of/every day
Effect: Completely destroys *all* daily affective quality of life
Clinical priority for relief of distress: Compelling and urgent

Advanced	**Desperate, almost unbearable**	**Most times and events**

Language: Thought of suicide *or* wishes to be dead/life not worth living *and* depression last more than hours
Effect: Destroys *much* of daily affective quality of life
Clinical priority for relief of distress: Compelling, may be urgent

Dominant	**Preoccupying**	**Most times and events**

Language: No enjoyment *or* future bleak or worries about everything and depression last more than hours or regularly
Effect: Greatly downgrades important parts of affective quality of life
Clinical priority for relief of distress: Cannot be ignored

Background	**Distracting distress**	**Occasional times or events, briefly**

Language: Any of the above *and* depression last *more* than hours
Effect: Moderately downgrades important parts of affective quality of life
Clinical priority for relief of distress: Interventions aimed at groups of persons rather than individuals

Minor	**Subduing distress**	**Fleetingly, few events**

Language: Cried/sad *or* less enjoyment than usual *or* regrets/pessimistic *but* depression lasts *less* than hours
Effect: Prevents affective quality of life from being really good
Clinical priority for relief of distress: Distress can be ignored and does not compel attention

Trivial	**Controllable distress**	**Fleetingly, few events**

Language: Only fairly happy or worries about something
Effect: Limits affective quality of life negligibly
Clinical priority for relief of distress: None

Positive	**Emotionally comfortable**	**Experiences happiness**

Language: No distress in this area and very happy
Effect: Affective quality of life enhanced
Clinical priority for relief of distress: Not applicable

Figure 15.3. Definitions of affective suffering levels.

depressive disorder and dysthymia, and with a scale of depressive symptoms (Gurland, Katz, Chen, 1997). However, diagnosis does not in itself provide a continuum of severity, and diagnostic qualifiers such as major, minor, or subsyndromal depression give little guidance in this respect. The companion axis of severity for DSM-IV diagnoses covers mixed subjective and objective elements and combined domains of living. There is a dimming of the distinction between depression as a diagnostic typology and depression as a degree-of-distress class, and between, on the one hand, depressive suffering itself and, on the other hand, depressive problems spreading over several domains.

Although a scale of depression does present a continuum of sorts, the salience for subjective severity is not explicit. A given scale score can be generated by one of many combinations of items and thus does not directly represent suffering even where the items in the scale are solely subjective. It follows that there is no meaningful rationale for examining relationships between physical disorder and any particular segment of a depression scale, with the exception of a threshold score corresponding to clinical levels of depression. The latter device forgoes the advantages of a continuum.

Concentration on a severity continuum of an apparently pure agent such as suffering appreciably clarifies the search for relationships with the physical side of the equation. Clarification emerges from separating the roles of disease and distress in a wide range of depressive disorders and states as causes, predictors, and sociobiological associations of physical consequences.

OBJECTIVE INEFFICIENCY IN PHYSICAL FUNCTIONING

For consistency with the reasoned paradigm laid out for the subjective, emotional side of the reciprocally interactive equation, it behooves us to find a parallel on the objective, physical side. To remain true to the ideal principles already proposed, the objective, physical actor–reactor would have to be a continuum reflecting graded severity and free of subjective admixture. What follows is a contemplative exercise in applying these principles to objective impairments in the elder's physical functioning.

Physical functioning is here taken to refer to the conduct of tasks that involve mechanical actions and are integral to the person meeting the demands of the environment. These tasks include preservation of independent living, autonomy in intimate acts, and meeting socially expected standards for task performance. Widely used inventories of such tasks cover transferring from prone to sitting to standing positions, feeding, bathing, toileting, dressing, grooming, mobility, cooking, shopping and other household chores, self-administration of medication regimens, communicating, telephoning, managing personal business, visiting, and work and leisure activities. By selecting areas that are central to the person's successful management of living, a salience to the emotional, subjective side of the interaction is primed.

Experienced observation has shown that certain functional tasks are impaired in a regular sequence and recover in the reverse order (Katz et al., 1963). Apparently, the functions are responding to an influence with a regular gradient of severity or/and have a correspondingly regular gradient of vulnerability (i.e., orderly progression must be shaped by an organization of either extrinsic or intrinsic determinants, or both). Our goal is to find the influence that marches in step with the hierarchy of functional impairments but translates into a continuum of severity and can relate to the emotional, subjective continuum.

A concept of functional inefficiency appears to meet the criteria we have set for an objective physical agent. In the proposed working model, functional inefficiency and impairment are referenced to the routine challenges of living: fulfillment of social and civic obligations, maintenance of the living unit, and dealing with self-care (see Figure 15.4).

Inefficiency is defined as increased (i.e., wasteful) expenditure of time and effort (i.e., energy) and/or resources to obtain unchanged or reduced functional output. The features of this continuum concept are potentially accessible to a variety of measurement strategies. Loss of efficiency is postulated to be initially evident in the component abilities (here called subtasks) that enable the execution of a function, evident next in the actual execution of a goal-oriented function (here called a task), and eventually evident in failed ability to execute independently that function or task. Thus, for each challenge posed by the environment, there is a continuum of efficiency–inefficiency. The continuum is readily divisible into four hierarchically organized stages based upon transformations into substantial shifts of manifest functioning at critical thresholds:

Stages 1: *Normal Efficiency.* Efficiency (expenditure of energy-time and/or resources per unit of output) of component abilities and the conduct and completion of tasks are all within normal range (e.g., the person can walk at least 11 level city blocks within the population norms for time and without evidence of nonessential demands on en-

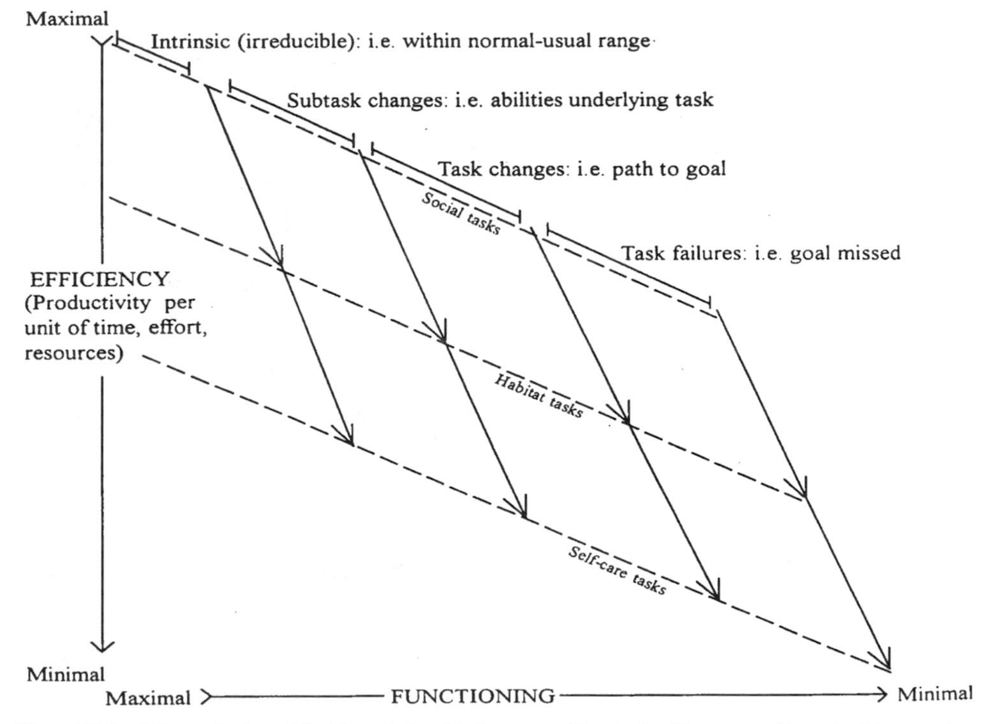

Figure 15.4. A theoretical model of the relationship between objective inefficiency and impaired functioning.

ergy, such as stumbling, limping, or holding on to wall, and without excess use of resources such as walking devices or personal assistance).

Stage 2: *Subtask Inefficiency. Components of a task* show a decline in efficiency (excess expenditure of energy-time and/or resources per unit of component output), but the task itself is conducted at a level that meets usual or normal standards (e.g., abnormal gait demands excess effort, but the person can walk 11 blocks within the normal range of time, without resting).

Stage 3: *Task Inefficiency.* Conduct of the *task as a whole* absorbs an excess expenditure of energy–time and/or resources per unit of task output, but the task is nevertheless completed without assistance (e.g., the person takes longer than normal and must rest at intervals, but can walk 11 blocks without assistance).

Stage 4: *Task Failure.* The given task is not completed, in part or whole, without the assistance of another person (e.g., person cannot walk without assistance).

Time and effort and/or resources per unit of task accomplishment are the main parameters describing inefficiency. Inefficiency is present from the earliest stages of the onset and progression of functional impairments. Psychological, behavioral, and material resources of the person and the demand characteristics of the task and its context set the thresholds for transitions from subtask to task inefficiency and from the latter to task failure. The content and sequencing of functional changes (deterioration or improvement) may vary between cognitively, physically, and emotionally determined disorders, and between objective and subjective dimensions of those changes.

The "inefficiency" model differs from that presented by the World Health Organization (WHO) International Classification of Impairments, Disabilities, and Handicaps (ICIDH) formulation for examining the progression from disease to impairment, disability, and handicap (Verbrugge, 1990) and from the various modifications of the WHO model employed in studies of aging (Kovar & Lawton, 1994; Johnson & Wolinsky (1994). The WHO model and its modified versions do not, as we see it, lay out the stages, much less a continuum, of progression of functional changes. Rather, those models emphasize relationships among causes and consequences revolving around functioning. They do not embody a concept of degrees of change in functioning (i.e., a continuum) leading logically from normal ranges through early changes, intrusions into the conduct of functional tasks, and, ultimately, to dependence. Functional limitation or its synonyms is left, ambiguously, to cover both poor conduct of the activities of daily living and outright inability (partial or complete) to complete a task. In those models, etiology (e.g., pathology) is given a linear place preceding functional changes, whereas etiological agents may in fact precipitate functional changes at any point in the continuum of severity, not excluding complete dependence. In other words, disease is made to appear as an early stage of functional impairment, which it is not; they too are distinct in nature. Disease is a pathological process, largely outside of social and conscious control, while functional impairment, at least at some stage of deterioration, involves goal-oriented tasks. Increasing intensity or extensity of the disease process does not merge on a continuum or at a particular threshold of severity, with intensity and extensity of functional impairment. Also, to some large extent, the indicators of "organ" or "system" impairment (e.g., symptoms and signs of medical disorders) as represented in those models are more readily translated into the indicators of etiology than into the early stages of functional changes. Another anomaly is that those models do not allow for a distinction (e.g., by measurement) between degrees of functional change and consequences such as dependence and handicap. Perceptions or realities of poor health or handicap are contingent consequences, not extreme developments of functional changes. Perception of poor health or social disadvantage may arise at any stage of the progression of functional deterioration and may even improve as the person adjusts to deteriorating function. Thus, these widely used models would not serve for the continuum of changes in functioning that we need for our further discussion.

QUALITY-OF-LIFE RELATIONSHIPS

Seeking to find and understand interactions between subjective (emotional) and objective (physical) agents is an exercise in connecting types and domains of qualities of life. The illustrative measures are of functional inefficiency on the one hand, and affective suffering on the other. Figure 15.5 shows a potential range of pathways from subtypes of the former to the latter. We have chosen for brevity to detail only the inefficiency subtypes. We have also not broached causes common to functional inefficiency and affective suffering, because temporal sequences would not raise an issue as to which of functional inefficiency and affective suffering is cause and effect; both would be effects with respect to the common cause(s).

As illustrated in Figure 15.5 for the inefficiency-suffering pathway, there are intervening variables that import mixtures of subjective and objective influences into the pathways connecting the purer extremes. This is a frequent finding in the literature (e.g., Vilhjalmsson, 1998). Figure 15.5 shows, for example, objective inefficiency in functioning may give rise to subjective awareness of changes in subtasks and tasks. Although deterioration in function may be objectively evident, the subjective extrapolation of that fact into an expectation of further decline in the future may be a potent connection to affective suffering. Moreover, the ideal causal directions represented in Figure 15.5 are open to the same assessment confounds that have been discussed for the pathway between the purer extremes.

RESISTANT AMBIGUITIES

Certainty about some important facet of the relationship between depressions and physical disorders (in any of their accepted or acceptable meanings) would form a base for further exploration. One starting point is to determine whether there is something about mood that alters function, and vice versa, or more particularly, whether there is connection between worsening of mood and worsening of functioning. Stating the nature of this inquiry implies at first sight that a cause–effect relationship may be discoverable. Detailed information about the cause-and-effect roles of depressions and physical disorders in a clinical, public health, or humane context ought to provide a basis for rational, effective, and early intervention. Nevertheless, we have emphasized the limits of certainty inherent in a cause-and-effect model as applied to complex conditions.

Our opening caution in the interpretation of cause–effect relationships based on epidemiological studies, including those with longitudinal designs, was that differen-

Deteriorations of functioning ⟷	Behavioral, psychological, and biological mechanisms ⟷	Affective suffering
	Loss of autonomy	
	Fear (despair)	
Task failure	Assistance (dependency)	
Task failure	Restricted activities	
Progressive inefficiency	Expectation of decline	
Task inefficiency	Current uncertainties	
Progressive failures	Future uncertainties	
	Loss of status	
	Exposure to indignities	
	Lowered self-esteem	
Task inefficiency or task failure	Awkwardness (humiliation)	
Task inefficiency	Effort (fatigue, distraction)	
Task failure	Discontinued plans	
Task failure	Vulnerability to actions of others	
	Chronic physical suffering	
	Loss of sensual gratification	
Task failure	Discomforts (disgusts)	
Symptoms	Discomforts (disgusts)	
	Loss of intimacy–love	
	Sense of helplessness	
Contextual reactions	Loss of resources	
Contextual reactions	Withdrawal of others	

Note: Even where the extreme poles of the connection between task inefficiency and affective suffering are relatively purely objective and subjective respectively, the intervening mechanisms may introduce a mix of subjectivity and objectivity.

Figure 15.5. Mixed subjective–objective pathways between physical functioning and affective suffering.

tials in recognition-detection thresholds for interacting domains or elements could give misleading results on the sequencing over time of the onsets of occurrence or change in the agents of interest.

One way that we considered, in depth, for looking at cause-effect relationships without the confounds that accompany the assessment of thresholds, was through relationships among severity continua of agents that might serve cause or effect roles. Relative changes along segments of the continua are made the substance of observation rather than transitions across selected points. Continua were proposed, then, as a means of escaping the confounds surrounding the recognition–detection of thresholds for operationally defined categories or for selected scale cut-scores. Temporal relationships among severity continua for each domain and element participating in change might have been expected to offer a foundation for interpreting sequences among potential actors and reactors, taking account of all degrees of occurrence or change. However, conceptual analysis suggested that the pace and degree of change along a segment of a given severity continuum could be altered by preceding changes in other interacting continua and thus be made more or less detectable. In addition, descriptive or assessment language might be richer or poorer, and therefore more or less sensitive, along various segments of one or other continuum, thus inconsistently magnifying or minimizing information on change along those segments of the continuum. For example, the prevailing professional culture is probably better at making fine discriminations in the milder levels of functional impairment than in the milder levels of depression. Moreover, instrument makers, and those who measure with those instruments, are immersed in this world of colloquial and professional languages, perceptions, and underlying values, any of which might calibrate the discriminations attained along segments of the continuum, so that observed and measured change are not absolutely comparable between severity continua of different domains or elements, nor over different segments of the same continuum. For example, vivid and obvious changes in suffering (e.g., depression, affective suffering) might be induced by subtle and unmeasured changes in task efficiency and thus appear to precede it or replace it as a precedent to task failure; or *mutatis mutandis* task failure may give the illusion of preceding suffering.

It is also difficult to maintain the feature of continuity in severity or progression when what might be potentially a smooth continuum of increasing degree of influence of one condition on another can be transformed into a series of discontinuous levels. One illustration of this phenomenon was offered in Figure 15.4, relating increasing inefficiency in functioning to subtask changes, task changes, and task failures. As the successive inefficiency thresholds for those changes and failures are passed, a transformation, usually a boost of strength, of the influence of function on emotional states is likely to occur. Correspondingly, though more subtly, as intensity and extensity of affective suffering increases, the meaning for the sufferer, and the implications for functioning, may surge at certain threshold values. For example, the health care profession may choose to intervene when affective suffering is sufficient to elicit the criteria for formal diagnosis; or family may become involved when intensity is such as to provoke suicidal threats; or afflicted persons may lose motivation when suffering pervades most of their days. It may be the case that underlying continua (e.g., inefficiency, suffering) do have an effect in complex interactive comorbidities, but this is overlain by transformations that are hierarchically organized discontinuities (e.g., task inefficiencies–failures).

Transformations that are the result of increasing strength of the cause may be mistaken for the onset of the cause (e.g., onset is recognized when increasing inefficiency results in task failure). A variant on the theme of the "threshold confound" is the seductive notion that one can muster a cohort of subjects who are free, at baseline, of one or more of the conditions under study. If that were a given, the sequencing of incident conditions would reveal the cause–effect relationships. However, we have pointed to potentially misleading observations arising from problems in detecting onsets, and here, we add that baseline readings are also vulnerable to errors of insensitivity in measurement. Moreover, baseline assessments typically do not take account of past events, or they have no reliable way of doing this. Measurement may register the absence of a particular condition at baseline, but that condition may have been present in that subject's recent or distant past. Empirical observations tell us that the presence of indicators (e.g., symptoms or signs) of a condition is a strong predictor of continuation or relapse of those indicators (see, e.g., Gurland et al., 1988, p. 131). It follows that indicators that are absent at baseline but subsequently emerge, have an increased chance of having been present in manifest or latent form prior to baseline. Thus two fallacies lurk in the shadow of the assertion that a cohort is selected to be free of specified conditions: (1) A condition of interest may be present even though its proxy measured indicator is not; (2) a condition of interest may be currently absent yet an influence on interactions through a chain of events it (i.e., the condition of interest) instituted in the past.

The need to separate objective and subjective elements within domains was also emphasized and attempted. However, as previously mentioned, although continua for inefficiency and suffering can help to anchor the extreme ends of the pathway between the objective, physical condition and the subjective, emotional condition, there are intrusions of critical transforming thresholds at which new and relevant phenomena can emerge. These thresholds were discussed for inefficiency and suffering, and illustrated (refer back to Figure 15.4) for functional inefficiency. Also, there are intervening variables that can introduce an uncertain mix of subjectivity and objectivity into the pathways of influence between objective elements in functional inefficiency and subjective elements in affective suffering (refer back to Figure 15.5).

Difficulties in separating subjective and objective elements were found to be compounded by imprecise language in information gathering and a tendency for hybrid subjective–objective influences to be at play. Even capacities such as subtask inefficiencies, which might appear to be objective and outside the subjective influence of value systems and goal orientation, may not be pure. Where a subtask is presented for measurement to a subject in a laboratory setting, it (i.e., the subtask) can assume the character of a task in that it is summoned into awareness as a challenge to be accomplished and is thus exposed to subjective influences. For example, a person, without discernible loss of efficiency, may be able to reach for and grasp common objects such as articles used in feeding or dressing, and may be completely unaware that his or her efficiency is reduced in grasping small objects without direct visual guidance but may become anxious when asked to perform the latter action in a laboratory setting. Moreover, subtask objective inefficiency may be accompanied by symptoms (such as pain or awkwardness) that are easily recognized even where not labeled as inefficiency. Task inefficiency, complaints of difficulty in managing tasks, and self-reports of task performance, all are blends of objective and subjective elements.

Among other potentially confounding matters, we encountered discordance between frequency of observations and speed of changes in the putative cause-and-effect agents. On the one hand, stages of progression can become folded into each other in calamitous circumstances, for example, massive stroke or heart attack, acute critical illness, accidents and operations, overwhelming pain, radical escalation of environmental obstacles, catastrophic emotional collapse, or sudden death. On the other hand, progression may be so prolonged that causal events are lost in the distant past, or effects lie latent in the far future.

Despite a concerted effort to eliminate confounding of subjective and objective influences, and to avoid the pitfalls of threshold phenomena for determining onsets where sensitivity and specificity of methods for detecting causes and effects may differ, these issues persist. We are also left with the vagaries arising from measures which entail a combination of domains or both subjective and objective elements, information lost in the past and in other asynchronies between the timing of events and that of information collection.

On the subject of complexity, if we have traveled in a circle, beginning with the puzzling open ends of findings addressing the cause–effect relationships of depression and physical disorder, at least it is not a full circle. In the course of this excursion, we have suggested means of clarifying concepts, measurement and understanding. This puts us in a position to face the next hurdle in this field of study, namely, the great complexity of connections that arise in comorbidity among aging persons.

In older persons, inefficiency, suffering, and their transformations and relationships are often the resultant of more than one disorder, a final common pathway of cognitive, physical, and emotional disorders accompanying aging. This resultant is also altered or predisposed by past lifestyle, inheritance, and experience; the current social and material environment; and future anticipations. This final common pathway can provide a logical link between complex disorders and their consequences for qualities of life, but this common ground of linkage is occupied by the aggregate complexities of its numerous components. Complexity of this intricate degree must be formed and continuously reformed by numerous if not infinite numbers of cause–effect microcosms. Given that this insight is accepted as persuasive, one need not look further to accept that the cause-and-effect model will inevitably defy definitive deciphering. To move on, we must find a route to useful knowledge about complex relationships that is not circumscribed by the tenets of an ironbound cause–effect model.

AN ARGUABLE RESOLUTION

Many of the dilemmas stemming from efforts to disentangle cause-and-effect (acting and reacting patterns) surrounding the relationship of physical disorder and depression could perhaps be solved by reframing the purposes of such inquiry. Failing a statement of bounded purpose by investigators, the default expectation is that discovery of cause-and-effect will add information on origins, course, and outcome of both physical disorder and depression. This information is supposed, in turn, to serve the manifold purposes of understanding, directing further research, risk assessment, and prognosis; developing and improving management and treatment, evidence-based selection of effective treatment; and setting standards for quality control. Justification for this encompassing position could only be found in the belief that there is one root truth to be extracted respecting a particular pairing of cause-and-effect, with all other influences being modifiers of that basic truth.

Tolerable Uncertainty in Complex Relationships

We hold that observations on temporal sequences are, in the field under review, generally not interpretable in a rigid cause-and-effect model for reasons already laid out. Comorbid physical disorder and depression interact to form a complex unity, a whole that is further shaped by past experience, future anticipation, and the living and material environment, including treatment. It is not possible in the attempt to designate a clear active or reactive role for interacting agents to isolate one agent, such as physical disorder or depression, without altering its nature and thus undermining the inferences that can be drawn for causes and effects. There cannot be a prime cause and main effect attributable to physical disorder or depression, since the two agents converge as irreducible wholes. Similarly, moderators of a complex interaction are also parts of the whole, not readily located on a path between cause-and-effect.

On the one hand, this view may be hard to swallow, because conditions that are essentially continua, when given arbitrary threshold values, project the illusion of entities that materialize or vanish at precise points of time relative to each other. On the other hand, by accepting that cause, moderator, and effect terminology is inappropriate in exploring interactions within a complex whole, one can escape from a frustrating analytic struggle and can then take a fresh look at the constructive gains to be obtained through reconstructing the purposes of the physical disorder–depression dialectic.

Rules of Evidence

The complex, reciprocally interactive model can be informed by findings from epidemiological research on the relationship between physical disorder and depression, including moderators and mechanisms. However, this model accepts that uncertainty will characterize that epidemiological information. It is sufficient that there be statistical evidence of association between the components in the complex, but inconsistency in strength or direction of association are expected. Either concurrent or longitudinal associations qualify for membership in the complex. A necessary feature of members of the complex is that, in addition to evidence of association, they must be credible as reciprocally interactive agents in that complex. Credibility is a judgment based on information from one or another of a wide range of sources: for example, biomedical, psychological or behavioral fields of study, serendipitous accident, or controlled interventions.

Organizing According to Purpose

Analyses of physical disorder–depression interactions become more productive when closely tied to purposes. For example, the fruits of analysis may be intended to allow risk predictions of the likely course and outcome of untreated comorbidity. Clinical or more formal risk-assessment algorithms are anchored in temporal relationships between profiles of the complex constituted by the domains and elements of physical disorder and depression, as well as the characteristics of the enlarged whole that includes time and context. It is not necessary to adduce cause-and-effect in calculating and acting on levels and type of risk of future events (we avoid the term *outcome* because of its implicit reference to cause-and-effect). Risk assessment can in turn guide the need for treatment, prepare the patient and family to cope with the future course of the condition, and provide benchmarks against which treated outcomes may be compared.

Another example of the productivity in choosing analyses that are keyed to specific purposes is noted in the testing and selection of treatment. Confronting established or new evidence (see previous comments on rules of evidence) of the reciprocating interactions stirring within the comorbid complex and its corollary determinants, it is reasonable to intervene in one or more of the interactive pathways without definitive proof of cause-and-effect. There is bound to be more than one possible target of intervention across and within the credible pathways of physical disorder, depression, time and context. *Need* for intervention action is brought into focus by epidemiological information on the frequency, distribution, and course of the complex web of conditions and context. *Credibility* of intervention is borrowed from the lore of clinical experience, interventions studies, and a wide variety of behavioral, psychological, and biomedical research strategies that attest to the mechanisms for influencing the course of the complex and its components. Prioritizing the repertoire of interventions will be guided more by familiarity with certain interventions, together with knowledge of their relative effectiveness, safety, side effects, and cost than by evidence of prime and secondary causes. There is something to be said in favor of evaluating treatment prospects for any and all interacting agents in a complex, chronic set of conditions, not just for the assumed originating cause.

It is worth expanding on the assertion that a spurious aura of prime status may surround the accolade of cause, with attributed effects apparently adding to the perceived importance of the assumed causal actor. This is seen, for instance, where advocacy for treatment of depression is fortified by claims that physical function impairments can be induced by the former condition. In part, this assertion builds on the assumption that causes offer the most potent target for intervening to prevent and relieve effects. We question whether the value attached to treating either depression or physical function impairment needs to be gilded by secondary consequences. Elsewhere (Gurland & Katz, 1997), we have laid our reasons for pressing the treatment of depression for alleviating subjective suffering itself, and the treatment of impaired physical functioning, because the latter is a central ingredient in impoverishing life. Where these conditions occur together as a complex interacting whole within a specific context, both conditions, and the whole, call for treatment. Public health or policy positions that pit one condition against the other in a competition for resources are political, not scientific or clinical realities.

Accepting and Working with Complex Wholes

In line with the position we have reached, we recommend that research attention be directed to describing interactive associations between any and all of the potential actors and reactors in the arena of physical disorder and depression, including subjective and objective aspects, and context. Temporal sequences, course, and outcomes furnish important information in qualifying these associations. However, we regard as futile and unnecessary attempts to determine the relative strength of causal pathways or to separate prime and secondary causes. Where associations are demonstrated between plausible actors, reactors, and moderators, it is useful to assume that an interactive, reciprocating complex is at work. Converting this knowledge into avenues for progressive research and for clinical applications depends on reviewing a wider scope of information than can be embraced by an epidemiological study.

Beyond epidemiological data on relationships between physical disorder and depression lies knowledge gained from the wide variety of sources we have outlined, as well as from a grasp of health service capacities for responsive care, reinforced by community and personal helping resources. Convergence of these bodies of fact and understanding, leavened by skilled judgment and human compassion (Katz, 1998), builds the foundation for evidence-based practice standards in the prevention and relief of comorbid physical disorder and depression. We posit that the leap from epidemiological findings to salient research on therapeutic interventions in comorbid physical disorder and depression (and possibly other comorbid complexes), will be more readily launched from the complex model than from the conventional cause–effect models.

Interactive comorbid conditions and their context form a complex. Complex situations often elude simple understandings and interventions. Methods of gathering, displaying, and interpreting information of this complexity should be developed that assist the decisions of those who would use the information for improving the quality of life of aging groups and individuals. These methods should not overstep what can be logically inferred from empirical findings, but should not shrink from summarizing information commensurate in scale to the complexity confronting decision makers. The engagement between information and decisions, if kept within their respective constructive limits, can lead to incremental enlargement of the effectiveness of both. These application processes are interactive, like the comorbid conditions they seek to address.

RESTATEMENT

It has been said that anything worth stating is worth overstating. Nevertheless, we wish to set limits to the interpretation of our stated position. Our remarks have been entirely directed at understanding and applying naturalistic information about complex interactive conditions in general and, specifically, the nexus of depression, physical disorder, and their time–space context. Within that frame of reference, we have found that the conventional use of cause–moderator–effect models, longitudinal designs notwithstanding, attracts confounds and ambiguities that diminish the usefulness of findings. We have therefore proposed ways of viewing complexity without definitive dissection of causal directions or relative powers of the interactive conditions.

Evidence of interaction is intrinsic to the complex of interest we have defined. Of course, interactions involve the roles of causes, moderators, and effects, but evidence of interaction does not require proven assignment of those roles. We have pointed to a frustrating level of uncertainty in assigning those roles from information gathered on complex conditions in naturalistic studies. Moreover, we have spelled out reasons for regarding those assumed roles to be a weak link in a search for mechanisms of interaction or ameliorative interventions.

Since we have no desire to be nihilistic, we have illustrated several alternative methods for extracting value from knowledge about complex interactions and their longitudinal patterns. In this, we seek to open a discussion and avenue of discovery that fully respects the unique nature of complex wholes.

REFERENCES

Aneshensel, C. S., Frerichs, R. R., & Huba, G. J. (1984). Depression and physical illness: A multiwave, nonrecursive causal model. *Journal of Health Social Behavior, 25,* 350–371.

Clark, D. C., vonAmmon Cavanaugh, S., & Gibbons, R. D. (1983). The core symptoms of depression in medical and psychiatric patients. *Journal of Nervous and Mental Diseases, 171,* 705–713.

Dorfman, W. (1978). Depression: Its expression in physical illness. *Psychosomatics, 11,* 702–708.

Gurland, B. J., Golden, R., Lantigua, R., & Dean, L. (1984). The overlap between physical conditions and depression in the elderly: A key to improvement in service delivery. In E. Aronowitz & E. M. Bromberg (Eds.), *Mental health aspects of long term physical illness* (pp. 23–36). Canton, MA: Watson Publishers International.

Gurland, B. J., Katz, S., & Chen, J. (1997). Index of affective suffering: Linking a classification of depressed mood to impairment in quality of life. *American Journal of Geriatric Psychiatry, 5,* 192–210.

Gurland, B. J., & Katz, S. (1997). Subjective burden of depression. *American Journal of Geriatric Psychiatry, 5,* 188–189.

Gurland, B. J., Wilder, D. E., & Berkman, C. (1988). Depression and disability in the elderly: Reciprocal relations and changes with age. *International Journal of Geriatric Psychiatry, 3,* 163–179.

Katz, S. (1988, January). *Practitioner's judgments that affect quality of life.* Kent Award Lecture at the 1997 meeting of the Gerontological Society of America. Reprinted as Stroud Center Working Paper No. 21.

Katz. S., Ford, A. B., Moskowitz, R. W., et al. (1963). Studies of illness in the aged. The index of ADL: A standardized measure of biological and psychosocial function. *Journal of the American Garroter Society, 185,* 914–919.

Katz, S., & Gurland, B. J. (1991). Science of quality of life of elders: Challenges and opportunity. In J. Birren, J. E. Lubben, J. C. Rowe, & D. E. Deutchman (Eds.), *The concept and measurement of quality of life in the frail elderly* (pp. 335–343). San Diego: Academic Press.

Kovar, M. G., & Lawton, M. P. (1994). Functional disability: Activities and instrumental activities of daily living. In M. P. Lawton & J. A. Teresa (Eds.), *Annual review of gerontology and geriatrics* (pp. 57–92). New York, Springer.

Johnson, R .J., & Wolinsky, F. D. (1994). Gender, race and health: The structure of health status among older adults. *Girandole, 34,* 24–35.

Ouslander, J. G. (1982). Physical illness and depression in the elderly. *Journal of the American Garroter Society, 30,* 593–599.

Paradiso, S., Ohkubo, T., & Robinson, R. G. (1997). Vegetative and psychological symptoms associated with depressed mood over the first two years after stroke. *International Journal of Psychiatry and Medicine, 27,* 137–157.

Salzman, C., & Shader, R. L. (1978). Depression in the elderly: l. Relationship between depression, psychologic defense mechanisms and physical illness. *Journal of the American Garroter Society, 26,* 253–260.

Schrader, G. D. (1997). Subjective and objective assessments of medical comorbidity in chronic depression. *Psychotherapy and Psychosomatics, 66,* 258–260.

Verbrugge L. (1990). The iceberg of disability. In S. M. Stahl (Ed.), *The legacy of longevity: Health and health care in later life* (pp. 55–75). Newbury Park, CA: Sage.

Vilhjalmsson, R. (1998). Direct and indirect effects of chronic physical conditions on depression: A preliminary investigation. Social Science Medicine, 47, 603–611.

IV

Summary

16

Physical Illness and Depression in Elderly Adults

A Summary with Implications for Future Directions in Research and Treatment

L. STEPHEN MILLER

INTRODUCTION

The formidable compilation of this edited volume was initiated to fill what was seen by the editors and publishers as a critical void in the geriatric literature–that is, to bring together into a single source an empirically based yet available, sophisticated yet readable, state-of-the-art collection of the relevant factors involved in the interface between physical illness (and its concomitant, functional disability) and depression in older adults. In so doing, the editors have assembled a discerning and insightful group of recognized experts in the fields of gerontology, psychiatry, and health psychology. These experts, all of whom are active and accomplished researchers in their respective areas, present a wonderful set of summary reviews of the most important and salient areas of physical illness and depression. What is more, they have consistently presented these reviews in the context of theoretically based conceptualizations of both depression and physical disability, and have argued their points with data-based evidence from their own and others' laboratories, research centers, and community and hospital-based clinics. The result is this handbook, a collection of comprehensive, well-integrated, contemporary views of this complex relationship.

The overall theme that runs through most of these masterful chapters is clear. The relation between physical illness and the development of depression in older adults is affected (most likely mediated) by additional factors. It is crucial to identify and understand this relation and these additional factors if we expect continued progress in the

L. STEPHEN MILLER • Department of Psychology, University of Georgia, Athens, Georgia 30602.

Physical Illness and Depression in Older Adults: A Handbook of Theory, Research, and Practice, edited by Gail M. Williamson, David R. Shaffer, and Patricia A. Parmelee. Kluwer Academic/Plenum Publishers, New York, 2000.

advancement of our theoretical, research, and practical knowledge of physical illness and depression in older adults. Furthermore, the unique role of functional ability in both depression and physical illness appears as a particularly important factor in understanding this complex relation.

The handbook is coordinated into the following broad categories: (I) Risk Factors, (II) Conditioning Variables and Outcomes, and (III) Diagnosis and Treatment. Each category is then represented by individual chapters to form a substantive body of diverse yet surprisingly cohesive perspectives on what are thought to be the most current and important variables within the field of geriatric depression and physical illness. I use this categorical scheme to present my own summary of the current state of the art and likely directions for future research and treatment as they pertain to each of the categorical areas. I interject my own opinions of these points of interest where they converge with one another and, at times, when they diverge. Finally, I present my overall opinion about probable directions for both research and treatment in the near future.

While it is my task in this final chapter to give a perspective to the sizable amount of information put forth in these chapters, these represent my own opinions and personal summaries. Although food for thought, these opinions cannot replace the thoughtful and detailed views of the original authors. They, themselves, offer the best summary of future research and treatment directions within their respective areas of expertise. Nevertheless, I offer the following in the hopes that it may act as a broad guide to readers' formulation of this complex interplay and, perhaps, add to understanding and appreciating work of the skillful experts who contributed to this handbook.

PART I: RISK FACTORS

The first section, Risk Factors, focuses on various physiological, psychological, and psychosocial factors that may place a person at risk for experiencing depression in late life. It is clear that there is a strong relation between depression in the elderly and physical morbidity. However, it is just as clear that there is no simple, one-to-one association between the multitude of identified risk factors and either depression or physical illness. The message that consistently emerges is the bidirectional nature of associations between physical morbidities, related functional disabilities, and late-onset depression. Risk factors appear to be related in multiple ways to both depression and to physical illness in an elderly population. This leads to the necessity of multiple research strategies to clarify those relations, including cross-sectional, longitudinal, integrative, and focused, or microanalytic, methodologies. Thus, while each chapter stands alone in presenting important applied information about specific risk factors for both physical illness and depression in older adults, there is considerable across-chapter convergence regarding recommendations to add to our knowledge of the etiology, impact and, I hope, amelioration of these often devastating processes.

In the first chapter focusing on risk factors, Martha L. Bruce provides compelling evidence from epidemiological and clinical studies that research to date has shown a real, robust relation between depression and disability. She is also quick to point out that despite multiple measures and varied samples, we have not explained much about the mechanism(s) of this relation. In fact, this can be seen as a driving force in the compilation of this handbook and a primary target for the chapters herein. Bruce suggests that part of the reason for this apparent lack of understanding is that current analytic approaches do not provide enough detail to test theoretically driven hypotheses adequately. For example, two competing hypotheses can be formulated: (1) Disability is a risk factor for depression, or rather, a sign of depression and its underlying etiology versus (2) depression is a risk factor for disability–whether as a direct stressful condition leading to decline, an indirect impact on disability, or a biological effect of depression on functioning. Each hypothesis has been shown substantially to evidence a directional, if not clearly causal, relation. Bruce suggests that to understand the mechanisms involved, we first must recognize the bidirectional nature of depression and disability, and elaborate our research beyond its current level. A critical step appears to be specifying further the components and various aspects of both disability and depression that overlap. She recommends the disaggregation of current measures of both disability and depression. Furthermore, analytic approaches looking at longitudinal data that focus on *changes* in both depression and disability over time need to be undertaken. Structural equation modeling (SEM) may be an appropriate tool. Similarly, repeated measures across short intervals may identify underlying causal relations between depression and disability, and elucidate details of temporal change, particularly if focused on mediating factors of the disability–depression relation. In fact, as Bruce recommends, combining both detailed and longitudinal repeated measures may be our best way to increase the chances of identifying additional mediating factors (e.g., self-neglect, eating habits, loss of coping). Finally, there is clearly a demand to increase directly the interface between treatment and research, in that active intervention studies that modify the relation between depression and disability are needed. Manipulating one aspect (e.g., treating depression) is likely to have an influence on the other (e.g., disability), either directly (utilizing antidepressants for depression and subsequently measuring level of disability) or indirectly (measuring antidepressants' effects on the function of persons with physical disabilities). In this way, interventions can be used as manipulative strategies to tease out underlying etiological mechanisms.

In a similar vein, the seemingly specific risk factor of cardiovascular disease (CVRF) for late-onset depression is assessed by Jeffrey M. Lyness and Eric D. Caine in Chapter 3. The focus is on presenting models of the interplay between late-onset depression and cardiovascular compromise as broadly defined. They critically review what appears to be the primarily unidirectional-based literature to date for two pathobiological models of the contribution of CVRFs to depression. Using these models as examples, the authors efficiently walk the reader through a critical assessment of the literature supporting (or contradicting) each paradigm, elucidating their view of the type of research than can and should be conducted to study relations between depression and physical disability (in this case, vascular disease). They provide two compelling examples of how a possible underlying depressive pathogenesis may result in the development of depression in late-life: (1) development of brain damage via small vessel cerebrovascular disease, and (2) production of functional alterations in nortriptyline (NT) systems. Their conceptualization mirrors research guidelines espoused by Bruce in the prior chapter. Tellingly, limitations identified in reviewed research of these two hypotheses for depressive presentation emphasize the difficulty in establishing causal links and call attention to the need for recognizing nonbiological roles (psychological/

psychosocial) in the interface of depression and CVRFs, as well as the possibility of bidirectionality of depression and physical illness relations. Lyness and Caine, in particular, support a theory-driven, model-building approach with appropriate accumulation of empirical data that will support, disprove, or modify a given model.

Gail M. Williamson, in her presentation of relations between pain, functional disability, and depressed affect, takes a clearly empirical approach in assessing the important variables in her discussions of the mechanisms that explain the consistent, but modest, association between subjective experiences of illness-induced pain and depression, exemplifying the need for looking beyond simple cause–effect relationships between physical illness and depression. Specifically, past literature has actively looked for mediators (such as various types of functional disability) of the relation between pain and depression and come up with sometimes contradictory findings. Williamson identifies a specific aspect of functional impairment—restriction of routine daily activities—and her own and others' cross-sectional and longitudinal data appear to converge in suggesting that pain and depression are most strongly associated in its presence. Through a systematic approach across multiple patient samples, she has evaluated a number of related hypotheses. First, activity restriction, which appears as a particularly necessary component of instrumental activities of daily living (IADLs), generally mediates the association between pain and depression. Second, this mediation is attenuated by increasing age, or more specifically, by habituation to chronic pain. This finding has implications for intervention. It implies that it is length of adaptation and, thus, habituation to chronic pain, rather than age per se, that is the important variable for why older individuals are less distressed than younger individuals by activities that pain restricts. Several research directions are implicated by these results. The need for continued empirical research regarding the exact role of activity restriction as it relates to the pain–depression association seems obvious. Williamson suggests traditional controlled manipulation of specific variables, singly, to look at their resultant impact on the nonmanipulated variables. Examples include evaluating the impact of increasing routine activities in the presence of pain; assessing the effects of increased activity directly on depressive symptoms; evaluating patient trade-offs on acceptability of more pain for greater activity; and identifying differences between these individuals. As do others, she correctly attests to the need for longitudinal studies following the impact of these variables over time, including their predictive ability. As Bruce advocated earlier, increasing the interface between treatment and research is both a natural and necessary progression. Here, several of the previous suggestions can be incorporated, including focusing on increasing activities while minimizing factors (internal and environmental) that restrict or hinder activity. Adding a focus on psychological/behavioral interventions aimed at pain management, depression, and a combination of the two may provide a worthwhile window into the exact mechanisms involved. Finally, Gail Williamson's work suggests that we may be able to predict which people are most susceptible to depression as a result of pain and resultant activity restriction (i.e., those with little pain experience, or those who expect to have little activity restriction).

Alex J. Zautra, Amy S. Schultz, and John W. Reich describe a quantitative and sensitive measure, the Inventory of Small Life Events (ISLE Scale), as a way of assessing everyday events. From their perspective, these events can be used to study the impact of both desirable and undesirable transactions on multiple areas of interest, including stressors, the environment, and personal cognitive appraisals. Such a measure allows theory testing of the impact of major and minor life events in older adults. What I find particularly compelling about this measure is its conceptual base (i.e., the focus on small events in the lives of elders). I believe this can assist us in our understanding of heretofore neglected aspects of stressors and desirable events in elders' lives as they relate to depression initiation, depression maintenance, and recovery from depression. This quantitative measurement tool provides a way to evaluate the role of everyday stressors on depression as well as the role of desirable everyday events as sources of recovery from depressive symptoms. Data from Zautra et al. suggest differential effects of everyday events in elders, depending upon the source of stress and depression. Everyday events were not important in depression due to bereavement of spousal loss. This finding has implications for current theories of depression and current life stressors in elderly adults, as well as our theories of grieving. These authors also found that elders are different in this way from younger adults, in whom everyday events *were* associated with bereavement as well as psychological distress and depression. Depression due to functional disability in elderly adults, however, was clearly associated with everyday life events, and the quality of those life events accounted for observed depression. These results, together, suggest interventions oriented toward coping with everyday life stressors for this group and, perhaps, focusing on areas such as loss of meaning and purposiveness in life following bereavement. Finally, positive or desirable events seemed to lower depression in disabled elders. Again, this has implications for theoretical views of depression, since it suggests that stressors alone are not the cause of depression, but, rather, may be associated with the lack of desirable events, perhaps akin to the social learning theories of Peter Lewinsohn (1981). These findings point directly to several research needs: (1) a greater focus on varied sources of depression and differing relations between those sources of depression and their impact on everyday events; (2) additional study of the impact of desirable and undesirable events on depression as a result of varying sources (e.g., loss of spouse vs. functional disability); and (3) the role of positive events and quality of life experience as a contributor to the mental health of elders. Treatment implications seem obvious. The data suggest that different sources of depression may be differentially affected by treatment methods. For example, while depression as a result of functional disability may lend itself to treatment strategies to increase desirable events, and thus behavioral-based interventions, depression as a result of bereavement may lend itself to other strategies, perhaps related more to existential issues. Clearly, these are empirical questions that can be tested.

Extensive literature exists on the impact of caregiving of disabled elders and in the final chapter of this section on risk factors, Jamila Bookwala, Jennifer L. Yee, and Richard Schulz present a comprehensive review of that literature. Through their critical review, the authors report advances over time in the rigor and methodology of this line of research, including a greater reliance on comparison groups, objective measures of psychiatric morbidity (e.g., diagnostic-based measures) and physical morbidity, and large population-based samples. However, they caution us regarding findings, since most studies still rely on cross-sectional data and lag behind in examining physical morbidity in caregivers. Nevertheless, through their review, it is clear that caregivers evidence significant psychiatric morbidity when compared to noncaregivers or community normative data. Poorer physical well-being additionally appears to be fairly consistently related to caregiving, both by self-rating and, in many cases, by objective measures of physical health. The authors conclude that there is likely a link between poorer physical health and psychiatric morbidity among caregivers. This indirectly suggests that caregivers need treatment intervention, as they are at risk for both psychological and likely physical morbidity because of the caregiving process. Still unanswered questions remain, including what kind of treatment might be best, and how and when do we know whether to implement treatment? In partial response, the authors suggest that knowledge of physical morbidity–psychiatric morbidity links over time, particularly regarding the temporal relationship and lagged pathways between the two, can give us clues. I am in agreement and echo their sentiment that only continuation of greater rigor, more objective measuring, adequate use of control groups, greater focus on physical health status as well as mental health effects, greater focus on the links between psychiatric morbidity and physical morbidity, and longitudinal studies on the deterioration or improvement of caregivers' health as the caregiving context changes will allow us to fully answer these important questions.

It is clear from this above that there are many factors across psychological, psychosocial, and psychobiological domains that increase a person's risk for experiencing depression, physical illness, or their combination in late life. It is also just as clear that depression and physical morbidity are intertwined in multiple ways, and no simple hypothesized causal relation between so many risk factors and either depression or physical illness can explain that relation. However, I do think that we can make a few conclusive statements. There appears to be a bidirectional association between physical morbidities (and related functional disabilities) and depression in late life. Risk factors are differentially associated with both depression and physical illness in an elderly population. The dynamics of these relations, while still not entirely explained, nevertheless suggest potentially predictable outcomes when carefully analyzed. Finally, treatment of both depression and physical disability are important obligations.

PART II: CONDITIONING VARIABLES AND OUTCOMES

The second section of this handbook is primarily aimed at the delineation of conditioning variables and their impact on resultant outcomes in depression and physical health in older adults. Specifically, the authors have chosen to focus on those variables that are most likely to intervene directly or indirectly in the relations between depression, functional impairment, and general physical health in the elderly population. Thus, the variables of interest are those that in some way qualify these relations. As in the first section, what becomes abundantly clear is the complexity of relations between variables associated with depression and physical illness. Each chapter describes a particular subset of important conditioning variables and provides compelling evidence for their respective roles. The final chapter in this section by Charles Lance, Adam W. Meade, and Gail M. Williamson offers a statistical methodology that is likely to increase our sensitivity in assessing and explaining the myriad of possible conditioning variables over time, that of latent growth modeling (LGM).

Katherine L. Applegate, Janice K. Kiecolt-Glaser, and Ronald Glaser present a conceptualization of the relation of older adults' immune functioning to depression and physical health, and as a possible mediating factor between the two. Through a concise literature review, as well as descriptions of their own studies, they provide evidence that immunological decline is associated with depression and stress, regardless of age. They address the elder literature and see a parallel but exacerbated relationship, as aging in and of itself negatively impacts immunological functioning because of a downward change in immune system abilities. Furthermore, different from other, younger groups, in elder persons, stressors often include caregiving and bereavement, both of which seem to be linked to poorer immunological functioning. This appears to be the case whether immune functioning is measured at the direct neuroendocrinological level or at the behavioral level in terms of compromised health habits. The negative effect of both depression and stress on immune functioning, combined with the negative effect of immune compromise on overall physical health, point to some very specific issues for treating elderly persons. First is the necessity to treat depression aggressively given its potential for physical health complications. Second is the potential utility of identifying persons undergoing stress (e.g., bereavement, caregiving) as a way of getting them additional family or community resources. Finally, there is the practical need for vaccination to protect against infection, given the poorer vaccine response of these individuals and their greater susceptibility to infection I believe that the compromising aspects of the immune system detailed by Applegate and her colleagues indicate aggressive treatment for these associated factors (i.e., depression, stress) critical to reducing subsequent increases in morbidity and mortality. However, continued prospective research on the *dynamic* aspects of these variables and their relation to immune system compromise is needed.

M. Powell Lawton examines relations among physical illnesses, depression, and end-of-life attitudes and behaviors. He first presents the concept of health-related quality of life (HRQOL) and a number of limitations to our current perspectives of quality of life (QOL). These limitations include artificial limitation of variables not readily defined as directly health related, a traditional lack of focus on positive or ordinary health, and what he perceives as a lack of accounting for individual variation in personal definitions of health status (i.e., they are social-normatively based rather than individually or subjectively based). Lawton follows this by hypothesizing the role and place of depression in that concept. His review of relevant literature reveals that depression is neither inevitable in the terminal phase of life nor a necessary motivation for avoiding life-sustaining treatment or ending one's life. And although QOL often erodes with declining health, he provides a convincing argument that many positive, non-health-related aspects of life strongly influence the judgments of seriously ill elders, depressed or otherwise, when thinking about life-prolonging treatment. To support this view, Lawton presents a model of quality of life and mental health broken into four sectors of QOL: (1) behavioral competence, (2) environment, (3) perceived QOL, and (4) psychological well-being. The first two are presented as objective and directly observable in terms of absolute or social-normative standards, and the latter two are presented as subjective, dependent on persons' own evaluations of their QOL or the lack thereof. The model proposes that these four sectors are different and variably appropriate indicators of QOL but stresses that none are all-encompassing. From this conceptualization of a broader view of QOL as it relates to health, Lawton describes a number of critical issues that I agree require additional consideration as factors involved in overall perceptions of health in elderly adults—all of which have not, as yet, been primary foci of research. These include the prevalence of depression as life nears its end, the relation between the seemingly accepted desire for maximal length of life and the character of end-of-life treatments by elders, and associations between depression and cognitions in the context of end-of-life issues. Broadening our view by including additional subjective/individual-based perspectives creates new possibilities for these factors regarding QOL effects on health. In an attempt to quantify the effects of these more dynamic factors, the author has hypothesized a possible mediator between QOL and end-of-life attitudes and behavior, which he calls Valuation of Life (VOL). The strength of proposing such an intervening variable is that it can be tested as a possible mediator of the impact of *both* positive and negative aspects of one's life (including perceived health and depression) on end-of-life attitudes. Nevertheless, longitudinal study of persons as they move from good health to poor health to terminal illness seems essential. Basic questions such as whether people become more depressed over time as significant illness intrudes into their lives, as well as questions regarding the impact of positive variables on end-of-life decisions, attitudes, and treatment decisions, remain unanswered and warrant further study.

Gail M. Williamson and David R. Shaffer present data supporting an Activity Restriction Model, which describes the mediating aspects of restriction of normal activity between physical illness and depression. In a particularly programmatic fashion, they support the Activity Restriction Model of Depressed Affect from multiple studies of physical-based major life stressors (e.g., cancer, amputation) as well as non-illness-related major life stressors (e.g., caregiving), and consistently show evidence for a strong mediating effect of activity restriction on main effect relations between these stressors and depressed affect. They make the convincing argument that functional disability and, in particular, activity restriction is the primary (though not the only) mediator between major stressful events and the development of depression. They argue that this mediating role is large and remains even after controlling for other known contributors to depression and, again, present convincing data. The success of their programmatic approach illustrates the increased potential explanatory possibilities when research methodology moves away from looking for simple main effects and instead attends to mediating and moderating variables in explanatory models of physical illness and depression. Such theoretical model testing appears to be a primary research tool that will be extremely helpful in explaining more fully the depression and physical health relation. What remains to be done is to expand research on the role of functional disability in general and activity restriction in particular as explanatory variables for the role of non-illness-related stressors and depression. Similarly, there remains a need for postulating still more theory-based questions aimed at identifying additional mediating/moderating variables that are either independent of functional ability or, perhaps, complimentary. At the level of treatment, the Activity Restriction Model suggests an active manipulation of functional ability to impact depression. Indirectly related to treatment, the Activity Restriction Model suggests predictive utility of understanding predisposing factors (illness-related or otherwise) for activity restriction to better target groups at risk for developing depression.

The final chapter in this section on conditioning variables and outcomes is a sophisticated methodological treatise by Charles Lance and associates. In keeping with the tenor of this section, they suggest taking a fresh look at the way we measure and detect developmental change over time in our longitudinal study methodological analyses. An initial review is presented of what they term the "state-of-the-practice" for analyzing longitudinal data in the literature on aging. The authors conclude that longitudinal data collection is crucial to understanding complex relationships. However, they point out what has been stated in the first section of this handbook–that while growing, there remain relatively few longitudinal methodologies utilized in aging research, relying instead on cross-sectional study. Lance et al. continue by discussing limitations of longitudinal studies that are being conducted, showing how many fall short by studying participants for relatively brief periods of time with few data-collection intervals. They review the actual analysis of longitudinal data in the aging literature, conclude that general linear modeling (GLM) techniques are primarily utilized, and follow with a critique of GLM-based techniques compared to an "idealized" set of criteria for measuring change over time. The pitfalls of GLM-based techniques, when used as the only way to describe change over time, suggest a need for altering our perspective. Lance and colleagues recommend a continued and even greater emphasis on longitudinal data collection, but with multipoint data intervals. They also recommend that we begin utilizing the more sophisticated statistical methodologies that are available and explain LGM techniques as a potential methodology when time points greater than two are utilized. They provide compelling evidence that LGM techniques may be closer to meeting their "ideal" criteria. While LGM may not be the most appropriate methodological tool in every case, it clearly has some advantages over more traditional general linear models, particularly when dynamic change is possible. As illustrated by the authors, LGM can measure change at the individual level of analysis rather than the typical aggregate-score level. The technique allows the presentation of alternative functional forms of change over time and assessment of parallel change across multiple variables simultaneously. Finally, LGM supports modeling individual-difference predictors of intraindividual change. I direct the reader back to their working example with an existing data set to illustrate these points in better detail. One of the great advantages of LGM methodology is its potential to provide a new perspective on individual change in response to treatment interventions, taking into account variability of change. This perspective may help us better conceptualize the real impact of an intervention and reduce the likelihood of inaccurate conclusions when the impact is dynamic over time (e.g., when data from one set of time points indicate a relation (e.g., Time 1 to Time 2) but data from another set of time points do not (e.g., Time 2 to Time 3).

Throughout this second section, the description of additional important variables and ways of analyzing their role in the relation between physical health and depression in older adults has been emphasized. Delineating these intervening variables enhances our understanding and, I believe, will continue to direct our research inquiries. Expanding research utilizing these already identified qualifying variables and characterizing still more possible variables will increase our overall understanding of the association between physical health and depression.

PART III: DIAGNOSIS AND TREATMENT

The final section of this handbook is the most applied. It deals with the multiple challenges that primary care physicians, mental health care professionals, and others face in properly diagnosing and treating depression, declining health, and their concomitants in older adults. Logically, application of the theoretically driven techniques and methodologies discussed throughout this handbook should be reflected in these last chapters, and that is what, in fact, is seen. While specific diagnostic and treatment issues are emphasized within each chapter, they are held together by a similar thread of reliance on empirical support and a knowledge of the complexity of the relation between physical health and depression. This is clear from Schulberg and colleagues' very applied chapter on comorbidity issues, through the direct treatment intervention evaluations of the other chapters, to the final discussion by Gurland and colleagues of the limits of trying to account fully for all aspects of causality when dealing with such a dynamic and complex interplay.

In the first chapter in this section, Herbert C. Schulberg, Richard Schulz, Mark D. Miller, and Bruce Rollman focus on diagnostic and treatment issues that globally pertain to physical illness and depression in older primary care patients and present a general theoretical model based on the relative effectiveness of control strategies (both selective and compensatory) used by older persons. These strategies are described as predicting loss of perceived control that, if ineffective, ultimately can lead to depression. Schulberg et al. identify specific assessment strategies that assist in differential diagnosis when physical signs and symptoms co-occur with mood problems. The authors do a straightforward and admirable job delineating the elements of treatment necessary to take into account the multicomponent and complex aspects of this combined physical and psychiatric presentation. They then describe the current treatment strategies of psychopharmacology and psychotherapy, and review their effectiveness. Their conclusion is that pharmacological and, likely, combined psychotherapy and pharmacological interventions are best, but also they point out that different classes of treatment vary in their mechanisms for achieving specific outcomes. This chapter serves as an excellent guide for formulating basic treatment plan decisions prior to implementation and gives helpful information regarding some of the factors necessary to address when assessing effectiveness. It is clear that treatment implementation currently has multiple problems and is not done at the level we might expect given data on availability of effective treatments. These problems appear to result from a combination of patient, provider, and health policy factors. Specifically, patient factors include negative attitudes toward mental health issues, concerns about the stigma associated with mental impairment, common misattribution of depressive symptoms to "normal" aging processes, impact of physical illness on the side effects of antidepressants, and sensory and cognitive impairment that can decrease compliance. Provider factors include problems for primary physicians in diagnosing depression when it co-occurs with physical illness, physicians' perceived lack of time, and inadequate mental health field support available to primary care physicians. Crucial health policy factors are identified, including a perception that artificially "carves out" depression exclusively for the field of psychiatric service, poor reimbursement for patient contact, and higher copayments by insurers for mental-health-related work. Despite these serious limitations toward effective treatment of depression in elders, the authors give us specific recommendations and directions for further study. They suggest that educating and training primary care physicians to recognize and treat depression, and to identify differences between physical illness and depression can and should be implemented. Certainly, more effective procedures are needed by primary physicians to identify older patients at high risk for experiencing a major depression. Schulberg and colleagues suggest a thorough physical exam and laboratory tests as primary rule-out steps, followed by explicit quantitative rating scales for the assessment and diagnosis of depressive symptoms specific to this older population. They further suggest that, regardless of etiology, treatment is necessary all the time given the relative safety and effectiveness of modern therapy regimens. Finally, they also point out the need for mental health professionals to recognize and be aware of physical illnesses that mimic affective disorders. As stated explicitly in the chapter and summarized here, continued refinement of assessment tools is necessary, as are continued refinement and availability of effective treatments. Overcoming barriers constraining the use of these strategies in everyday practice is critical.

The next three chapters in this section, while maintaining a reliance on empirical findings, focus to a greater extent on the application of diagnostic and treatment factors. Myron F. Weiner and Ramesh Sairam consider the relation between clinically diagnosed major depressive disorder (MDD) and the specific disorder of Alzheimer's disease (AD). A well-organized brief review of the pathophysiology of both MDD and AD includes recognition of several differentiating factors by summarizing areas of convergence (e.g., physiological changes, including neurotransmitter effects) as well as divergence (e.g., imbalance of different neurotransmitter systems, protein disturbance prevalence in AD) in the likely biological etiology of the two disorders. Behavioral similarities (e.g., cognitive complaints) and differences (e.g., developmental course, chronicity of mood changes) are elucidated. Weiner and Sairam focus on research concerning the cognitive and functional impact of MDD on AD and vice versa. They also discuss predisposing elements of MDD as a risk factor for AD and vice versa. Again, here, as in earlier chapters on conceptualizing the etiology of depression and physical illness, the importance of recognizing bidirectionality is apparent. Several points are clear in their discussions. MDD in late life appears to be a risk factor for AD, but mechanisms are not well understood. MDD exacerbates cognitive dysfunction and functional ability of persons with AD, but treatment seems to ameliorate those portions of cognitive dysfunction and functional disability related to MDD. Furthermore, there appears to be a "greater risk associated with depression earlier in life[,] suggest[ing] the possibility of a critical period in physiological development during which depression can predispose to the later development of AD" (Weiner & Sairam, p. 242, this volume). Findings from their research indicate that MDD is probably overdiagnosed in AD, especially by caregivers. It may be that the overlapping symptoms, particularly misattributing cognitive impairment as depression, are involved, as these are more likely a result of AD. In a promising vein, the authors give evidence to support the effectiveness of treating MDD in persons suffering from AD with conventional pharmacotherapy and, perhaps, specialized psychotherapy. However, one of the larger hurdles in this treatment (as well as in outcomes research for treatment efficacy) is differential diagnosing, separating MDD symptoms from AD symptoms in a reliable manner. The authors propose an operationalized set of criteria for diagnosing MDD, specifically in AD, geared around a symptomatology cluster that minimizes overlap. By focusing on aspects of depressive features clearly separate from or in excess of those found in AD, clearer delineation may be obtained. These are primarily greater chronicity of depressed mood and multiplicity of depressive symptoms seen in MDD as opposed to AD. I agree with Weiner and Sairam that broader, multiassessment strategies for both research and treatment are needed, as are better validated assessment tools specific to MDD within the AD population. Finally, and in support of earlier chapters in this handbook, the need for continued theoretically driven study relating MDD, cognition, and functional ability to AD remains.

Like the previous chapter's authors, Charles F. Reynolds, Mark D. Miller, Benoit H. Mulsant, Mary Amanda Dew, and Bruce G. Pollock focus on treatment issues as they relate to geriatric depression. Adopting a medical model, Reynolds et al. view geriatric depression as a chronic, relapsing illness that requires proactive treatment. After presenting basic information on the global burden and prevalence of depression, they relate it specifically to aging and depression, with age as a positive weight. This is followed by a brief review of evidence that depression in later life increases mortality from both illness and suicide. The main thrust of the chapter is the pharmacotherapy of late-life depression, focusing on the tricyclic antidepressant, nortriptyline, selective serotonin reuptake inhibitors (SSRIs), paroxetine, and their comparison and combination with interpersonal psychotherapy (IPT). Data comparing nortriptyline and IPT from their 10-year NIMH study in Pittsburgh are presented for acute and continuation treatment, as are data from their research and clinical trials of paroxetine. Specific details are given for their medication-clinic treatment protocol, along with relevant data from that protocol using paroxetine through acute, continuation, and maintenance phases of treatment. This detailed description offers a useful model of applied research methodology in comparing varying treatment strategies. The authors make a strong, data-backed argument that probably the most effective current strategy for elderly depressed individuals, especially those older than age 70, is continued long-term treatment from acute through maintenance phases using a combined approach of pharmacotherapy and psychotherapy, with appropriate augmentation pharmacotherapy as needed for poor responders. However, they point to the need for cost-effectiveness assessments of maintenance treatment to tease out which elderly depressives are suited for each particular therapy regimen (i.e., monotherapy of pharmacology or psychotherapy, or a combination of the two). Thus, while providing a treatment protocol with empirically based evidence for its effectiveness, they urge continued and expanded studies focusing on maintenance therapy, particularly with respect to the, most likely, better tolerated SSRIs. Only further cost-effectiveness studies that break down the components of combined therapy intervention strategies (e.g., pharmacotherapy alone, psychotherapy alone) will tell us whether it is "worth" the additional costs, particularly in terms of maintenance therapy. What I find most heartening about the data presented by Reynolds et al. is the convincing argument that maintaining remission in geriatric depression *will* reduce mortality. As we gain a greater understanding of appropriate treatment interventions for depressed elders, I believe we can improve treatment and compliance, lower the level of recurrence and relapse, promote quality of life, and increase overall cost-effectiveness.

In the next-to-last chapter of this section on diagnostic and treatment issues, a decidedly psychological perspective is presented. Based on Snyder's Hope Theory, Ellen J. Klausner, C. R. Snyder, and Jen Cheavens focus on measuring and, especially, treating hopeless thinking and related suicidal behavior in depressed elders. They present a group-based intervention utilizing a goal-focused model of hope. After a brief description and explanation of Snyder's Hope Theory, the authors relate it to the development of depression, particularly in older adults. The components of Hope Theory include movement toward a goal, motivation and determination (agency), and planning and problem solving to attain the goal (pathways). From this perspective, the authors discuss the literature on hopelessness and depression in general and as it relates to treatment-resistant depression, suicide, and feelings of hopelessness. They additionally briefly review the literature on positive effects of psychotherapy on older adult depression. Their primary purpose is to describe a goal-focused, group psychotherapy (GFGP) methodology utilizing Hope Theory as an intervention technique. Data from a modest sample of geriatric depressed patients provide results of a successful application of this technique compared to a reminiscence-based group psychotherapy previously shown to have mild effectiveness in depressed elders. The target group consisted of persons, who, while achieving some remission with either medication or psychotherapy, continued to report symptoms of anxiety, hopelessness, suicidal ideation, psychomotor retardation, or decreased daily functioning. GFGP treatment was associated with decreased disability scores and increased hope scores. For illustrative purposes, a case study of one participant in the GFGP is described, detailing successful treatment outcomes. Besides being another excellent example of potentially effective treatment for geriatric depression, this research raises optimism for the use of group-based psychotherapy. Given the symptom-resistant nature of the population in this study, it may be useful for long-term maintenance of symptom alleviation. The data call for expansion of the sparse literature currently available on controlled treatment trials of psychotherapeutic techniques with depressed older adults and support larger replications across other groups, more debilitated groups, and additional comparisons to other treatments.

The final chapter in this section tackles one of the stickiest questions for which this handbook was compiled, that is, the difficulty in delineating causal effects from such a multivariable, multifaceted relation. Barry Gurland, Sidney Katz, and Zachary M. Pine discuss the cause-and-effect bases of current longitudinal studies of physical illness and depression, and question whether this is the best way to study such a complex phenomenon. They convincingly point out multiple ambiguities that arise from cause-and-effect-based studies that limit dynamic and inseparable relations between physical illness and depression, in effect, artificially separating them. In their place, they provide a model of complexity independent of definitive findings of causal directionality, including the need for delineating relative power of interactive conditions between physical illness and depression. To this end, they offer a model that uses the interaction (or "complex") itself as the point of study. In many ways, Gurland and colleagues represent the "yang" for the "yin" of the rest of this handbook. They caution us about the limitations of demanding, or even expecting, a "prime" cause and a "primary" effect ever to be fully realized in our understanding of the physical illness–depression relation. But they do not leave us with such a pessimistic "comeuppance." Instead, they offer a new, perhaps more dynamic, approach to studying the whole of this complexity without resorting to the need for complete determinism of cause–moderator–effect relationships. They suggest the need to focus more on treatment of this "whole" complex, rather than an arbitrary cause (whether depression *or* physical illness), and to describe interactions and associations without the constraint of cause and effect being forced upon components of the whole complex of interest. It is certainly a perspective that deserves careful thought and, in my opinion, one that, in its own way, further explains the, at times, seemingly unexplainable complexity of this relation.

CONCLUDING REMARKS

It has been my intention to summarize the major themes and principles that have already been so eloquently presented in much greater detail throughout this comprehensive handbook. Through briefly reviewing each chapter and commenting on those areas that I feel are most critical to our understanding of the complex relationship of depression to physical illness so carefully presented within these pages, I have tried to emphasize areas of both continuity and divergence. At the same time, I hope the reader comes away from this final chapter with at least a sense of the multiple paths available for future research, improvement in diagnosis, and hopes for more effective interventions and improved public policy.

In this handbook, the authors have supplied us with their own perspectives on the relation between depression and physical illness as it particularly relates to an older adult population. They have shown us varying but sophisticated and well-thought-out conceptualizations of the relative impact of mediating and moderating variables. They have given us examples of the complex nature of the relation between depression and physical illness for this population and have included theoretically based explanations for aspects of this complex association. They have presented empirical data supporting these relations and have converged as a group on a complex but likely bidirectional relation. They have provided heuristic models, useful not only in terms of understanding at a conceptual level the possible role of multiple factors in the development of both depression and physical illness but also have provided empirically testable models. Furthermore, specific guidelines and recommendations for the applicability of these multiple perspectives have been presented and should be used as the starting place for continued study.

The next few years will continue to show the dynamic, changing level of our understanding of physical illness and depression in an older population. From the findings presented in this handbook, I believe that we are well on our way to narrowing the gap between conjecture and true understanding, between wishful thinking and effective treatment.

ACKNOWLEDGMENT: Manuscript preparation was facilitated by support from Grant No. AG15321 from the National Institute on Aging (G. M. Williamson).

REFERENCE

Lewisohn, P. M., & Arconad, M. (1981). Behavioral treatment of depression: A social learning approach. In J. F. Clarkin & H. I. G. Eager (Eds.), *Depression: Behavioral and directive strategies.* New York: Garland.

Author Index

Subject Index

ISBN 0-306-46269-9